Introduction to Facility Management

Hester van Sprang

Bernard Drion

First International Edition

Noordhoff Groningen

Cover design: Shootmedia, Groningen

Any comments concerning this or other publications should be addressed to Noordhoff Uitgevers bv, Afdeling Hoger onderwijs, Antwoordnummer 13, 9700 VB Groningen or via the contact form at www.mijnnoordhoff.nl.

0 / 20

ISBN: 978-0-367-72386-6 (hbk)
ISBN: 978-9-001-75255-2 (pbk)
ISBN: 978-1-003-15459-4 (ebk)

NUR 801

Preface

If you ask ten experienced facility managers what facility management entails and what their priorities are, there is a good chance that you will get ten different answers, especially if those facility managers come from different industries, countries or continents.

Facility management contributes to the success of organisations, but to quote the late world-famous soccer player and coach Johan Cruijff: 'you only see it after you've figured it out'.
Therefore, this book explains the subject of facility management in all its diversity in understandable language, with quotes, cases and many practical examples. First and foremost, to students but also to everybody else with interest in this professional domain.
The field of facility management is changing. That is why this book devotes a great deal of attention to trends and developments in society and the effect they will have on the work environment and thus on the work of facility professionals. An example of such a trend is the integration of facility management (FM) with human resource management (HRM) and information management (IT). Because of these developments, we have devoted a separate chapter to the future of facility management.

As we hope that we, too, will continue to learn, we invite you to send us your comments, discussion points or suggestions by e-mail.

We wish you much pleasure in reading.

June 2020
Hester van Sprang, Deventer (The Netherlands),
hester@meerwaardehuisvestingsadvies.nl
Bernard Drion, Meer (Belgium), b.drion@telenet.be

For the translation of *Basisboek Facility Management* into this first edition of *Introduction to Facility Management*, we have gratefully used the expert comments of the members of the sounding board group: Wil Gooskens, Harry Reinders, Johan Offringa, Pieter le Roux and Jaap Wijnja, as well as the sounding board group of the third edition of *Basisboek Facility Management*.

We want to express our immense gratitude to Lynne Ball and Dick van Velzen (translators) and the excellent support of our publisher Petra Presscher.

We are also indebted to the (FM) experts who shared their insights in opening columns: Peter Ankerstjerne, Jacqueline Bakker, Michiel Bakker, Susanne Baslev Nielsen, Diane Levine, Elizabeth Nelson, Jens Schlüter and Nicolas White.

Moreover, we thank Joost Andela, Charlotte de Block, Matthijs Gallé, Brigitte van Gompel, Lisa Hut, Thomas Klein Bramel, and Loesje Ophuis for their contribution to the 'Young Professionals in FM'; and the professors Joop van Duren (Zuyd University of Applied Sciences), Brenda Groen (Saxion University of Applied Sciences), Frans Melissen (BUAS) and Mark Mobach (Hanze University of Applied Sciences) for their contribution to Chapter 7.

Last but not least, we like to thank Merel de Ruijter, Nynke Visser, Thom Wernke and Michel Tobé for their contribution to 'A day in the life' in Chapter 6.

Index

good
vibes
only

1
What is Facility Management?

- What is facility management?
- What are the responsibilities of the facility manager?
- What is the position of facility management within an organisation?
- What are the roles of the facility professional?
- How has the profession developed over time?
- What is the added value of facility management?
- What determines the added value of facility management?

◀ Through the eyes of the facility manager:
'You couldn't tell, but this is an office. Furnished as a meeting place, where, naturally, you can also work individually. As a facility manager, you serve the organisation by stimulating the community spirit with a hospitality-like concept.'

1

Expert Column

Facility management, meeting the demands of a continually changing environment

Facility management is the integration of people and business processes of an organisation with its physical workspace impacting the triple bottom line of economy, equity, and environment. Meeting this demand is critical to the energy efficiency of the built environment, reducing greenhouse gas emissions, increasing the economic productivity of buildings, and providing efficient workplaces.
Business activity and the workplace infrastructure available greatly influence organisational workplace strategies. How people interact with their physical work environment can offer insights into what is happening within organisations through spatial and resource allocation and the interrelationships of staff and visitors within a facility.

The world of work is changing and so too is the nature of the workplace and the field of facility management. The function, design, location and operations of facilities have undergone a significant transformation in recent years. This process is continuing as the world of work is increasingly global, networked, distributed and virtual due to exponential advances in technology, a more integrated global economy, and requirements for sustainability solutions. Such change requires the facility manager to react quickly and decisively to the demands of a shifting environment.

We know that the workplace of the future will be much different than it is today, just as we can easily observe the radical changes in how and where we worked in recent decades. But how much different will work be is the big question, and how do we deal with such change? Seeking these answers will require different ways of thinking, new tools and openness to new ideas. This includes the design of the workplace to reflect the impact of emerging technologies like artificial intelligence, the internet of everything, robotics, and a cultural revolution in the way we work. With the increase in facility management degree programs around the world, this book shows students how facility managers support business leaders who are constantly challenged to delivering higher value more efficiently.

Diane Levine, MCR
Executive Director, IFMA Foundation

1.1 Introduction

Describing clearly what a dentist or a bicycle mechanic does is much easier than explaining what the work of a facility manager involves because all over the world, a dentist and a bicycle mechanic do more or less the same things. The latter cannot be said of facility managers. Facility management occurs in all kinds of forms and all sorts of ranks and positions: from a concierge at a secondary school to vice president Real Estate at a multinational. Moreover, the term facility management is far from being commonplace in all countries. In Switzerland, for example, the discipline is called object management. In the United States, in addition to the term facility management, the terms property management and operations and maintenance management are used. In France and the United Kingdom, the discipline has undergone a name change. The British Institute of Facilities Management has changed its name to The Institute of Workplace and Facilities Management because it believes that workplace management better indicates the scope of the discipline. The French professional association ARSEG has replaced the term facility manager (director/responsable des Services Généraux) with work environment manager (director de l'Environnement de Travail). In this book, we use the terms facility management, facility manager and facility professionals.

What the facility manager is responsible for, and how the business unit facility management is set up differs per country, per industry and even per company. You can be a facility manager in the service of an organisation, but you may also work as such through a hired commercial service provider. Others work as consultants, specialists or independent entrepreneurs. Moreover, the term, let alone the profession, is far from universally known. Just ask facility managers if they recognise that they always have to explain what their job entails at parties and other occasions.

The so-called elevator pitch is an excellent way to simplify this explanation. Suppose you meet a very important person, from whom you would like to receive a rewarding assignment, in the elevator. You only have thirty seconds to get that person interested in what you can deliver before the elevator arrives at its destination. What you say in that short time can also serve as an answer to the question of what you are studying or which profession you are in:

'I make sure you can do your job better.'

You bet that'll spark the interest. Try it if you like.

A well-known anecdote showing the importance of supporting processes is the following. In 1962, President Kennedy announced that the United States would put people on the moon and bring them back safely. Not long after that, Kennedy went on a work visit to NASA, where he saw a cleaner walking with a broom. He interrupted the tour, walked up to the man and said, "Hello, I'm John Kennedy. What are you doing?" The cleaner replied, "I'm helping to get a man on the moon, Mr President."

And he was right. After all, whereas in the past the facilities department was mainly seen as a cost driver, it is now clear that the facilities function has a significant impact on employee satisfaction, the organisation's image and the degree of sustainability of its operations, to name but a few. More and more studies are showing the effect of workplace designs on employee productivity. For example, the layout of the work environment can stimulate informal knowledge sharing in organisations. Visibility and proximity lead to spontaneous encounters and spontaneous encounters lead to knowledge exchange. From this point of view, coffee, tea and lunch facilities can contribute to the productivity of employees. The quality and the location of these facilities then become new issues.

Until the 1980s, every size organisation had a technical, housekeeping or internal department. These departments were responsible for building installations, maintenance, cleaning and catering management. The organisational unit in question was supportive, almost submissive. At that time, the employees involved usually wore workwear, whereas nowadays this only occurs functionally, for example, for reasons of recognisability and hygiene (see section 3.10). Awareness grew slowly but surely that a relatively large amount of money was spent on these services, without the organisation having any control over them. Therefore, a need arose to combine the household, technical and internal services into one business unit. The underlying idea was that integration would allow for more control and thus would make the costs and quality of the services better manageable. This supportive organisational function became known as facility management.

The term facility management originates from the United States, where the first facilities association was founded in 1981: IFMA, International Facility Management Association. The multitude of definitions of facility management illustrate the comprehensiveness of the discipline.

> (IFMA) Facility management is a profession that encompasses multiple disciplines to ensure functionality, comfort, safety and efficiency of the built environment by integrating people, place, process and technology.

> (EN 15221) Facility management is an integration of processes within an organization to maintain and develop the agreed services which support and improve the effectiveness of its primary activities.

> (EN-ISO 41011) Facility Management (FM): an organisational function which integrates people, place and process within the built environment with the purpose of improving the quality of life of people and the productivity of the core business.

These definitions show how diverse the view of facility management as a discipline is, as a reflection of the development of the profession over time. The IMFA's definition places the emphasis on maintaining the quality of the built environment through an integrated approach to facility support, at the lowest possible cost (efficiency). From this perspective, facility management is likely to be seen as a cost factor.

The definition in EN 15221 emphasizes supporting the effectiveness of the primary process, a first step in thinking about the added value of facility management in achieving organisational goals. However, it is so broad that this definition can also apply to the fields of human resource management (HRM) and information and communication technology (ICT) without modification.
The definition in the ISO 41011 is consistent with thinking in terms of the added value of facility management, as it makes the connection between the FM-process and the added value that this process provides, for people and the organisation, highlighting the contribution of facility management to the success of organisations. Furthermore, ISO 41011 does not distinguish between types of facility organisations. Whether you work for a facilities department of an organisation (under its management) or a commercial supplier, under this ISO standard all facilities parties are treated as providers. This definition is therefore the most suitable for this book.

1.2 Facility Management in the organisation

We distinguish between primary functions and support functions in a company. Facility management is often one of the support functions, as are Human Resource Management (HRM), Legal Affairs, Financial Affairs (FA), and Information and Communication Technology (ICT). In a hospital, performing surgery is one of the primary functions, but cleaning the operating room - sterilising it - is part of the support services. Facility management can also be one of the core activities of a company. For example, in the leisure industry (e.g. hotels, bungalow parks, amusement parks and aqua parks) as well as commercial facility providers (such as security companies, caterers, event agencies, and cleaning companies).

Support functions

In short, facility management is the business unit that is responsible for optimal work, accommodation and care environment - both physical and virtual. By physical, we are referring to the built environment (office building, hospital, bungalow park, factory, school, etc.), including its layout. In addition to these buildings, a different work environment is becoming increasingly important, that is any other place where people can work. Information technology (IT) enables us to work almost anywhere at any time: at home (the so-called 'second workplace'), at a coworking place, in the car and the train, even in the countryside. Employees can increasingly be productive in that 'other' work environment thanks to laptops, tablets and smartphones. This set of remote locations outside the company's core working environment is what we call the 'third workplace'.

1.2.1 Facility Management in the organisational structure

An organisational structure expresses the distribution of tasks, powers and responsibilities in an organisation, and the mutual relationships are regulated. Which structure suits an organisation depends on its strategy and environment. Mintzberg, an organisation scientist, distinguishes three levels in a company, which connect to the pyramid in figure 1.1: top management (strategic level), middle management and support functions (tactical level), and the executive core (operational level).

Organisational structure

Three levels

An organisation is often depicted as a pyramid; this is also the approach taken in this book. Why a pyramid? A pyramid has the characteristic that it narrows towards the top. The base is vast and deep, while ultimately the top is reduced to a point, indicating that in general, there are far fewer people working at the strategic level than at the operational level. The arrows in figure 1.1 illustrate the direction of the focus. The top management mainly looks at the external environment, market trends and future developments, and decides upon the policy. Tactical management focuses primarily on the inside, on how the organisation should be structured. This level translates strategy into objectives (preparation). Operational management concentrates on implementation at the process level, on the actual provision of services and the provision of resources (performance). Facility professionals support all levels of an organisation. However, customers manifest themselves differently at every hierarchical level of the company. Therefore, different interests and areas of attention are at play at every level. Let's take a closer look.

FIGURE 1.1 The hierarchical levels in organisations

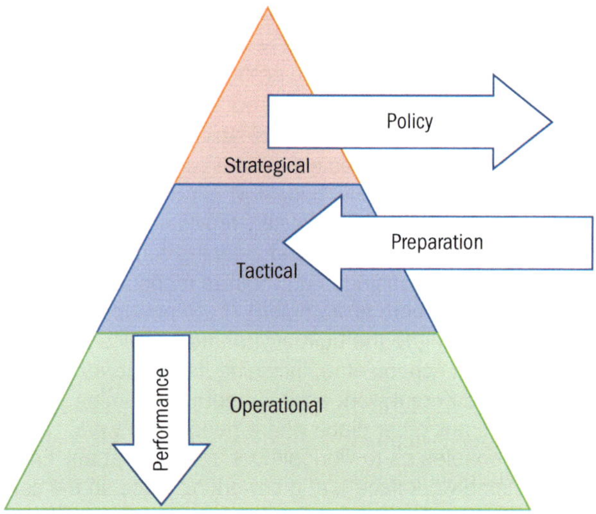

The strategic level of the organisation

Strategic organisation level

The strategic organisation level is the highest level of the organisation: it includes the Board of Directors and similar functions, also referred to as the senior management or C-suite. Here, *policy* is determined, i.e. the strategy and objectives of the organisation are aligned with market developments. This level is responsible for all strategic decisions taken. Let's suppose that a leisure park company decides to increase its market share among the well-to-do elderly. Based on market research, it will then opt for a new park concept: very luxurious bungalows on estate-like grounds. The facilities policy must be consistent with this development, meaning, for example, that all services such as reception, security, green services, cleaning and maintenance of the bungalows have a quality level that corresponds to the expectations of the intended guest. And because the audience group consists of senior citizens, adjustments will also have to be made to the terrain and the bungalows to ensure that guests with a physical or mental disability can indulge in recreational activities. An elaboration of the facilities policy is to determine how luxurious or sober the service package should be, at what price and for which client. You could compare this with deciding the service level of a hotel, expressed in stars: one star indicates the lowest level and five stars the highest level of quality.

The tactical level of the organisation

Tactical organisation level

The tactical organisation level represents the divisions that are one level below the strategic level. Here the policy - for example, the decision to develop a specific type of bungalow park - is translated into *actions to be taken* for the establishment of the estate park. This includes acquiring the site, conducting research into the applicable safety requirements, designing prototypes for the bungalows, the design of the services, the product mix, in-house or outsourced work, up to and including the design of the work processes and the deployment of personnel. It is at this level in the facilities organisation that the services are set-up.

The operational level of the organisation

Operational level

All activities, actions that are necessary for the actual running of the bungalow park belong to the operational level. The same applies to support activities. Activities on this organisational level involve the *performance* of activities such as handling the reception, maintaining the site, supervising the swimming pool and offering catering facilities. The operational management relates to the question of how the services are provided, by whom and when. The facility staff member, whether employed by the organisation itself or by a facility provider or a subcontractor, always has to deal with two stakeholders: the internal customer (or the (other) employees in the primary process of the company) and the external customer (patient, visitor or client of the organisation). The intensity of the direct contact between the facility staff, the internal customer and the external customer differs per industry.

Stakeholders

Examples

In the average office-based organisation, the facility staff member mainly has contact with the internal client (colleague, the staff member involved with the primary process). For example, those at the facility service desk, in the company restaurant or during the cleaning process. The contact with the external customer (the visitor) is less intensive; it mainly takes place via the intercom at the gate, at the reception and during the serving of coffee or a meal. The employee in the primary process does have intensive contact with the visitor. Figure 1.2 depicts the nature of these relationships.

The healthcare industry provides a different setting. There, in addition to the employees in the primary process, FM-employees also has an intensive contact with the external client (Figure 1.3). Think of the patients and visitors that make use of a hospital's restaurant. At the information desk, too, there is a great deal of interaction between facility management staff, patients and visitors. Room service is also part of facility management, which involves contact with the patient when serving meals and drinks and keeping the patient's room clean. Moreover, volunteers who work as hosts who guide visitors or show them the way are, in most cases, managed by facility management.

FIGURE 1.2 The intensity of customer relationships of operational facility management staff in office-based organisations

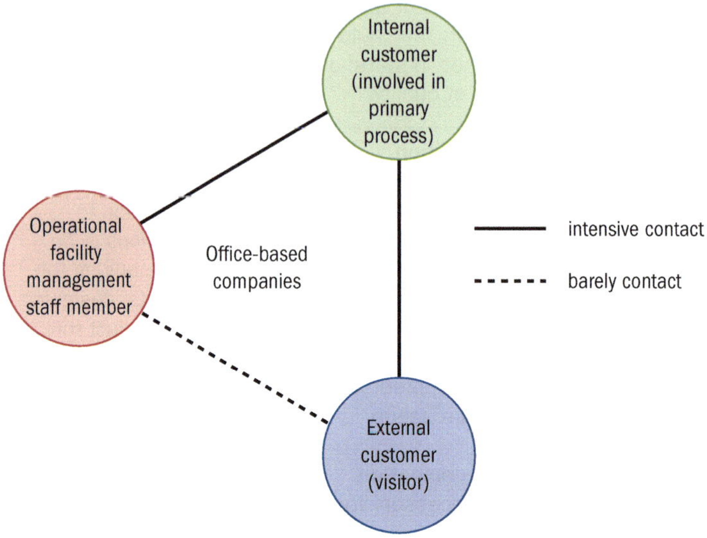

The hospitality and leisure industries offer another variation. Although the name for the facility activities is different in this context (e.g. housekeeping or food & beverage), these are in fact, facility operations. In these

FIGURE 1.3 The intensity of customer relationships of operational facility management staff in healthcare

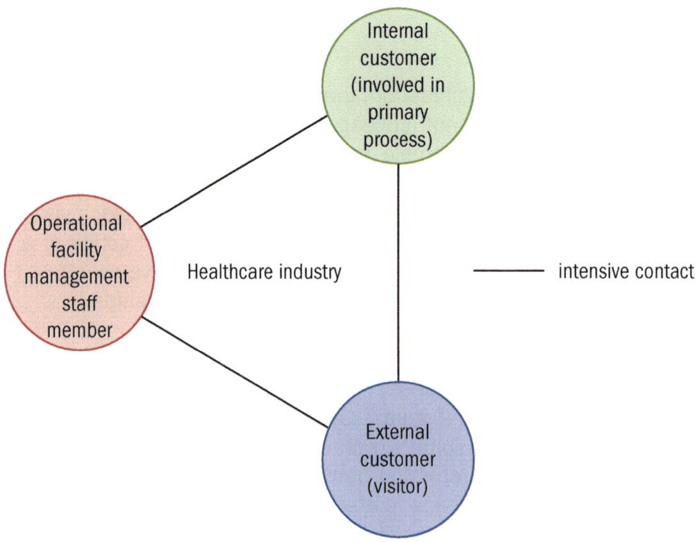

industries, the core activity is the provision of hospitality services to external clients. The number of internal clients is limited. For this reason, the executive (facility) staff is mainly in direct contact with the external client. The intensity of a contact in Figure 1.4 also applies to commercial providers of facilities services.

FIGURE 1.4 The intensity of customer relationships of operational facility management staff in the hospitality and leisure industries

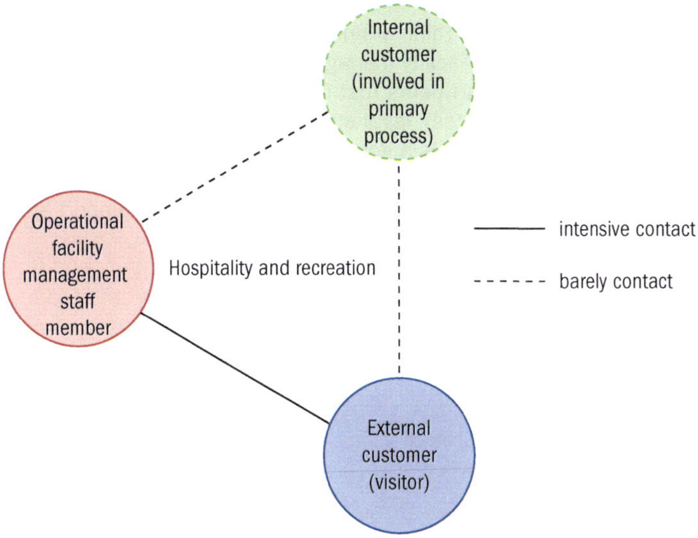

Table 1.1 summarises the levels of an organisation and the associated FM-focus and activities.

TABLE 1.1 Organisational levels in relation to activities

Organisational level	Focus	Activity	Underlying issues
Strategic level	Policy	Defining the facilities and the services	What does the market/ client want, and how do we respond to it?
Tactical level	Preparation	Specification of the product range, the offer, the staffing, sourcing and procurement.	What is achievable at what price? Who is responsible for it, when and where?
Operational level	Performance	Execution of services and supply of resources (e.g. office supplies)	How do we carry out the offer at the required service level for the agreed price with a high level of customer satisfaction?

Policy

In this context, policy involves aligning the organisation with external developments, setting the course and is long-term in nature. Strategic facility management determines how to translate the strategy into a strategic plan for the facility management organisation and how to add value for the organisation.

Preparation Preparation is the translation of the agreed strategy for the benefit of the internal organisation. This can also mean the outsourcing of activities. In that case, the provider determines how it organises the services within the set conditions.

Performance Finally, the performance is the execution of activities needed to implement the agreed strategy. The responsibilities for execution may be outsourced.

Mintzberg divides the support functions into operational and organisational support (see figure 1.5). These functions directly or indirectly serve the functional core. Mintzberg calls this the 'technical staff'; responsible for making the necessary preparations and for controlling the core process, i.e. the production of the company. Think of financial control, planning, training and research & development. Facility management is part of organisation-oriented support. Similar departments that fall into this category are among others: HRM (or human resources), legal affairs, marketing and public relations (PR). This classification is a general classification, which does not apply everywhere in the same way. For example, IT will be a comany-wide activity that is both a support function for the operational core support and a support function for the organisation.

FIGURE 1.5 The position of supporting functions in the organisation - inspired by Mintzberg

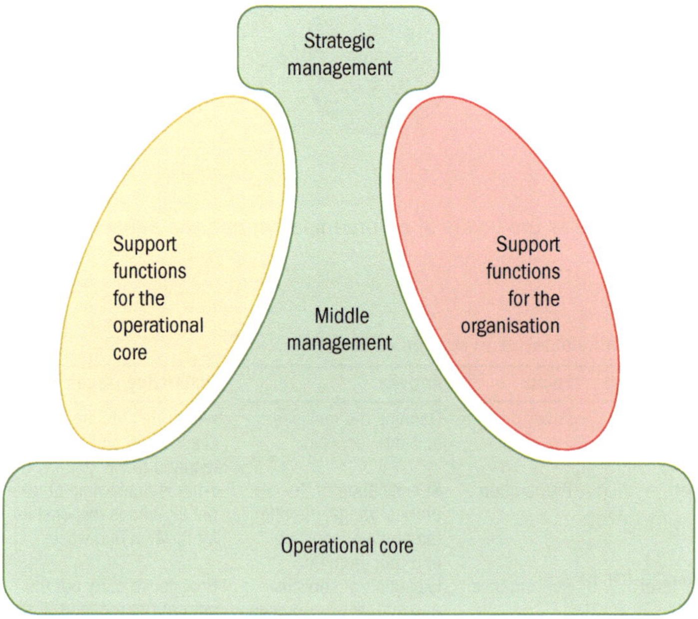

Figure 1.6 shows an example of an organisation chart from an educational institution. In this organisation chart, the executive core is divided into eleven Vocational Study Programmes. Supporting services are: Communication and Marketing; Facilities, Housing and IT; Financial Affairs, Student Administration & Control; Human Resources; Education and Quality Assurance. As can be deduced from these functions, the task of these

organisational support services is to create the conditions for the implementation of vocational education and training. The Board of Directors is also supported: by the Administrative Secretariat and Legal Affairs and its representative bodies (Central Student Council and the Workers Council).

FIGURE 1.6 ROC Twente organisational chart (2019)

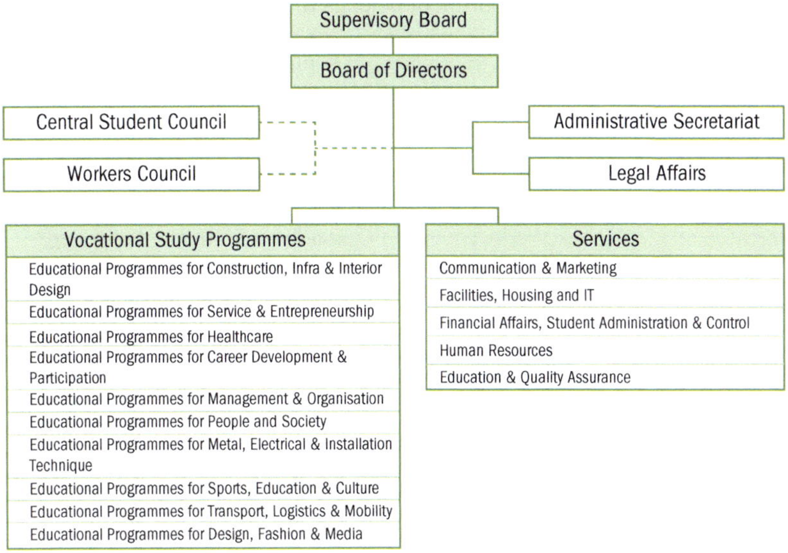

In large companies, it is customary for facility managers not to have direct access to the board of directors. The reason for this is usually that the board makes decisions about billions of euros, while facility costs amount to a fraction of this. Take Air France KLM, for example. Here, most of the costs are for aeroplanes and kerosine. In comparison, the facility costs are almost negligible. In such a case, it is logical for the facility manager to be accountable, for example, to the financial officer (Chief Financial Officer or CFO) or the person responsible for personnel and internal affairs (Chief Operating Officer or COO). Still, facility management issues will be scheduled regularly on the management agenda. After all, only if a facility manager is involved in organisational developments at an early stage, can s/he make an optimal contribution and create added value from a facility management perspective.

The challenge for facility professionals is to come up with creative solutions that contribute to the organisation's success. Above all, a good understanding of the core activities and stakeholder's needs is required. Only then can costs be balanced against customer satisfaction and quality. But keep in mind: the most appropriate decision is not always the cheapest solution.

In some organisations, the facilities department (often referred to as the facilities function) is structured as an executive core, as can be seen in the organisation chart of the Deventer General Hospital (Figure 1.7). As Figure 1.3 illustrates, the intensity of the contact between facility staff and patients is high, and the influence of facility support on the quality of care is substantial. Therefore, facility staff have a significant impact on the perception of hospitality by patients and their families and hence on patient satisfaction. Also, cleaning is an essential service for the prevention of cross-contamination and the development of resistant hospital bacteria that can seriously threaten the health of patients. This is probably the underlying reason why the facilities department belongs to the executive core.

FIGURE 1.7 Deventer General Hospital organisational chart (2019)

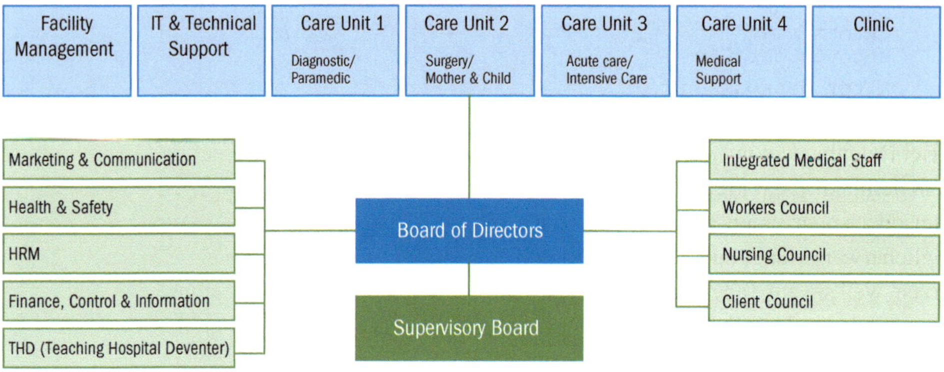

1.2.2 The customer

Like in society, in facility management, the customer ultimately determines what is to be delivered. At least, that is how it should be. It is also the customer who assesses the facility company, whether consciously or subconsciously. The customer, therefore, is essential to everyone active in the facilities industry. You could say that if the facility professional maintains a good relationship with the customer, then everything will be fine. But it's not that simple.

Figures 1.2 to 1.4 may give the impression that there is only a distinction between internal and external customers. However, as Figure 1.8 illustrates, reality is more complex with a diversity of (internal) customer groups. The centre of the model depicts the hierarchical layers in the organisation (comparable to Figure 1.1). To the left, you can find the organisation with its core activities. On the right side, the business-support processes are shown (see figure 1.5); with providers for the supply (part of) the facilities services on the far right.

Defining customer, paying customer and benefitting customer

What matters now is the enumeration 'client', 'customer' and 'consumer or end user'. These are also referred to as defining customer, paying customer and benefitting customer. A simple example shows the difference between these. Imagine that the employees of an organisation want to have a choice between freshly grounded coffee, Wiener melange, cappuccino and hot chocolate. These are the benefitting customers who use the coffee supply

FIGURE 1.8 EN 15221 Facility Management model

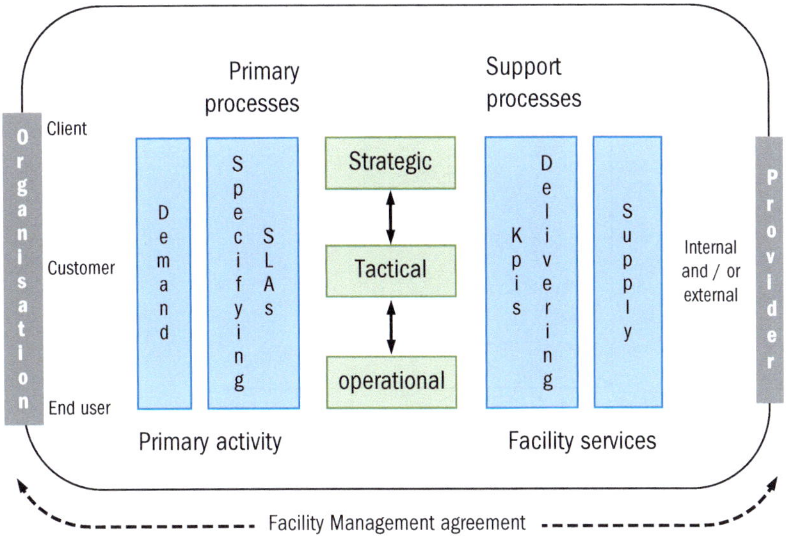

daily. However, the invoice is not settled by these users personally, but by the department managers (the paying customer). Logically, a more luxurious assortment is more expensive. Whether or not the organisation finds these costs acceptable can only be determined by the paying customer, not by the benefitting customer. The paying customer will base his decision on the guidelines set by the defining customer. It is therefore important for every facility professional to reach agreements at the right level in the organisation and with the right people: the paying customers. In Chapter 4, we will discuss this so-called 'stratification' of the customer in more detail.

1.2.3 Stages of development of the FM-organisation

As you have already understood from section 1.2.1, there are various ways to design a facilities organisation. These are called the principal forms. There is no standard, no best or worst solution. By design, we mean the placing of functions in the organisational chart. The position of the facilities company in the organisational structure determines to whom the facility manager is accountable, which largely depends on the management philosophy of the entire organisation. Various layouts are possible, depending on what the organisation wants to emphasise.

The classification used in this section is in line with the development of many facility management companies (see Figure 1.9):

- in-house sourcing;
- the directing organisation;
- the demand management organisation.

The development stages are not a normative classification, in the sense that a demand management organisation would be better than in-house sourcing. It is about finding a suitable management model in the context of the organisation.

1

FIGURE 1.9 Development stages of the Facility Management function

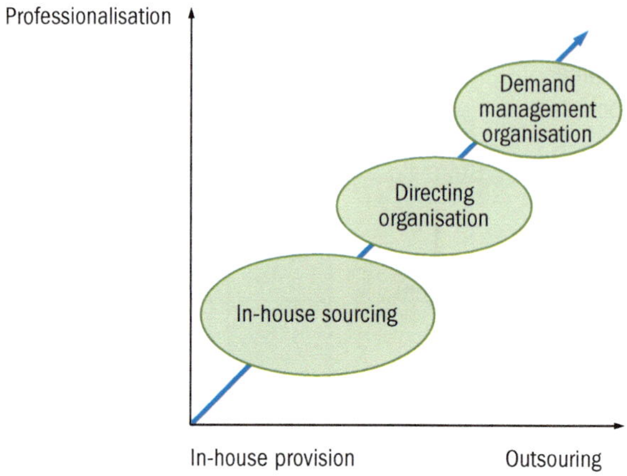

In-house sourcing

With in-house sourcing, the operational services are carried out and managed by the facilities organisation. Non-profit organisations mainly use In-house sourcing. However, due to increasing market forces (and the associated need for flexibility) and the need to reduce costs, more and more attention is being paid to outsourcing in these industries as well. Organisations that prefer in-house sourcing mostly work with a budget set by the management, which may be charged to the users (departments). Characteristics of in-house sourcing are:

Budget

- extensive facilities organisation with low outsourcing rate;
- focus on operations;
- various (internal) contracts based on effort (see also section 3.3.9).

Directing organisation

When more attention is given to the quality, flexibility and cost of service provision, organisations will begin to wonder whether it is wise to continue with in-house sourcing of services. Market parties may have knowledge and expertise that the organisation does not have at its disposal. Another reason may be that external providers can provide the same service cheaper or with more flexibility. If so, outsourcing will not be limited to a single function but will encompass a fair share of the facilities services on offer. In this situation, operational services are no longer carried out and managed by the facilities organisation itself. Instead, external providers step in to take care of operational FM-activities. Subsequently, the question arises as to how a facility manager can keep a grip on the facilities organisation. S/he has to translate the customer's demand into an appropriate range of services, purchase these in the market, and implement and manage the contracts. The latter is particularly important. Attention needs to be paid to contract management, to ensure that the organisation will receive what has been agreed to, and at the agreed price. Characteristics of a directing FM-organisation are:

Contract management

- a compact facility management organisation
- focus on matching supply and demand
- some contracts based on performance (or output) specifications (see also section 3.3.9)

'While service providers are ready to offer Total Facility Management, customers are still very hesitant.'

— FM-Trendreport Belgium 2019, p. 32

Demand management organisation

The more suppliers, the more contracts, contacts, invoices, overlaps and gaps. Managing contracts and consulting with contact persons is labour-intensive and demanding. Facilities providers have found the following answer to this issue: integrated facility management (IFM), also known as total facility management (TFM) or integrated facility services (IFS). In IFM, all services are outsourced to a single external supplier that acts as a partner in facility services delivery. An integrated contract with one supplier is easier to manage than the multitude of contracts that comes with working with single suppliers. However, the downside is the strong dependence on this one supplier. Another provider is not easily found and the impact of switching providers is very high, as it will be noticeable in the entire facilities organisation. Multinationals in particular have opted for this far-reaching form of outsourcing. This choice is often part of a global facilities strategy, whereby a provider is sought that can provide facilities services in all the countries in which the multinational is active.

Integrated facility management (IFM)

Multinationals

In this service model, the management-operational split is high (see Figure 1.10). Processes are outsourced at the tactical level (structuring) and operational level (performing), and in some cases even partly at the strategic level. The supplier can determine *how* to organise the facility services, provided that it complies with the specifications. Characteristics of a demand management organisation are:

- minimal in-house facilities management organisation;
- focus on the definition of demand;
- cooperation with one or more partners in a long-term relationship (partnership).

Degree of outsourcing

The preceding sections describe in-house sourcing, the directing organisation and the demand management organisation. Figure 1.10 shows an illustration of the management-operational split of these principal forms. The degree of outsourcing (shown in blue in the figure) is rather small for the first type of organisation. The in-house FM-organisation provides most services. In the second principal form - the word says it all - the facilities company directs the services outsourced. The degree of outsourcing is higher than with (predominantly) in-house provision. And in the case of the demand management organisation, the facilities company acts purely as a client - and it even leaves the control to the contracted facilities providers of services. The demand management organisation has the highest management-operational split.

Management-operational split

The number of organisations with dominant in-house sourcing will most likely decrease in the coming years. The market trend of increasing outsourcing of services (which means that an organisation is shifting towards the direction

1

FIGURE 1.10 The relationship between sourcing and the principal forms

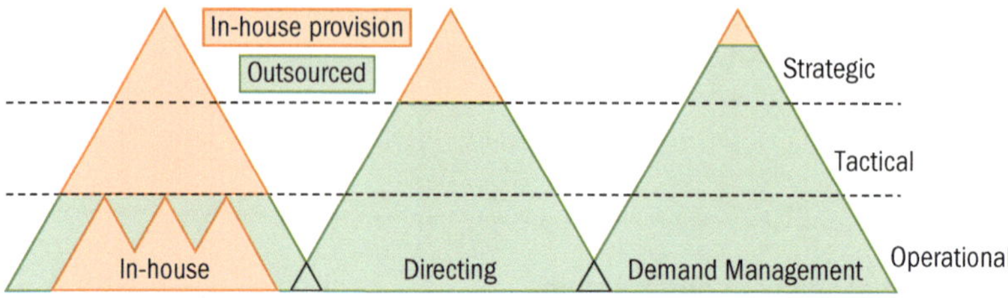

of demand) is a driver for this expected change. From the Global Facilities Market Report 2018, it appears that the average outsourcing percentage (measured among 41 countries) is more than 50. 11.5% of outsourcing takes place in the form of an IFM contract. In 2019, the average outsourcing ratio in the Netherlands was 64% (FMN/TG, 2019). A comparison of FMN/TG's periodic facilities market surveys shows that the average outsourcing rate in the Netherlands increases by approximately 1% per year. An opposite development is visible in Belgium, in 2018 48% of the FM-market was outsourced, which is a decline of 4% compared to 2016. However, this seems to be a temporary setback: in 2022, the expected average outsourcing rate in Belgium is 53% (IFMA/PROCOS Group, 2019).

1.3 The roles and responsibilities of the facility professional

Facility professional

Now that you know what the field relates to, we can look at the tasks and responsibilities of the facility professional. What exactly do facility professionals do?
Facility professionals create a fit-for-use environment. They do this by connecting space, technology, services, processes and people. Facility professionals offer an appropriate experience within the set means, to create added value for the organisation (LOOFD, 2018).

Areas of focus

This description contains several keywords that we will explain below.

Fit-for-use means suitable for the intended use, and tailored to the wishes and needs of the organisation and its users. Fit-for-use refers to:

- Environment. This book assumes that facility management always takes place in a physical or virtual environment that involves a form of organisation, and in which people provide professional services. Just about the only thing excluded is the private environment (people's homes and immediate surroundings). However, care homes, which are private domains for residents but are part of a healthcare environment facilitated by a professional organisation, are exceptions to this rule.
- Space; concerns the built, virtual and sometimes also temporary (work) environment and layout of space.
- Technology; concerns the whole of technical structures (building and facilities).

- Soft services; involves the complete range of facility products and services.
- Processes; facility management supports the core functions of a business or institution through the organisation and integration of space, technology and services in activities that add value.
- People; the (end)users and the various stakeholders of the organisation, as well as the facilities staff. People and their performance are the focus of facility management.
- Experience. The overall perception of the (end)user is always taken into account when developing a service or environment.
- Resources. The resources include the budget, but also space, knowledge and skills that are available to provide the service.
- Added value; refers to the impact of the FM-services on the performance of the organisation. For example: by creating a work environment that supports the productivity of employees or by a hospitality concept that strengthens the organisation's image. For a long time, the discipline revolved around cost savings. Now that we understand the impact of the workspace on the functioning and wellbeing of people, facility management is increasingly seen as a driver of success.
- Organisation; a person or group of persons who have their function with responsibilities, authorities and relationships to achieve its objectives (ISO 41011).

Added value

To achieve fit-for-use, the facility professional must take into account:
- the internal client (who performs professional activities, the worker);
- the external client (e.g. the visitor, patient, hotel guest);
- soft services;
- the physical environment (hard services);
- the third workplace and virtual environment;
- the matching of supply and demand and its sourcing; and
- procurement of facility resources.

The areas of attention briefly described above do not apply equally to all industries. Both the industry and the stage of development of the facilities function influence the main focus areas of facility management. Table 1.2 shows industry-specific examples of internal customers, external customers, services, physical environments and third/virtual workplaces.

Industries

Stage of development

TABLE 1.2 Industry characteristics in terms of customers, services, the physical environment and third workplace

	Examples of internal customers	Examples of external customers	Examples of characteristic services	Physical environment	Third workplace
Hospital	Doctors Nurses Assistants Laboratory staff Volunteers	Patients Visitors	Nutritional service Hospital logistics Parking services Room service	Hospital Clinic Pharmacy Laboratory	Outpatient clinic
Bank	Board of Directors Management Investors Bank employees	Private accounts Business accounts	Property management Security Money transport Banqueting	Headquarters Affiliates ATMs	Home office Car Customer location

1

TABLE 1.2 Industry characteristics in terms of customers, services, the physical environment and third workplace (continued)

	Examples of internal customers	Examples of external customers	Examples of characteristic services	Physical environment	Third workplace
University	Board of Directors Deans and professors Researchers and teachers Student assistants	Students	Audio-visual service Catering Library	Campus Laboratory Auditorium Lecture hall	Home office
Industry	Head of a workshop Mechanics Assembly staff	Wholesale Consumers	Logistics Site maintenance Security	Production hall Warehouse Office building	
Ministry	Directors-General Policy officers Spokesmen	Citizens Business world	Driver service Banqueting Messenger service	Department building	Home office
Nursing home	Management Geriatrician Care staff Physiotherapist Volunteers	Patients (clients) Relatives Other visitors	Cleaning Meal supply Linen service Maintenance of auxiliaries Reception	Apartments Restaurant Therapy room Day-care Mortuary	Residential supported living
Municipality	Mayor Aldermen Municipal Secretariat Officials	Inhabitants of the municipality Local business	Public service desk Meeting service Greenkeeping	Town Hall Municipal Archive Council chamber Wedding venue	Home office Customer location
Bungalow park	Reservation desk	Guests	Reception service Food & beverages Park management Pool maintenance	Bungalows Swimming pool Playground Supermarket	
Hotel	Hotel manager Administration Marketing	Leisure guests Business guests Event organisers	Porter service Valet parking Room service Banqueting	Lobby Meeting centre Hotel rooms	
Museum	Conservators Museum Shop Education Department Marketing Fundraising Volunteers	Visitors (groups and individuals) Exhibitors Sponsors	Climate control Guards Security Transport Catering services	Halls Storage space Restoration workshop Coffeeshop Museum shop	
Retail	Store Manager Shop staff	Shoppers	Security Stocking Cleaning Building management	Retail property Warehouse Headquarters	

Roles

Facility professionals have many roles within organisations. For example: facility manager, facility employee, purchaser, contract manager, project manager, service coordinator, supplier, entrepreneur and consultant, to name but a few. It is essential always to be aware of the current role. We distinguish between four main FM-professional roles:

1 Facility manager. Facility managers oversee hard and soft facility services (see Chapter 3). As a rule, this is an internal role. Still, organisations can also outsource this role, as you can read in the vacancy text for a facility manager in the box below.

Facility Manager for the Utrecht Region at Dolmans Group

Job description
Taking care of the customer is is the thing for you. You like to motivate and enthuse a team, but at the same time, you dare to deliver criticism when necessary. You know better than anyone that you, as Facility Manager, are an extension of our client's core business.
You can work on a strategic, tactical and operational level. You are ultimately responsible for all services provided at different locations of our client in the Netherlands. Also, you are the point of contact for our client, all team members, and supply partners.
You are familiar with the trends and developments in the facilities market. You see commercial opportunities and respond to them by introducing innovations. Besides, you monitor the quality of the service and the corresponding budget and present clear reports. You lead and coach a team of Service Hosts and Service Desk employees spread across the various locations in the Netherlands and focus on personal development.
Despite the many distractions, you manage to create and maintain an overview. You are motivated to excel and take pride in exceeding expectations.

Experience and background required
- You have a degree in Facility Management from a university of applied sciences.
- You have at least 3 to 5 years of managerial experience in a similar role.
- You have substantial financial expertise and can compile reports.
- You have the communication skills to communicate with stakeholders at different levels easily.
- You have an in-house emergency response certificate or are willing to obtain one.
- You are proactive and have a great sense of responsibility.
- You have a driver's license.

2 Specialist. Facility professionals can operate as specialists for a specific (spatial or technical) facility or one or more services (such as internal communication). Specialists are either employed by the organisation or hired through an external party. It depends on the issue at hand and the activities of the organisation, whether specialist knowledge must be available at all times or whether there is an occasional need for it. An example of a facilities specialist is a senior buyer (see the job offer in the box below).

Senior Buyer Indirect categories, London, United Kingdom
A leading FMCG brand is seeking an ambitious, senior buyer following internal promotions in the procurement function. The procurement portfolio focuses on Indirect Categories, such as FMCG's Facilities Management category. You will be working across a range of categories such as Facilities Management, IT and HR, and across the full end to end procurement process - from request for quotation and tender negotiations, through to contract and post-contract supplier management. Your responsibilities include taking autonomous ownership of £Multi-Million spends and developing sourcing strategies across a range of indirect categories, particularly HR, Facilities Management and IT. You are also responsible for managing key suppliers, ensuring that they are continuously meeting service level agreements. We value building long-lasting, commercially successful relationships.

Experience and background required
- Experience of the end to end procurement process, including contract reviewal and supplier management;
- Strong facilities management procurement experience including sub-categories such as cleaning, chemicals, security, health & safety and pest control;
- A CV demonstrating outstanding cost savings achieved within one or more Indirect categories, with evidence of leading strategic tender activities;
- Excellent communication skills; a confident negotiator who can speak with stakeholders at all levels;
- Strong academics; degree educated (or equivalent) and Chartered Institute of Purchasing and Supply (CIPS) preferred.

3 Advisor. Advisors, or consultants, are often involved in the design and organisation of the overall facility management (tactical/strategical level). The consultant may do this work as an employee or as an external contractor. In general, the facility manager engages an advisor for projects that are complex in nature, require specific knowledge, or that occur so infrequently that the experience is lacking in the facilities organisation. Many consultants also offer (facility) auditing services, for example, as a mystery guest for checking reception or hospitality services. Next, we will show a job offer for a facility consultant. Advisors who deal with strategic issues, often have a master's degree.

Senior Workplace Consultant at CBRE Amsterdam
As a Senior Workplace Consultant, you will operate in an enthusiastic and professional team that deals with (strategic) housing advice for end-users. You are the point of contact for Dutch and international clients, most of whom are office-based. You advise them in making strategic housing decisions. In practice, this means that you translate your client's organisational goals and business processes into a specific housing need and suitable office concept. For this purpose, you have a wide range of knowledge at your disposal, from advice on workplace consultancy to change management. You are familiar with the latest innovations in your field and that you include them in your advice.

Experience and background required
- You have completed a university or higher vocational education.
- You have at least five years of experience in business-to-business consultancy in the field of office accommodation.
- You have a broad knowledge of office concepts and innovations.
- You have experience with change management concerning workplace innovation.
- You have a good understanding of numbers, an analytical mind, and are capable of developing strategies.
- You have excellent communication skills and can present yourself well.
- You can work both independently and in a team.
- You have (near) native Dutch and English language skills, both written and spoken.

4 **Entrepreneur.** The entrepreneur can take on all previous roles, but 'at one's own expense and risk'. This means that an entrepreneur runs a financial risk with the entrepreneurial activities s/he undertakes. Still, any profit is for him or her. We do not have a job notice for entrepreneurs since they are self-employed.

The division into these four main groups is not complete. One finds facility management graduates in a multitude of functions, such as project manager, account manager, policy officer, lecturer or researcher in facility management.

1.4 The development of the FM-domain

Facility management is a relatively young field that is continually evolving. How these developments take shape, vary from country to country and from region to region. For example, Figure 1.11 shows the considerable differences in the FM-market size per continent (Global FM market report, 2018). The markets of North America, Europe and Asia/Pacific are many times larger than those of South America, Africa and the Middle East.
Over the past decade, we have seen substantial growth in the European market for facility management. Research shows that the size of the facilities market (including real estate) amounts to an average of 5 to 8% of gross domestic product (GDP) (depending on the country and the maturity of the discipline in that country). **European market**
Various sources from both Europe and the United States indicate that the average workplace costs around 10,000 euros per year in facilities support. Based on these figures, the European facility management industry (including real estate management) employs approximately 40 million FTEs per year (EuroFM, 2017).

Another indication of the maturity of the market is the outsourcing rate (see section 1.2.3). In general, the outsourcing percentage increases as the market matures. Moreover, the share of integrated facility management (IFM) increases with the professionalisation of the discipline. More and more providers of real estate management also offer facility services, and

1

FIGURE 1.11 The global FM-market (in billion dollars)

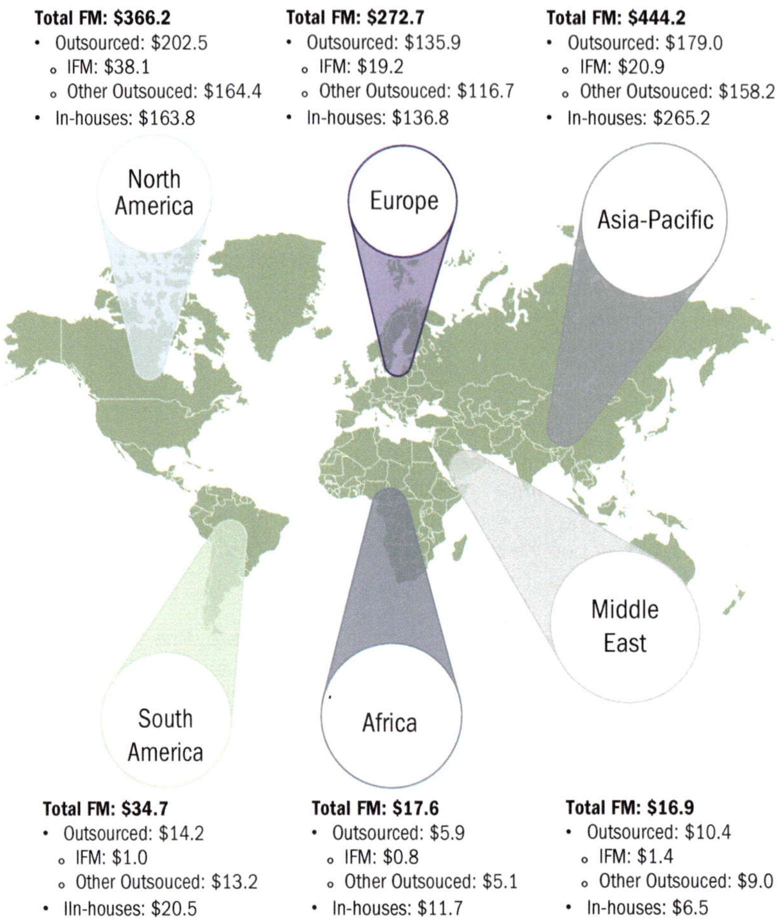

facility service providers from origin add real estate management to their package. This results in a blurring demarcation of the fields of facility management and real estate management in several countries.

Approach

In addition to the differences in the development phase or the degree of professionalisation of the discipline, differences can be identified in the approach to facility management. These differences have a cultural origin but are also related to the genesis of the facility management profession in the country concerned. In the United States of America, some areas (like the West coast, especially Silicon Valley and the Bay area) are leading the market in both hard and soft services. At the same time, in other areas, the profession is lagging behind somewhat. In the general opinion, the USA is slightly ahead of Europe when it comes to hard FM-services (the building and the building-related infrastructure). In the USA, facility management has developed as a specialisation in the field of architecture. Europe is said to be leading in FM-soft services. However, here too, the development of the profession significantly differs per country. Some examples are that in

Germany, the emphasis is on building management and technology, while in the Netherlands, soft services with a strong focus on the customer or user experiences characterise the profession. The United Kingdom is one of the frontrunners in innovative forms of outsourcing, such as Public-Private-Partnerships. In the Scandinavian countries and the Netherlands, there is still a great deal of focus on location- and time-independent working ('new ways of working') and related workplace concepts such as coworking spaces. In this context, a user-centred approach to the further development of services is common. The underlying assumption is that if a person's well-being is cared for, he or she performs well. The focus on 'healthy buildings' and the development of the WELL Building Standard also fit in with this trend.

TABLE 1.3 The focus in Facility Management in the Netherlands, from approximately 1990 to the present day

	Focal point	Characteristic
1	Building	Until 1990, building managers and janitors were mainly responsible for the accommodation. The household department took care of, amongst others, the canteen and cleaning.
2	Building installations	Companies realised that by efficiently and automatically managing their climate installations (HVAC); they could save energy and reduce costs.
3	Outsourcing	Companies increasingly focused their attention on their core business activities. The common understanding was that outsourcing services to specialised external companies would lead to cost savings.
4	Computer-Aided Facility Management	A starting point for improving efficiency is gaining insight into quantities (such as square metres of office space and the number of workplaces) and occupancy rates, to generate management information. CAFM enabled this.
5	Procurement	Because much money was involved in outsourcing services, and European tendering came obligatory for (semi)public institutions, specialised buyers became in demand.
6	Ratios	One of the ways to assess the costs incurred for services is to compare them with those of other companies. To be able to compare apples to apples, standardised ratios were developed.
7	Quality	However, costs do not exist in isolation but relate to the service level. Quality was increasingly being considered as part of the focus.
8	Ergonomics in the workplace	There was a growing awareness that working conditions (ergonomics) have an impact on absenteeism rates.
9	Experience	In addition to costs and quality, customer satisfaction became an important indicator in the assessment of the service provider.
10	Office innovation	Powered by advanced IT and stimulated by the increase in traffic congestions, employees started working from home. Private offices were replaced by interchangeable and shared workstations. This change in functionality often leads to a reduction of floor space.
11	Marketing	The image of the organisation became important, as a result of the emerging 'experience economy'. Companies aimed to create a work environment that matched their corporate branding.
12	Hospitality	At its heart, it's all about people. If employees and visitors feel well treated, they will also perform better or give an order sooner. Hospitality became a spearhead.
13	Added value	Facility processes should add value to the processes they support. Slowly but surely, the question 'What does facility management cost?' shifted to: 'What does it yield?'
14	Healthy offices	More and more attention is being paid to the effect of the physical environment on the well-being of people. The healing environment concept is permeating the office environment in the form of healthy offices, with the underlying goal of increasing productivity.

TABLE 1.3 The focus in Facility Management in the Netherlands, from approximately 1990 to the present day (continued)

	Focal point	Characteristic
15	Sustainability	Together, building and construction are responsible for 39% of all carbon emissions in the world, with operational emissions (from energy used to heat, cool and light buildings) accounting for 28%. [en](World Green Building Council, 2019). Facility management plays a pivotal role in making business operations more sustainable, both in terms of buildings and installations and in terms of reduced energy consumption and waste.
16	Smart facility management	More and more buildings are generating enormous amounts of data. Linking data within a building (smart building) or between buildings (smart city) offers many new possibilities for data-driven facility management. Internet-of-things also creates challenges, for example, in the field of (cyber) security and privacy.
17	Risk management and business continuity management	Climate change increases the risk of extreme weather and thus, the risk of damage to or disruption of the business process. In addition, companies have to operate at a time of geopolitical instability with an increased threat of attacks and new technology creates new risks for business operations. The facility manager is responsible for safety and security on-site and for taking measures to be able to continue the organisation in the event of calamities.

Table 1.3 illustrates the shift in focus in the Dutch facilities market over time - however, it is indicative and only intends to give an impression of the influence of trends and developments in the field.

Inter-nationalisation

It is expected that the facilities market will become increasingly standardised. One of the drivers of standardisation is globalisation. The internationalisation of facility management is progressing rapidly. No more than five globally operating real estate consultancy companies advise many multinationals about all their subsidiaries all over the world. These consultancy companies use a limited number of office design concepts that are very similar. This creates a worldwide 'standard' for office design. Like fashion, office concepts are rapidly adopted by others.
The facility multi-service provider ISS is active in almost fifty countries around the globe. As a multi-service provider, Sodexo is also a global player - active in 72 countries. The Compass Group should also be mentioned here, albeit as a single service provider in the foodservice industry: present in around thirty countries. More and more of these (formerly) single service suppliers (providers of one type of service: either cleaning or catering or security) are focusing on the cross-border provision of integrated facility management (IFM). Among IFM's customers are many multinationals that are active in many countries. The European tendering procedures (mandatory in public procurement), also contribute to the internationalisation of facility management.

Norm

Another factor that contributes to harmonisation is the development of standards or norms. A standard is a document that provides requirements, specifications, guidelines or characteristics to ensure that materials, products, processes and services are fit-for-purpose. A norm thus ensures (inter-)national standardisation. There are two types of standards: technical standards and management system standards.

Technical standards

Technical standards provide support for the organisation to define and administratively organise components at various levels in the same way. Examples are the ISO 41011 and ISO 41013 (that replace the EN 15221-1) on facility terms and definitions. Another example is the ISO 41012 Facility management - Guide for sourcing and drawing up agreements replacing the EN 15221-2. The purpose of this type of standards is an unambiguous definition of all aspects of the field of study, which, for example, makes benchmarking and professionalisation possible.

Management system

A management system is the set of interrelated or mutually influencing elements of an organisation to establish a policy and objectives, plus the processes to achieve those objectives. The components of a management system include the organisational structure, roles and responsibilities, planning and execution. The scope of a management system may consist of the whole organisation or sub-areas in the organisation. The system can also relate to one or more disciplines. Management system standards provide a methodology to structure and optimise the management system of the organisation. A fundamental management system standard for facility management is the ISO 41001, which was published in 2018.

ISO 41001

The ISO 41001 describes the structure of a facility management system. Before the introduction of ISO 41001, there was no management system standard for facility management. This standard applies to all types of facility organisations, as it does not distinguish between internal facility organisations and external facility providers. Moreover, the standard is independent of culture, legislation and regulations, profit, non-profit or not-for-profit, public or private. For the standard, there is only one facility management organisation and one client organisation (the demand organisation). As a result, every facility management organisation, by definition, is seen as a provider organisation and is subject to the same rules. However, the standard only applies to facility management organisations that are a separate entity (organisation, business unit or department). Small organisations, where the management or the secretary performs the facility role, are not covered by the standard.

In addition to the ISO 41001, other standards are decisive for facility management, such as:

- EN 15221-3 Facility Management - Part 3: Guidance on quality in Facility Management (European standard)
- EN 15441-4 Facility Management - Part 4: Taxonomy, Classification and Structures in Facility Management (European standard)
- EN 15221-5 Facility Management - Part 5: Guidance on Facility Management processes (European standard)
- NEN-EN 15221-6 Facility Management - Part 6: Area and Space Measurement in Facility Management (European standard)
- EN 15221-7 Facility Management - Part 7: Guidelines for Performance Benchmarking (European standard)
- ISO 22001 Food Safety, Hazard Analysis Critical Control Point (HACCP) (global standard)
- ISO 22301 Business Continuity Management (BCM) (global standard)
- ISO 26000 Guideline for corporate social responsibility (global standard)
- ISO 31000 Risk management - Guidelines (global standard)
- ISO 90001 Quality management systems - Requirements (global standard)

It goes without saying that it is still essential to adapt services to local customs, for example, concerning working hours, eating habits and manners. On the other hand, social and economic mobility has been greatly enhanced by the free movement of people and goods in Europe. Increasingly, facility professionals are dealing with people from different cultural backgrounds, both clients and employees. Facilities associations can play an essential role in developing an international approach that matches local customs.

1.5 The positioning of Facility Management

A very decisive development for the facility management field is clarifying the contribution of facility management to the success of the organisation; the so-called added value of facility management. Thinking about added value is not new in itself. As early as 1985, Porter introduced the model of the value chain. This value chain (see Figure 1.12) provides insight into the value that the primary and secondary (supporting) activities of an organisation add to the customer. In this model, facilities activities are part of the infrastructure and procurement (both secondary processes).

Value chain

FIGURE 1.12 Porter's value chain

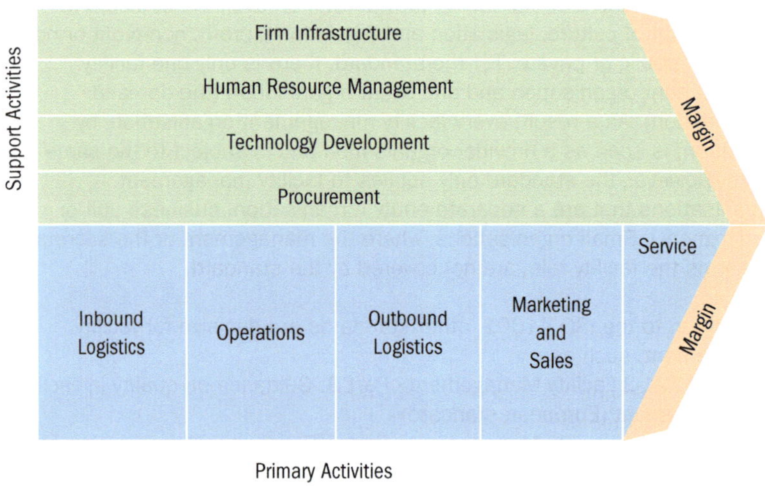

To know where and how they can create added value, facility managers need a good understanding of the primary business processes. In this, the focus of the facility manager is enabling these primary processes.

As Table 1.3 illustrates, the focus on the added value of facility management is relatively new. If facility professionals can demonstrate the added value of facility management, thinking about facility management broadens from efficiency to effectiveness. In other words; the added value of facility management used to be to save cost. However, it is now clear that facility management is a success catalyst.

Success catalyst

An example: cleaning a building costs money. The direct result of the cleaning service is a clean building. In the past, clients usually considered how much money they wanted to spend on a clean work environment. Cheap services were seen as an indicator of the success of the facilities company. If the client includes the indirect result of cleaning in this assessment, it becomes a different story. The indirect effects of cleaning are employee satisfaction, extending the lifespan of materials and the desired image of the organisation ('everything under control'). The question thus is no longer: 'Can we decrease costs?', but: 'What is the desired level of cleanliness, and how can we achieve this with the available resources?'. In other words: the cleaning quality and the positive effects of cleaning are assessed in relation to costs.

Indirect effects

Another example: the work environment costs money. These costs can be expressed in costs per workplace or annual costs per employee. If an organisation reduces the footprint of a workplace, the costs per workplace and employee will decrease. If the facility manager focuses exclusively on the accommodation costs, the introduction of an open-plan office concept is a successful operation. However, if s/he looks at the goal of the work environment, the optimal support of employees in their work, another image emerges. It may be that people are less able to concentrate because they do not have sufficient opportunities to isolate themselves. Disturbance reduces employee productivity. In that case, there is no question of a successful intervention at all. There is an excellent saying: 'Penny wise, pound foolish'. It means that someone is thrifty when it comes to small things and details but loses sight of thriftiness in the bigger picture. Insight into the relationship between savings and reduced productivity leads to a different perspective: the value-added. This insight drives the paradigm shift from a cost centre to a hospitable success enabling division.

Goal

A better understanding of the effects of facility management also changes the perception of the role of accommodation. Nowadays, accommodation is seen as a business resource instead of a necessary 'roof over your head'. Some examples:

Strategic resource

- The building, with its specific appearance characteristics, is a means of expressing the identity of the company: transparent, modern, flexible and open, or somewhat hierarchical, reliable and traditional.
- A building contributes to attracting and retaining staff. In a company with an attractive and pleasant work environment, people will be happy to come and continue to work here.
- Buildings can contribute to cultural change. For example, if a company wants more cooperation between specific departments, openness and the exchange of knowledge between employees can be stimulated through the use of space.
- The physical environment influences people's well-being, for example, through the indoor climate. Smart design can stimulate (or 'nudge') healthy behaviour, for example, the use of stairs instead of the lift by placing the elevators out of sight.

A well-known model that illustrates the added value of facility management for the organisation or client is the FM value map by Per Anker Jensen (2010; see Figure 1.13). In this model, facility management can provide added value in the areas of satisfaction, cost control, support for productivity, increasing the reliability (stability) of the organisation, increasing the adaptability (flexibility), and supporting the organisational culture. Facility activities also have an impact on the environment the organisation operates in. These may include economic impact (e.g. by purchasing from local suppliers), social impact (e.g. by creating employment), spatial impact (through a choice of location, buildings and parking facilities) and environmental impact (CO2 emissions, waste, local warming, local generation of renewable energy). This model illustrates that the added value is the *impact* of the facilities provided (the basic range and additional supply). Added value relates to the effect of facility management on core business performance, albeit mostly indirect.

FIGURE 1.13 The FM-Value Map

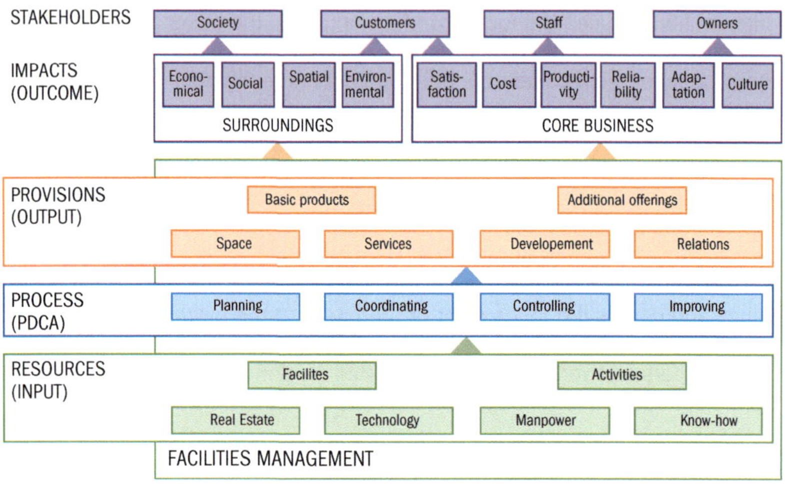

Value proposition

The added value of facility management for an organisation depends on the value proposition that the organisation (client) uses to distinguish itself from its competitors. Treacy and Wiersema (2002) identify three value propositions:
1 operational excellence;
2 product leadership;
3 customer intimacy.

An organisation will use all three strategies, but to stand out from the competition, it will want to excel in at least one of them. The dominant value proposition determines the added value of facility management for an organisatio - because for a successful implementation, alignment of the internal organisation with the value strategy is essential.

Operational excellence

The value strategy of operational excellence is about the competition on price. In this type of organisation, the emphasis is on the most efficient and effective deployment of processes, people and resources to keep costs low. Examples of organisations that use this strategy are Ikea and McDonald's. The added value of facility management, therefore mainly lies in controlling costs, increasing productivity, controlling risks and increasing flexibility. Thorough risk management reduces the risk of (costly and process disruptive) calamities. The more flexible the design of the (work) environment and service processes, the faster and cheaper they can be adapted to changing demand.

Product leadership

The value strategy of product leadership revolves around a highly distinctive (innovative) product, as at Nike and Google. Organisations that use this strategy are continually looking for improvements and innovations in products and services to be able to distinguish themselves from the competition. Creativity and innovation are cornerstones for this strategy. The added value of facility management for this type of organisation mainly lies in the area of increasing customer satisfaction and supporting innovation, productivity, culture and image. Innovative organisations are always looking for talented employees. Thus, the image of an organisation is essential to attract the right people. Attention to employee satisfaction is vital to ensure that people are happy to work for the organisation. The building, the workplace concept and the hospitality concept all convey cultural values. The productivity and creativity of people are influenced by the layout and ambience of the workplace. If healthcare organisations have product leadership as a value strategy, the healing environment is an outstanding added value of facility management.

Customer intimacy

The strategy of customer intimacy focuses on building a sustainable relationship with the customers, in which a continuous response is given to changing needs and wishes. This strategy then creates added value by offering customised and semi-customised solutions and acting as a partner in the development of products and services. Organisations must maintain the relationship with the customer to keep track of the market and to recognise trends and developments in good time. Examples of companies characterised by this strategy are Ahold Delhaize and Phillips. In any case, the added value of facility management in this type of organisation is to support the image and increase customer satisfaction and flexibility of operations. Depending on the values that are important to the organisation's customers, other values will also play a role, such as sustainability.

Customer excellence

Some organisations combine customer intimacy with operational excellence to customer excellence. This value strategy links customer focus (customer intimacy) to efficient operations (operational excellence). In particular, organisations involved in lean management apply this value strategy (see Chapter 5).

● www.facto.nl, 29 januari 2019

Best Practice Award 2018

Cleaning company Gom and Ibis Styles Haarlem City Hotel won the Best Practice Award. According to the laudatory opinion of the jury, the Ibis Styles Haarlem City Hotel and Gom create working conditions that 'set an example for the hotel industry'.
This award shows that it is possible to do things differently. At the beginning of this year, André Aaij signed the Code Responsible Market Conduct. His vision is that attention to people always pays off: 'The result is high job satisfaction and development opportunities for employees and a very clean hotel with a good occupancy rate. There is no difference between in-house

staff and hired staff: all employees regard each other as colleagues. Cleaning employees notice that they have added value and feel like an important part of the team and the process'.

1.6 Structure of this book

Coherence

The chapter classification of this book is based on a scheme that will be explained step by step. We have chosen a cone shape to define the coherence of the chapters. The basis of this cone is formed by the organisation, with the human being in top position (figure 1.14a). The mission, vision, values and brand (image) of the organisation form the context for facility management to take shape (Figure 1.14b). Several intermediate segments, or disks, are needed to achieve an optimal relationship between people and the organisation in terms of facilities. Figures 1.14a to 1.18b show a step-by-step closing of the 'gap' between humans and organisations from the bottom up and from the top down. A lateral and top view of the cone is shown at each step.

FIGURE 1.14a Lateral view

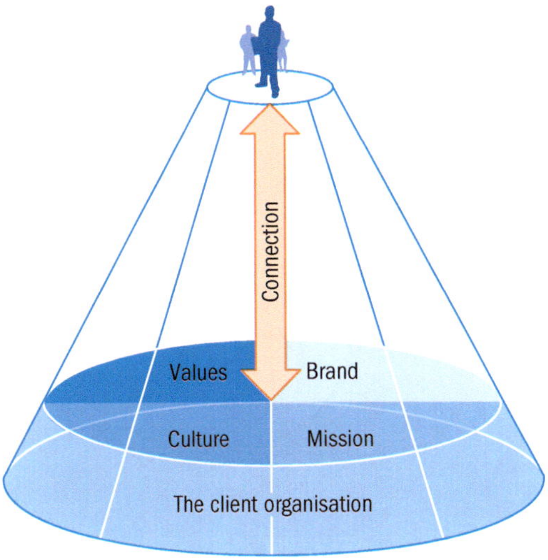

FIGURE 1.14b Top view

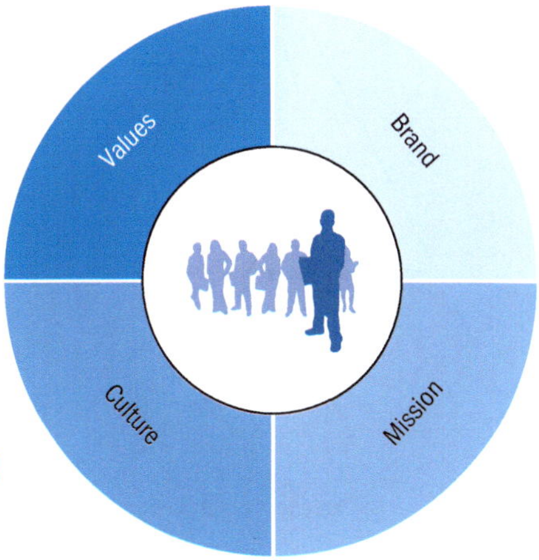

Facility management is about people in relation to the organisation to which they are affiliated. People are the focal point. The underlying assumption is that people will only make a useful contribution to the organisation if they are committed to it.
Every organisation has its distinctive characteristics, expressed through a brand, mission, culture and core values. The more people recognise themselves in these characteristics, the more loyal they will be to the organisation.
As far as the work environment is concerned, facility management deals with the relationship between users, whether they are employees, visitors or suppliers, and the organisation. A work environment is defined in the broadest sense of the word, including the (residential) care environment and the recreational environment.
Facility management is highly context-dependent. To properly match supply and demand, it is crucial that the facility manager knows the organisation well and that the facility company or supplier also has a clear profile of itself. This is what it is all about:

1 Values. Core values express what is essential for the organisation in the ethical field.
2 Brand. This is the desired image of the organisation, identity, service or product with which it distinguishes itself from competitors.
3 Mission. This is the proposition of an organisation, usually starting with: We want ...', like Apple's: 'We want to make the best computers in the world.'
4 Culture. This is the common set of standards, values and behaviours shared by the members of the organisation.

FIGURE 1.15a Lateral view

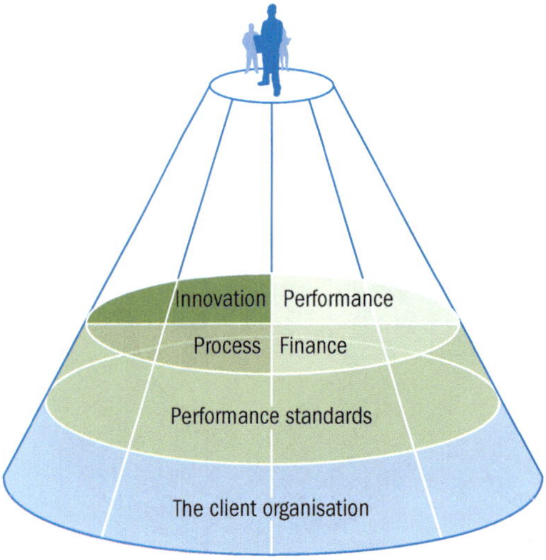

FIGURE 1.15b Top view

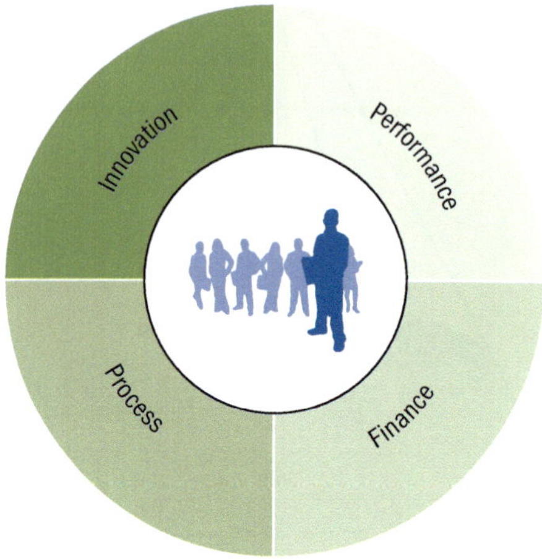

In addition to the characteristics mentioned above, every organisation has conditions that need to be met in order to continue to exist. This ring represents the four perspectives of the balanced scorecard: customer (performance), finances, processes, and learning and innovation. The balanced scorecard is a strategic planning and management system that organisations use to align the day-to-day work with their strategy. Every organisation will have to deliver a certain level of performance to stay in business. Besides, of course, the finances must be in order, and all processes must run efficiently. To remain competitive, the company must innovate.

1 Process. This concerns the joint activities for the creation of a product or service - to make a positive contribution to the organisation's success.
2 Performance. All stakeholders and departments within the organisation must deliver a specific result, to jointly guarantee business continuity.
3 Finance. Commercial companies must make a profit, while non-profit organisations (such as the government) must remain within budget. Commercial facilities providers must make a profit to be able to guarantee the continuity of their business.
4 Innovation. The enormous technological developments are forcing every company to keep a close eye on these developments and to adapt to them with innovative solutions where necessary.

FIGURE 1.16a Lateral view

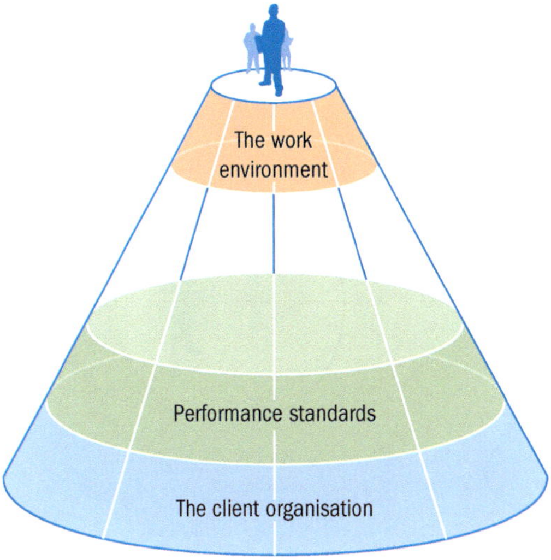

FIGURE 1.16b Top view

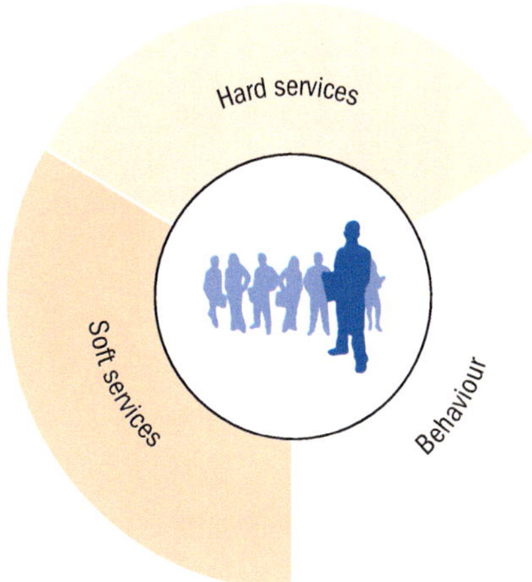

A person experiences the organisation through the work environment, among other things. There s/he perceives the behaviour of other people, the associated *hard services* and the user-related products and services, which we call *soft services*.

1 *Hard services.* These are services related to the built environment, such as new construction, renovation, refurbishment, building-maintenance, and maintenance of building installations.
2 *Soft services.* These include catering, cleaning, security and printing.
3 *Behaviour.* Behaviour determines how the work environment is experienced and vice-versa.

1

FIGURE 1.17a Lateral view

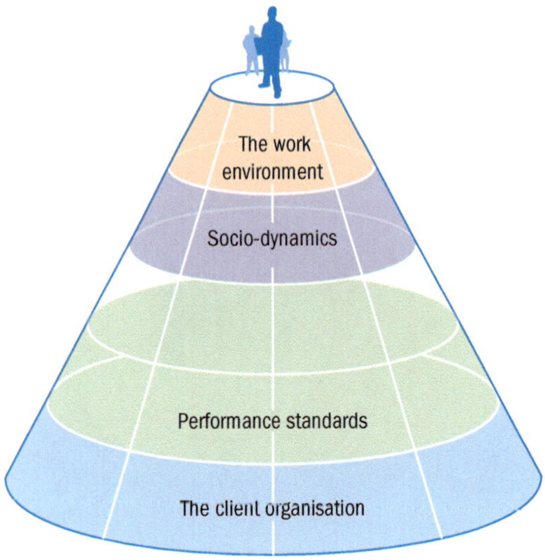

FIGURE 1.17b Top view

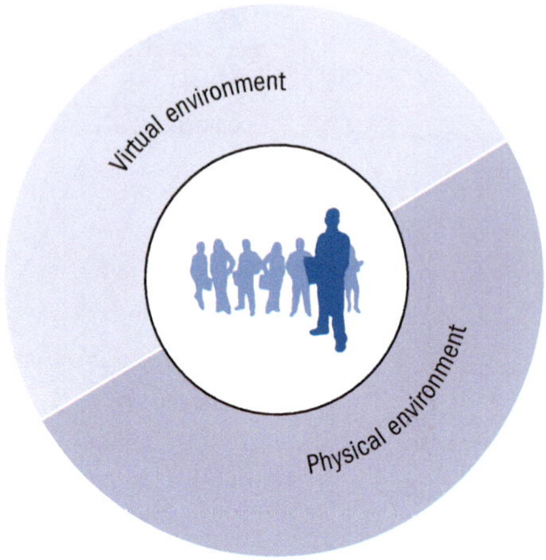

Due to far-reaching social developments, the work environment is subject to enormous changes. These changes concern the physical and virtual work environment. The latter can be anywhere if one can work with tools that are connected to the internet, such as laptops and smartphones.

1 Physical. This concerns the built, physical environments, such as the building and its layout.
2 Virtual. Virtual refers to everything that takes place outside the physical work environment and in which information technology is used (such as the internet, the cloud, videoconferencing).

FIGURE 1.18a Lateral view

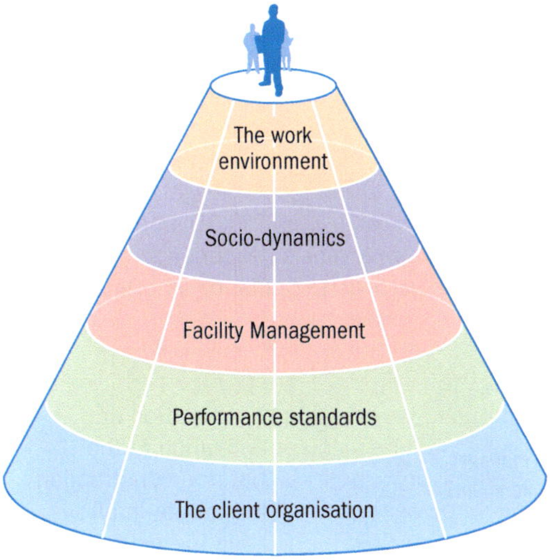

The field of facility management is concerned with adapting to the changing work environment to maintain an optimal relationship between the guests and the organisation.

FIGURE 1.18b Top view

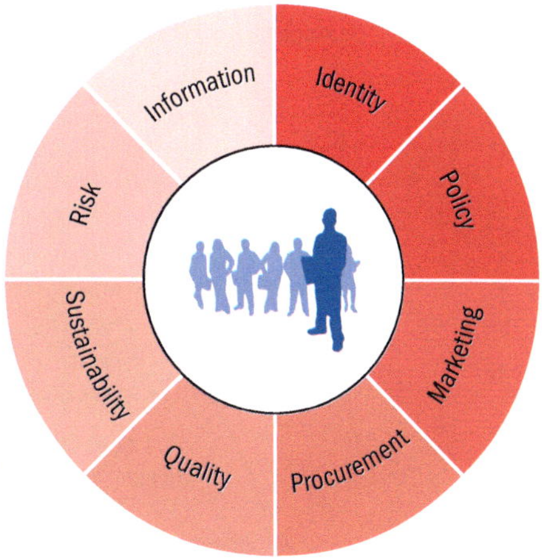

This segment of the model contains the issues that the facility manager deals with at strategical level. Identity has to do with the branding and public relations of the organisation. Information management plays a vital role in maintaining control and being able to manage. By marketing, we mainly refer to the relationship with the guest and the management of his expectations. The policy also includes account management. The other topics in the cone are self-explanatory.

Figure 1.19 summarises figures 1.14a up to and including 1.18b, and relate to how the chapters are organised. A question reflects the core of each chapter.

FIGURE 1.19 Chapter structure and reading guide

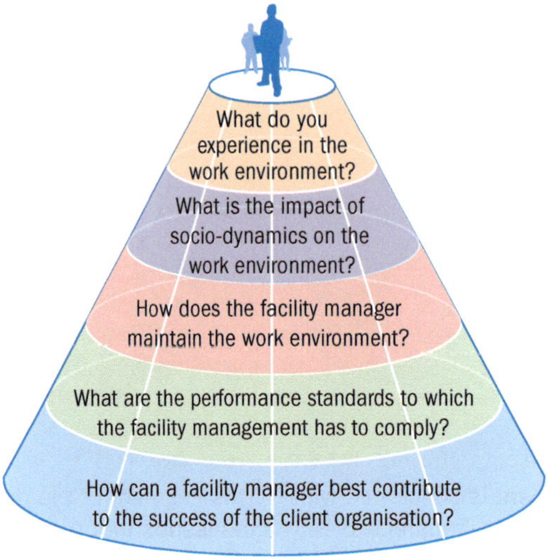

The questions in the figure above are discussed and answered in this book. We will start with the key question: what is facility management? This question is the focus of this first chapter.

Facility management is highly influenced by social developments. This is why we believe it is essential to discuss the dynamics this brings about before we consider other aspects of facility management in more detail. The focus of Chapter 2 is *the influence of socio-dynamics on work and the work environment*.

In Chapter 3 you will find an explanation of the *perception of the work environment*. Hard services, soft services, and attitude and behaviour are topics that are discussed in this chapter, noting that hospitality is becoming increasingly important.

Chapter 4 deals with *the functioning of the facilities organisation*, related to the essential question: how does facility management maintain an efficient and productive work environment? The main functions of facility management are discussed.

Because the client organisation determines the conditions for facility management, we devote a great deal of attention to this in Chapter 5 *Quantifying facility management*. A facility manager is only taken seriously by the board when s/he is in control, in other words: when s/he steers on the relevant dashboard or scorecard parameters.

Facility management is a relatively new area of expertise. The developments in the past until now are discussed in Chapter 1, but further professionalisation does not stop there. How the discipline may develop in the future is outlined in Chapter 6 *The future of facility management.*

Finally, in this data-driven era, we believe students must develop skills in research. Research agendas and current research by Dutch universities of applied sciences with the relevant professorships are therefore discussed in *Chapter 7 Facility management research*.

Young Professionals in Facility Management

Lisa Hut, Customer Success Director at Measuremen in The Netherlands
FM Graduate of the Year 2012 Hanze University of Applied Sciences, FMN Bachelor Student Award 2012; Finalist Student Poster Competition at EuroFM EFMC Conference 2013

Where do you want to be in 5 years?

Where do you want to be in 5 years? I never had my answer ready. After successfully graduating, I decided to find a job in Amsterdam. Lucky enough, my first job interview was a success. I started working for a start-up with global ambitions to improve workplace environments by collecting insights about the workplace.

I started as a project coordinator, and I loved working for all different corporate customers. The company expanded very quickly to other parts of Europe and inherently, my professional network was growing fast. Within three years, I was managing our operations in Amsterdam, London and Brussels. In 2016, my bosses trusted me in setting up shop in Australia. I became responsible for a widespread of activities such as Business Development, HR and Finance. Within a couple of months, our Sydney office was a fact.

I enjoyed every second of this journey and learned so much about different work cultures. Australians most likely do business with you if you come recommended by somebody else. Thankfully, there are a lot of network opportunities and Australians are very open to meet new people. For this reason, I chose to become a member of the FM Australia Association; it readily connected me to the right people.

With the Sydney office still operating, I moved back to our head office in Amsterdam. A new challenge awaited me: setting-up a new department to drive customer success in our meanwhile scale-up company. Life is unpredictable, so start journeys you never dreamed of starting. What will the next five years bring? Who knows!

Summary

- Facility management is the management of all services related to the work environment that are necessary to support people in their efforts to add value to the organisation. The work environment is a broad concept; it includes the office environment, but also the (residential) care and leisure environment.

- Facility management supports the primary processes of the company (in-house FM-division) or the primary processes of a client (consultants and commercial providers of facilities services).

- Organisations from various industries, with their specific primary processes, require different styles and approaches to facility management.

- In organisations, we distinguish between strategical, tactical and operational levels. The strategical level is responsible for determining the 'policy', the tactical level for the organisation of processes ('preparation') and the operational level for 'performing' activities.

- There are several ways to organise a facilities organisation: in-house sourcing, directing organisation or demand management organisation. It is not a normative classification, in the sense that a demand management organisation would be better than in-house sourcing. It is about finding a suitable service model in the context of the organisation.

- The role of the facility manager and what s/he is responsible for varies from industry to industry and from company to company. The four major roles of facility professionals are manager, consultant, specialist and entrepreneur.

- Various standards define the scope of the discipline. A standard provides for (inter)national standardisation, which avoids conceptual confusion. Two essential standards for facility management are the European standard EN 15221 and the global ISO 41000-series.

- Facility management can provide added value for organisations in the areas of increasing (customer) satisfaction, controlling costs, supporting productivity, increasing the reliability (stability) of the organisation, increasing adaptability (flexibility) and supporting the culture. FM-activities have an impact on the business environment: economic impact, social impact, spatial impact and environmental effects.

- Facility management is a relatively new field of expertise that is still being developed. How these developments take shape, varies from country to country, and from region to region.

- Internationalisation of the field of facility management:
  - There is a growing number of multinational suppliers.
  - There is an increasing number of multinational customers.
  - Facility management and real estate management are increasingly converging, stimulated by providers who offer these services in an integrated manner.
  - Real estate advisors working for multinational clients have led to more or less standardised (global) workspace concepts.
  - Standardised concepts leave room for local flavour.

- The position of the facilities organisation in the company determines the influence of the facility manager. The facility manager can add value to the organisation by thinking and acting strategically - in line with the organisation's strategic course.

- The value strategy of the organisation (client) determines the added value(s) on which facility management will focus. Treacy and Wiersema identify three value strategies: cost leadership (operational excellence), product leadership and customer intimacy. Some organisations combine operational excellence and customer intimacy in one term: customer excellence.

Questions and Assignments

1.1 Explain why it is difficult to describe the field of facility management in a few sentences. Choose three words that, in your opinion, represent the core of facility management.

1.2 In some organisations, facility management is not a supporting function but a primary function. Explain this and give an example.

1.3 Within facility management, everything is aimed at the provision of services to the customer organised in the best possible way. However, there is often confusion about whom we mean by 'the customer'. Give two reasons for this confusion.

1.4 Search the Internet for at least two organisation charts. Show where the support services (such as facility management) are positioned. And do you notice anything else? Motivate your answer.

1.5 A facility manager of an office organisation is asked the following two questions: 'What does a good cup of coffee cost' and 'What does a good cup of coffee yield'? Which of these two questions do you find most interesting? And why?

1.6 One of the main tasks of a facility professional is to create a 'fit-for-use' environment. Take the school in which you are studying or the environment in which you work as an example, and describe to what extent the facilities are geared to the wishes and needs of the users.

1.7 In day-to-day professional practice, facility managers are faced with the challenge of finding a balance between quality, costs and customer satisfaction. Explain this based on an example of your own choice.

1.8 The facility management field has become increasingly international over the years. Give two reasons for this development.

1.9 At which organisation would you like to be a facility manager: the head office of a tech company, a large municipality, a leading academic hospital; an event venue, or the internationally operating caterer Sodexo? Motivate your choice.

1.10 Suppose you are a facility manager in a hospital. Who decides whether the services you provide have added value?

NEW YORK

2
Social dynamics that shape work and work environment

In this chapter, we will answer the following questions:
- Why do facility managers focus on more than just the day-to-day operations?
- How can scenario planning help facility professionals to gain a grip on the ever-changing future?
- What changes are shaping our society?
- Which social developments influence how facility managers work?
- Which social developments influence the work environment?
- In what way do social developments influence the work of the facility professional?

◄ Through the eyes of the facility manager:
'Nice example of building transformation. A monastery was converted into a university building (BUAS, Breda), turning the former inner garden into a beautiful multifunctional roofed atrium.'

2

Expert Column

Social dynamics that shape the way we work and our work environment

As the nature of our world evolves from an industrial base, through service and into an information society, so our workplaces, work practices and communication processes are fundamentally changing in character. Such New Ways of Working bring with them new risks and challenges for organisations and their FM and Workplace Management teams.

The shift away from traditional work styles is reflected in the rise of the so-called 'gig-economy', which is characterised by high levels of self-employment and temporary work. It is expected that by 2025 more than one-third of the workforce is comprised of freelance workers. This is influenced by factors such as the influx of millennial workers, urbanisation, globalisation, mobility and new technologies. If we are to ensure that both our workplace as well as our workforce can accommodate these new trends, they must be fully understood and properly assessed so we can design and align our FM and workplace practices accordingly.

The new forms of corporate structure and organisation arising from the dynamics of internationalisation, often combined with technical advances and digitalisation, are having a direct impact on the workplace. The way we work has changed with the advancement of mobile technology, and thus the workplace strategy will increasingly include a combination of the "office"; coworking locations; public places; and working from home.
Flexible working not only engenders a greater need for worker mobility but also brings about new communication, leadership and design challenges. Future focus will be on creating a holistic workplace experience that supports the brand, culture and strategy of the organisation. This requires a shift in how we think about our physical environments as we move away from the traditional focus on cost towards a return on investment on people.

Peter Ankerstjerne
Global Facility Management & Employee Services Lead at JLL and Vice-Chair of IFMA's Board of Directors

2.1 Introduction

In the past, the emphasis in the work environment was on providing accommodation and facilities services. When the Industrial Revolution began - around 1850 - office buildings were constructed in which employees worked between 9 a.m. and 5 p.m. every working day. From then on, buildings became a synonym for the work environment.

Nowadays, this apparent connection between work and building is getting weaker. An example: decades ago, every account holder would go to the bank for money. Banks had branch offices in every town, and this was the place where bank employees would work. Gradually this has changed. Routine work was automated. For service purposes, banks started offering consultations at the home of their clients, also in evening hours. Nowadays, most people arrange their banking affairs via the Internet and use video calls to consult their advisors. It is sporadic for customers and bank employees to meet in person. More and more, banks are closing branch offices.

This development is exemplary for the trend towards online and remote consumption. Increasingly, even traditionally complex services can be ordered and delivered remotely (whether or not by smart devices). Just think of the purchase of a car. One doesn't have to go to the dealer anymore, because even those can be ordered relatively quickly via the importer's website with the help of the car configurator. In the healthcare industry, medical specialists are already experimenting with remote diagnostics and the use of health data from apps. Even for drawing up an official act as a will or a cohabitation contract, you no longer need to go to a notary. In some department store chains, you can arrange this online (although for signing you still need to go to an affiliated notary).

Change

Every area of our lives seems to be changing continuously and at an accelerated pace. A well-known expression that reflects the dynamics of our time is: change is the only constant. As a result of this, the work environment is also strongly subject to change, which is a real challenge for facility professionals. Accommodation and service concepts need to adapt smoothly to changes in use or demand. Therefore, facility professionals must have knowledge and understanding of external developments that shape our future. In this chapter, we discuss the impact of socio-dynamics on work and the work environment and its consequences for facility professionals (see figure 2.1).

Virtual work environment

Mixed reality

In addition to the physical work environment, the virtual work environment is becoming increasingly crucial to the facility manager. More and more, the distinction between physical and virtual will fade, resulting in a so-called mixed reality. An example of this is the HoloLens that projects holograms into the physical world. In addition to the physical environment, the facility manager has to immerse himself in the virtual environment, i.e. in all activities that take place outside the office, hospital, university building and the like.

FIGURE 2.1 The focus of this chapter

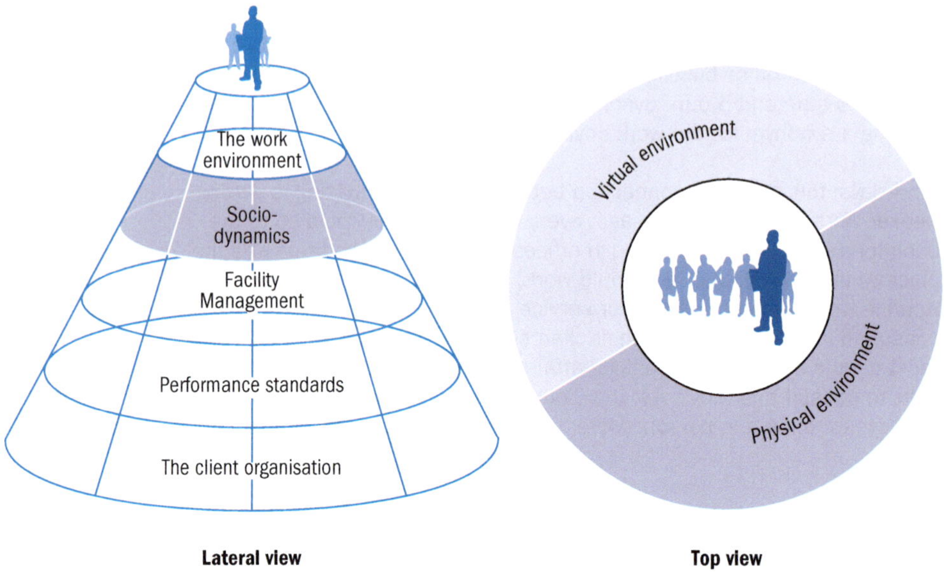

2.2 The importance of scenarios

Facility managers can no longer afford to limit their focus on day-to-day business operations. A facility manager who only deals with 'running the shop' loses sight of developments that will affect him or her in future.

FIGURE 2.2 The difference between the assumed course of events and the actual development

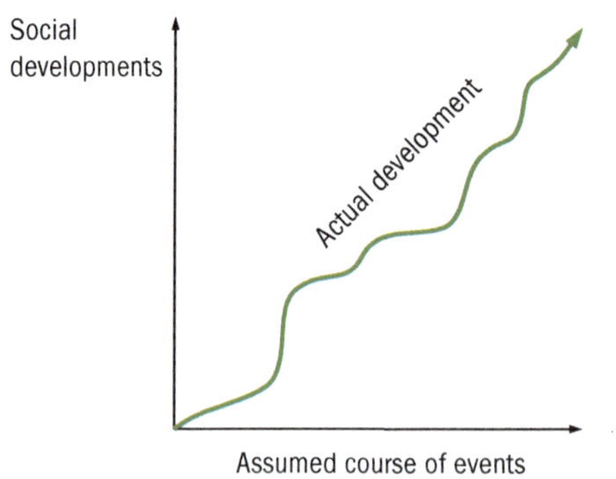

'Running the shop' means only managing and coordinating the facilities department based on agreements made with the organisation about service provision. In itself, there is little to be said against this. Still, at a time when so many social changes are taking place, it is necessary to monitor the market and social developments closely. Only then will you be able to adapt to changing circumstances pro- actively.

Figure 2.2 shows 'running the shop' on the horizontal axis and future developments on the vertical axis. Somewhere between these two, the actual events take place.

Future developments

2.2.1 Predicting the future

By now, you might be wondering if you need a crystal bal, because how else can one predict the future? Still, there are current developments (trends) that make reliable predictions possible. Viable, alternative and not improbable futures can be distinguished based on topical issues with the help of scenarios. The chance that one of these scenarios will become a reality is not unlikely if these are based on a thorough analysis of relevant developments.

Scenarios

Many decisions taken by the facility manager have a long-term effect. The decision to move, for example, and thus to lease or acquire a new building, has an impact that can last ten to twenty years. Service contracts can also have an extended validity period: three to five years. The longer the period, the higher the chance that changes in society or the organisation will make the conditions in the contract or arrangement redundant or invalid. The probability for this to happen is increasing as we live in a time of exponential changes. Exponential indicates that we are not only dealing with ever-increasing changes but that these changes come in more rapid succession. Like an inflatable balloon expands; first slowly but with increasing speed, until it transforms into a real hot-air balloon (see figure 2.3). For example, the first industrial revolution began around 1600 and lasted about 150 years. The second industrial revolution lasted only 100 years and the third only about 60 years. Now we are in the middle of the fourth industrial revolution, and there is already talk of what the fifth industrial revolution will bring us.

Long-term effect

Exponential changes

2

FIGURE 2.3 Exponential times

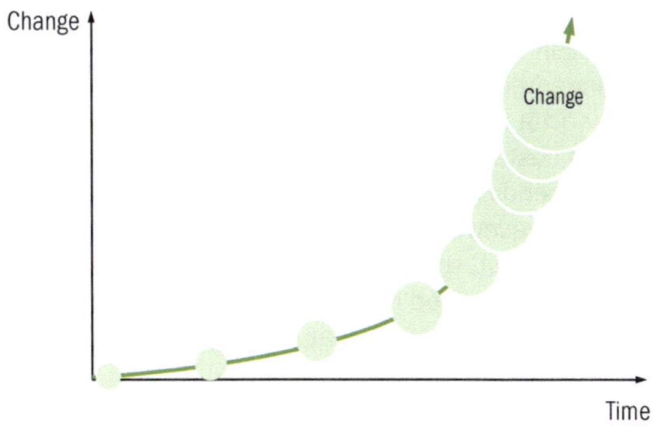

2.2.2 Socio-dynamics

The ever-faster expansion and growth depicted in Figure 2.3, indicate dynamics that also influence the work of the facility manager. Three generally observable (social) trends are increasing complexity of social systems and relationships; an increasing speed and availability of information; and a world that is evolving as a network society.

Dynamics

Examples of dynamics

Below, we will sum up some more examples:

- For younger people a world without the Internet is unimaginable, but the Internet was introduced only a few decades ago and yet it is an integral part of our everyday life. According to Internet World Stats, by mid-2019 almost 57% of the world's population will have access to the Internet. The Internet has radically changed the way and the speed with which we communicate. For example, in the Netherlands, the number of mail items sent is decreasing by an average of 10% per year. For business purposes, too, most communication is done digitally. For facility management, this means that the mailroom often digitises the few physical mail items that are still received by organisations instead of being delivered in postal rounds. Another effect is the lessened need for physical storage space. Nowadays, a locker is sufficient to temporarily store personal belongings.
- Nor can we do without Google, it seems, even though it was only established in 1997. Who would buy an encyclopaedia from a bookshop? Knowledge can be accessed quickly and everywhere. People are no longer tied to a fixed place to have access to the information.
- Social media have drastically changed our communication and will continue to do so. Networking has taken on a completely different meaning. For example, social media have had a great deal of influence on the political upheavals, among other things because opponents of regimes can connect and organise themselves. And in elections, social media appear to have a significant influence on voting behaviour. Just think of the scandal about the use of data from Facebook profiles by Cambridge Analytica (which in the meantime has been abolished) for the election campaign of President Trump. Without social media, these drastic changes would probably not have taken place at all. Companies also have to deal with the power of social media. They have become more vulnerable to negative publicity because messages can be shared very quickly. Customer service employees address this risk by actively monitoring reports on social media and by proactively reaching out to dissatisfied customers. They also establish active contact with potential customers: as soon as you visit a company's website, you are asked if you want to chat with an employee. Finally, social media have enriched the communication channels of companies with customers. You can now even take out or cancel insurance policies by sending a WhatsApp message.
- The amount of information we receive is growing enormously, but it also disappears very quickly. In the past, you had to have instant knowledge to be able to perform, which was learned by heart through frequent repetition. Nowadays it is much more critical that you have the skills to find information quickly and that you can assess the quality (or truthfulness) of it. The broad accessibility of information thus affects professional roles. For example, that of the advisor: where previously knowledge was the 'product', advisors now often offer guidance in (optimisation of) complex processes. Think of tendering procedures, supervising reorganisations and project management.
- In 2019, the British Bank Standard Chartered compiled a top 10 of the largest economies in 2030. Seven of these are now emerging markets. According to their forecast, China and India will become the largest world economies, ahead of the United States. If we are to believe this forecast,

'The pace of change has never been this fast, yet it will never be this slow again.'

— Canadian Prime Minister Justin Trudeau in his speech at Davos, February 2018.

the global market will be shaken considerably in the next ten years. As we describe in section 2.4, the economic climate has a direct impact on the goals and budget of facility management. Of course, this forecast was before COVID-19 pandemic. At this point in time these market developments are all but certain.

- Until the 1990s, we only had fixed telephony at our disposal. Nowadays, most people above the age of 10 possess a smartphone. While smartphone hardware is minimising in size, its capabilities and applications are expanding rapidly. In section 2.6, we will examine the influence of technological developments on facility management in more detail.

It is clear that organisations (and professionals) continuously need to adapt to changing circumstances. As has already been described, scenarios that outline possible futures are useful in this respect. They can function as a framework for thinking about future policy and can help to develop a vision and policy objectives. Scenarios can also be used to predict whether policy or strategy will have the desired effect. However, we agree with Prime Minister Trudeau: the future remains uncertain and unpredictable.

Scenario planning

Scenario planning

It takes four steps to arrive at useful scenarios:

1 The first step is to map out the current trends and developments. The DESTEP method is a useful tool to link up trends. This method describes, in a structural way, trends in the demographic, economic, social, technical, ecological and political-legal fields. For organisations that are directly influenced by globalisation, it may be wise to extend the DESTEP with the I for Internationalisation to DESTEPI. Several trends that influence society at large are described for each DESTEP factor from section 2.3 onwards. Some effects of globalisation are discussed under the social trends.

2 The second step is to identify the main driving forces. The question is which of the trends in today's society will determine the world view in the future. Driving forces are megatrends. They determine a large proportion of the other trends and can have a major impact on society. As the probability of each force plus the impact varies enormously, the course of each driving force is difficult to predict.

3 The third step is to create a scenario matrix and determine four perceptions of a future society. From the driving forces (step 2), the two most important ones are chosen for the axis. Each axis represents a progression from minimal to maximal change or an opposite development, for example, small-scale versus globalisation, increasing individualisation versus increasing social cohesion or recession versus economic growth. The combination of the axis results in a matrix with four quadrants. Each quadrant represents a future scenario. What, for example, would society look like if there was an economic upturn in

combination with far-reaching globalisation or a small-scale local economy? Or a combination of recession with globalisation or local economic development?

4 The final step is to interpret the scenarios in a business context. This step helps to determine what a facility professional can do to be prepared for these scenarios. For example, by making an impact analysis that assesses the probability and impact.

Social developments

You will probably only start developing scenarios when taking on strategic facility management. For now, it is essential to be alert to social developments that affect the work of the facility manager and to recognise them. In the following sections, we describe the DESTEP factors and give examples of the impact that these developments may have on facility management. We end this chapter with a detailed example of the steps from the scenario planning process in section 2.9.

2.3 Demographic factors

Ageing

Ageing society is one of the most critical demographic developments, which is not limited to Western countries. Japan, for example, also has to cope with an ageing population. The United Nations (UN) predicts that developing countries will also have to deal with an ageing population in the long run. The EU will go from having 3.3 working-age people for every person aged over 65 years to only two working-age persons in 2070. In the USA, the effect is even stronger. According to the US Census Bureau, in 2035, the elderly population will outnumber children for the first time in the country's history (www.census.gov). At present, one in six Dutch people are over 65 years of age; by 2050, this will be the equivalent of one in four. The population structure has a direct effect on the composition of the labour force, and thus also on the socio-demographic characteristics of both the company's staff and the clients of facility service providers. Many factors influence the ageing society trend.

In general, an increasing level of prosperity is related to a reduction in the number of births per thousand inhabitants. A declining birth-rate leads to a shifting demographic balance. How come the world population still grows, you might wonder?
First and foremost, because the death-rate is declining more steeply than the birth-rate. The average life expectancy in Europe has risen to over 78 years for men and over 83 years for women. The average life expectancy in the USA is over 78 years for men and women combined.

Migration balance

Another factor resulting in local variations is related to migration. In most western countries, population growth is the result of a positive migration balance (more people enter than leave a country). More (young) immigrants also mean that in time more children are born. However, the migration balance is a very uncertain factor. The Netherlands Environmental Assessment Agency (Planbureau voor de Leefomgeving), for example, uses two scenarios in its' Future Study on Prosperity and Environment for the Netherlands in 2030 and 2050'. In the Scenario High (which assumes a high economic growth of 2% per year and a high population growth) the population will continue to grow in the coming decades. By 2050, the population is expected to increase by about 2 million compared to 2018 due to a positive migration balance. In Scenario Low (which assumes moderate economic growth of 1% per annum and limited demographic growth), the migration balance and natural growth will be lower. This scenario foresees a population decline after 2030.

Furthermore, urbanisation has an even stronger local effect in demographics. Urbanisation is the proportion of the total national population living in areas classed as urban. Throughout history, cities have attracted people from rural areas; the so-called urbanisation. These days more than half of the world population lives in urban areas - although the level of urbanisation varies across regions. Latin America, the Caribbean and North America, for example, have urbanisation levels around 80%. In Europe, 72% of the people live in cities, towns and suburbs. Asia and Africa are expected to reach a 50% urbanisation-level in 2020, and 2035 respectively (www.ec.europe.eu). China already reached an urbanisation rate of over 60% in 2019 (worldometers.com). Urban structures vary across the world. A less concentrated urban structure characterises Europe compared to, for instance, China or the USA. The worlds' largest cities are mostly found in Asia, with Shanghai (over 24 million inhabitants), Beijing, and Karachi (both over 18 million) in top-3 positions in 2018 (worldatlas.com). In 2019, only 7% of the EU population lived in cities of over 5 million inhabitants, compared to 18% in China and 25% in the USA.

Urbanisation

Most cities aim for growth, under the assumption that this leads to economic activity and economic activity leads to wealth. Cities compete with each other by a thematic branding of their identity. For example, New York – 'the city that never sleeps'; Rotterdam - 'make it happen'; Wageningen – 'city of life sciences'; and Groningen - 'nothing beats Groningen'. These themes give rise to specific wishes and needs, which facility management must take into account. Hago, one of the largest cleaning companies in the Netherlands, published a white paper on the future of cleaning. They state that this specialisation of cities can lead to a further increase in scale if you arrange FM per town. Companies could unite and jointly purchase services, such as park management or security, and make use of shared restorative and workplace facilities on publicly accessible building plinths. Area-oriented facility concepts can contribute to the desired branding of the identity (see also Chapter 6).

A side-effect of urbanisation is a reinforced decline in population in rural areas. Areas that suffer from this are called urban decline areas. The government wants to keep areas with less growth habitable, for example, by maintaining facilities such as shops, library, music school, theatre and social meeting places. Facility management can play a vital role in maintaining the quality of life in urban decline areas. For example by profitably exploiting multifunctional community centres, such as a Kulturhus. Kulturhus is a Scandinavian concept and stands for a versatile community centre in which participants have a shared vision about the role of the facility in the village or the district. Joint management and programming are the cornerstones. Within a Kulturhus resourceful connections are made between participants and local clubs (see website Kulturhus Sølvberget). Maintaining the facilities in smaller towns is part of the mission of a Kulturhus.

Urban decline areas

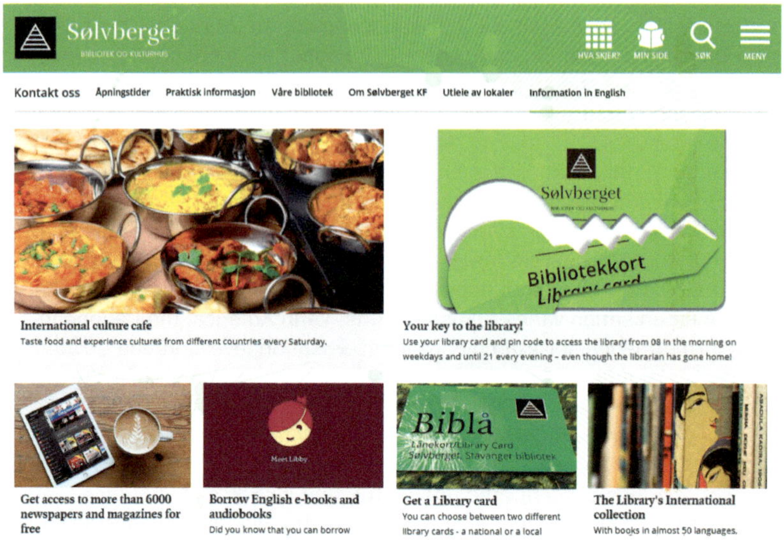

Website Kulturhus Sølvberget, Norway

A careful mix of services will create a welcoming meeting place for residents that can be exploited profitably by a multitude of tenants and users. A facility manager and his team ensure an attractive hospitality-concept, a good mix of user functions by attracting the right tenants, maintaining the quality of facilities, and therefore satisfied stakeholders. Another example of a functional mix that contributes to liveability is the provision of student housing in (partially vacant) nursing homes. Students can rent a room for reduced rent, in return for helping the care organisation with activities for older residents.

Population structure

And there are more effects of the changing population structure (the distribution of the population among age groups). For example, the decline in the number of young people leads to decreasing demand for services for this age group. In the Netherlands, nurseries and primary schools are already experiencing these effects. More and more schools are closing or merging because they have too few pupils to remain independent. After primary schools, secondary education and eventually, middle and higher (vocational) education will be affected. When schools merge or close, empty school buildings are left behind. Redevelopment of school buildings is an issue that facility professionals in this industry are already working on.

Vitality

As a result of medical science and better access to health care, people are getting older and older; hence, the introduction of higher retirement age. Given the vitality and thus the longer employability of employees, this seems the logical measure, although this also makes demands on the work environment and the organisation of work. Think of acoustics in open workplaces in combination with a decreasing hearing function among employees aged over 40 years. Conducting a conversation in an open-plan office space, therefore, requires more energy and reduces concentration. The Center for People and Buildings (CfPB) in Delft (The Netherlands) studies how the work environment best supports the labour productivity and satisfaction of older employees. One of the findings is that providing a suitable work environment is an essential factor for maintaining labour productivity and job satisfaction throughout a person's career. A comfortable work environment, in which you can enjoy working together, contributes to work motivation, job satisfaction and the vitality of people. Therefore, HRM and facility management are to join forces in promoting the health of employees, users and customers (fit for life).

Finally, the population structure (in combination with economic growth) determines to what extent the labour market faces a shortage of skilled staff. Organisations have to compete with attractive employment practices. They, therefore, pay more attention to employment conditions related to the work environment, facility facilities and flexible working arrangements. This is the so-called (war for talent). The population structure also influences the number of generations that together form the workforce. The more generations in the workplace, the greater the diversity in customer needs to which facility managers must offer a suitable solution (see section 2.5).

Diversity

2.4 Economic factors

Economic growth

Economic developments alternate between periods of growth (boom) and periods of contraction (bust). Due to demographic developments (see section 2.3), future economic growth in Europe is expected to lag behind developments over the past few decades. The slowing population growth and the ageing of the population are the main reasons for an economic slowdown. According to Eurostat (2019), labour force participation rates will slightly increase to approximately 80% in 2070 - mainly due to increased participation of women and older workers due to an increase in the retirement age. However, this is not nearly enough to compensate for the effect of the ageing population. The total supply of labour in the EU is expected to fall by almost 10% between 2016 and 2070. Robotization may compensate for some of this decline. However, none of these scenarios takes a pandemic like covid-19 into account. We will face its enormous impact on the economy for years to come.

The facility manager always feels the effects of the economic climate. In times of crisis, the emphasis is on cost savings and efficiency improvements. Reducing overheads, increasing flexibility, shortening the supply chain and outsourcing are essential themes in economic tight times. If the economy picks up and the government and companies have more money to spend, there will be room for other values. These values include improving customer satisfaction, supporting the organisational culture, improving the image, and improving sustainability. It is the challenge for facility professionals, especially in times of economic downturn, to make clear what the added value of facility management is, other than cost control. For example, supporting the productivity of employees, guaranteeing the continuity of business operations through sound risk management, or increasing the flexibility with which accommodation and services can move with the growth or decline of the client's activities. Social dynamics and increased uncertainty have an impact on the duration of contracts. For example, previously, long-term leases of five years or more were customary for leasing accommodation. Nowadays contracts with a term of two years or even less are in high demand. The average duration of contracts for the purchase of facilities services is also decreasing, and it is common practice to include clauses that enable interim adjustments to the contract agreements.

Knowledge economy

2.4.1 The knowledge economy

Several countries in Europe have a so-called knowledge economy. They cannot compete on low wages, but they can compete with smart technology and new solutions. In these economies, labour is seen as an essential investment that can lead to innovation and productivity growth, more than a cost factor.

Knowledge is a means of production for companies, but it is in the minds of and thus owned by- employees. To support the sharing of knowledge and innovation processes, the facility professional must have insight into these somewhat elusive 'knowledge production processes'. The facility manager must thus have an understanding of how people work together and how they arrive at knowledge sharing. Only then can he offer an environment and service concept that provides the right support. An example: research shows that a workplace concept in which people are completely isolated is counterproductive to knowledge sharing. But before you think it's a good idea to have all the knowledge workers work in an open-plan office, you have to understand the following: knowledge development takes place in phases, where sharing with others (interaction) is alternated with individual reflection and internalisation (working alone). For the latter, concentration is a prerequisite. And what supports concentration best? Visual and acoustic privacy. Or rather: the possibility for employees to isolate themselves, even if only intermittently.

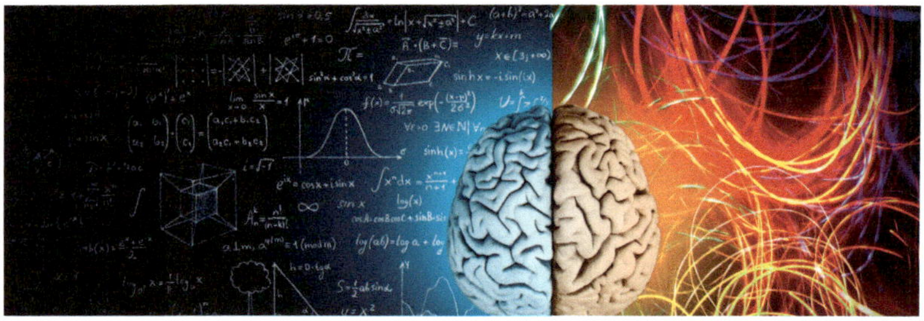

Sharing-economy

2.4.2 The sharing economy

A relatively new phenomenon is the sharing economy, in which immediate availability (use), rather than possession (ownership), is the new standard. This trend is a result of the unpredictability of the future, as property (especially of buildings, because property represent a high value) can be a financial risk in a period of economic decline.

Buildings as a service (BAAS)

In urban areas in particular, sharing of possessions is on the rise. A clear example is buildings as a service (BAAS). In this, the provision of office space is a service, instead of a product. Commercial real estate companies offer carefree, flexible, and full-service office accommodation solutions to companies. For short-term use of work- and meeting spaces, international parties such as Regus, Knotel, WeWork and Hana offer workplace and meeting facilities for a fixed (monthly) fee in short-term rental contracts. A third phenomenon is coworking. Coworking involves a diverse group of people who don't necessarily work for the same company or on the same project, but that work alongside each other, sharing workspace and resources.

Coworking spaces are offered by countless providers of coworking spaces, offering a total workspace concept in a business-to-consumer market. The shift from ownership to use has significant consequences for many areas of expertise. Hierarchy and property are connected. As soon as the importance of ownership of property decreases, the need for a traditional hierarchical structure reduces and power relationships shift. In the new economy, social capital has become a deciding factor. Cooperation within the supply chain and with all stakeholders is becoming increasingly important in this respect (see also section 2.4.3).

Coworking spaces

● D&B The Facility Group BV

What can facilities learn from the sharing economy?

1 Consider a multi-tenant environment. Proximity makes it easier to share knowledge and experience with like-minded companies.
2 Share your facilities such as company restaurant, reception and meeting centre. Not only does this enable you to save space, but it also means that you spend less time in control. You can use the money and time you can save to optimise the service level or the workplace concept.
3 Use technology to balance supply and demand. With the help of internet platforms, the match between supply and demand will improve. By implementing a direct-feedback system in your facility services, you can monitor customer satisfaction at all times and receive continuous input about the wishes and expectations of your guests.

2.4.3 The networked economy

In a networked economy, information technology is the dominant driving force, providing direct access to an abundance of knowledge and information. The dynamics of our society force organisations to use different combinations of expertise. Economic growth arises from the quality of the connections that an organisation can establish with its partners because these contribute to the agility and innovative power of organisations. Information technology connects companies and individuals all over the world. These connections create new economic structures: the so-called network organisations. A characteristic of a network organisation is that it can cooperate with like-minded companies. It takes a different way of thinking, aimed at exploiting opportunities for realising added value for all parties involved. An example of this is co-creation with suppliers and customers in the development of new products and services.

Network organisations

The teams or parts of a network organisation work - more or less - autonomously. They are not so much managed but facilitated. Increasingly, businesses rely on independent workers for projects outside core activities: the so-called 'gig economy'. The gig economy refers to a free market system

Gig economy

in which temporary positions are common and organisations contract with independent workers for short-term engagements. The term 'gig' is a slang word meaning 'a job for a specified period of time'. It is typically used in referring to musicians. Examples of gig employees in the workforce could include freelancers, independent contractors, project-based workers and temporary or part-time hires.
Descriptions of network organisations and networks of individuals (communities) can be found in Chapter 6. Here we also describe the role of facility management in this type of organisation.

2

2.4.4 Mobility and accessibility

Mobility issue

A social subject that is closely related to economic development is the mobility issue. As a result of urbanisation, cities across the world are facing increasing traffic volumes. Unfortunately, without adequate infrastructural facilities to accommodate these. It is not only passenger transport that is rising; freight transport is also growing. International transport in particular benefits from the growth of (global) trade. As a result, the volume of traffic jams (length x duration of the traffic jam) is increasing, especially around large cities. Congestion is thus inevitable. Long traffic jams lead to a reduction in the number of working hours resulting in considerable losses to the economy.

● www.tomtom.com

According to research by TomTom (manufacturer of GPS mobility solutions), India has the dubious honour of having four cities in the top-10 world rank of congested cities. Moscow is Europe's number-1 capital when it comes to traffic congestion; its runners-up are Istanbul, Kyiv, Bucharest, and Dublin. In North America, Mexico City leads the ranks, followed by Los Angeles, Vancouver, New York and San Francisco.

RANK BY FILTER	WORLD RANK	CITY	COUNTRY	CONGESTION LEVEL	
1	1	**Bengaluru**	India	**71%**	
2	2	**Manila**	Philippines	**71%**	
3	3	**Bogota**	Colombia	**68%**	↑ 5%
4	4	**Mumbai**	India	**65%**	0%
5	5	**Pune**	India	**59%**	
6	6	**Moscow region (oblast)**	Russia	**59%**	↑ 3%
7	7	**Lima**	Peru	**57%**	↓ 1%
8	8	**New Delhi**	India	**56%**	↓ 2%
9	9	**Istanbul**	Turkey	**55%**	↑ 2%
10	10	**Jakarta**	Indonesia	**53%**	0%
11	11	**Bangkok**	Thailand	**53%**	0%
12	12	**Kyiv**	Ukraine	**53%**	↑ 7%
13	13	**Mexico City**	Mexico	**52%**	0%

Top-10 cities suffering from traffic congestion

Longitudinal trend analysis shows that the time spent in traffic jams has increased considerably over the years. It is expected that this trend will continue. In 2019, on average, traffic congestions took up 29% extra travel. This is no longer limited to rush hours but throughout the day. And not only in mega cities only but also in small villages.

You may think public transport is the solution. However, in many places, public transport suffers from a lack of capacity during rush hour, despite a sharp increase in the number of trains per hour and the use of longer trains. Some take the measure to increase ticket prices for rush-hour journeys (instead of discounts outside rush hour) to encourage people to travel at other times. In the Netherlands, the Dutch Railways and major educational institutions are exploring the possibility of shifting the start time of schools until after the morning rush hour to relief traffic. No need to say that this will have consequences for the facility services at these educational institutions. Starting later means continuing for longer, for example, by scheduling some of the lessons in the evening. As a result, occupation and use of spaces will shift. As a consequence, opening hours of the reception and restaurant facilities will have to change, menus will have to include dinner, the closing round of security will take place at different times, and cleaning may be moved from the evening to the early morning.

In general, reduced accessibility is a problem for employers. Employees who regularly travel for work spend more time commuting, which is at the expense of productivity. However, increasing travel time is also a limiting factor when it comes to recruiting new staff. On average, people are willing to commute for up to one hour one-way. The more delay people experience due to traffic jams, the shorter the distance they can cover in one hour. Traffic congestions lead to smaller geographical recruitment areas for companies. To compensate for this, working from home and more flexibility in working hours are increasingly accepted. This, in turn, has consequences for the facility manager: for example, choice of location, opening hours, highly variable occupation of workstations, and the mobility management itself. Fleet management (lease cars) and parking facilities (parking spaces, bicycle sheds and recharging facilities) are among the responsibilities of the facility manager. Because employees generally prefer the fastest option to travel, the means of transport (or a combination of these) may differ per journey. Employers respond to this by granting employees a mobility budget. This budget can be spent as the individual employee see fit. Alternatively, employers issue a mobility card with which the (lease) car can be parked on a P&R site, followed by travel by train, hired a public transport bicycle or a public transport taxi. By providing insight into the social costs of car use, the facility manager can advise the HRM department on a greener mobility policies.

Recruitment areas

Mobility policy

2.4.5 The circular economy

Another factor that will influence our economy for the next decades to come has to do with the Climate Treaty. To limit global warming to an average of 2 degrees Celsius, we need to reduce greenhouse gas emissions. There are binding European agreements to reduce greenhouse gas emissions by 40 per cent by 2030 compared to 1990. This is an enormous reduction that can only be achieved by radical measures. Many European countries work towards a transition from a linear to a circular economy (see section 2.5). Research by the European Environmental Agency (EEA) indicates that in 2019, 21 out of 32 responding EEA member countries already support circular economy initiatives.

Climate treaty

A circular economy is an economic system based on minimising the consumption of virgin materials by reusing products, components and high-quality raw materials. Companies must contribute to this by reducing their ecological footprint and taking responsibility for waste streams. In an ideal and ultimate circular economy, waste no longer exists at all (cradle-to-cradle principle). Countries use regulation and market-based instruments mainly for recycling, energy recovery and waste management. At the same time, eco-design, consumption and reuse are targeted with softer policy instruments such as information campaigns and labels.

Ecological footprint

In a circular economy, supplying companies retain ownership (of lamps or floor finishes, for example). Customers no longer pay for the product, but its use (pay for performance). The manufacturer is responsible for the functioning of the product and benefits from the longest possible trouble-free service life, easy maintenance and the ability to replace parts. Prolonged ownership for manufacturers leads to different design choices that have a positive influence on the lifespan of products. Industrial examples of pay for performance schemes include lighting (pay per lux); tires (pay per kilometre); solar power purchasing agreement (pay per kilowatt-hour); office equipment (fee per page). This will change the business model of facility management suppliers and will also affect facility procurement.

The facility manager plays a vital role in limiting CO2 emissions because a substantial part of the CO2 emissions is related to user behaviour and thus to the energy efficiency of the building and its installations. The facility manager also has a pivotal role in the field of waste management. After all, s/he establishes contracts with waste processing parties and has a decisive say in whether or not to separate waste flows. Facility managers can use their buying power to enforce recycling.

- www.ceguide.org

PwC's uniform take-back program

PricewaterhouseCoopers (PwC) has six different suppliers in the United Kingdom who work at their various locations, including security, hospitality, catering and cleaning. PwC provides uniforms to suppliers working on site, to present a professional image. An estimated 60% of all textiles in the UK currently go to landfill - the UK reuses only 9% of the 39.2 million items of corporate wear. At PwC, under the old system, individuals determined the fate of corporate uniforms at end-of-life, typically municipal waste systems. The textile waste is estimated to be around 3,000 items or 1 ton per year. Some challenges PwC faced on thinking about changing the system: PwC logos must be removed when disposed of at end-of-life (for security reasons), and multiple suppliers need to be involved in a program that involves relatively small volumes.

Solution
PwC identified a specialist textile recycling company which sorts uniforms so they can be retained at their highest value:

- some will be sold on to secondhand markets, some turned into industrial rags and some shredded for use as automotive insulation.
- Buttons are also removed and sold back into the fashion industry, where they can be used in new manufacturing.
- All corporate logos are removed and destroyed for security and brand protection.
- PwC engaged their suppliers in a take-back program so that items are returned when new ones are distributed. Suppliers are also involved in setting up systems to measure the volume of waste textiles and estimate their end destination.
- PwC also provided instructions to supplier staff on how to extend the life of their uniforms through proper care.
- Program costs are only 5-15 pence per person-year, which covers the cost of shipping uniforms to the specialist recycling provider.

Key Benefits

- Supplier employees were pleased to participate in the program as they felt they were contributing to an environmental solution.
- Complete certainty of where uniforms are going and that all logos are correctly removed, avoiding impersonation of PwC personnel.
- Energy, water and greenhouse gas emissions reductions, by avoiding the need for new clothing or industrial cloths.
- Avoiding uniform disposal to the landfill saves approximately £100 per ton for local authorities.

2.5 Social developments

Social developments relate to how people live and how people live together. Over time, people's values and opinions change. Just think of how society was organised when your grandparents were young and how it is organised now. The society in which your grandparents grew up had many more ranks and standings, with the accompanying outward characteristics and possessions. In countries with low or middle incomes, people often work longer than in wealthier countries. Since 1990 average working hours have dropped sharply in Japan and most European countries. Asia seems to be the exception to this rule. In South Korea, in 2018, a law was passed that limits the maximum number of hours that employees can work. Instead of 68, the South Koreans are 'only' allowed to work 52 hours a week. In many countries, there are no legal limits at all. For example, the United States, like many Asian countries, have no maximum limits. In the Middle East, too, eight out of ten nations allow working more than sixty hours a week. However, excessive work is not without danger. In Japan, they even have a word for death by overwork: karoshi.

Another phenomenon is the rise of work on a part-time basis. In absolute terms, since 2007, part-time employment has grown in Europe, while full-time employment has declined. However, far more women than men work on a part-time basis. Furthermore, part-time work is not equally popular among countries. In Central and Eastern European countries, part-time work remains a marginal phenomenon, even among women (see figure below). In contrast, Western countries have embraced it much more widely. A clear outlier is the Netherlands where three-quarters of women work part-time, but also one-fifth of men, almost three times as many as the European average. The Netherlands also has one of the lowest shares of involuntary part-time workers (for example as a result of unemployment).

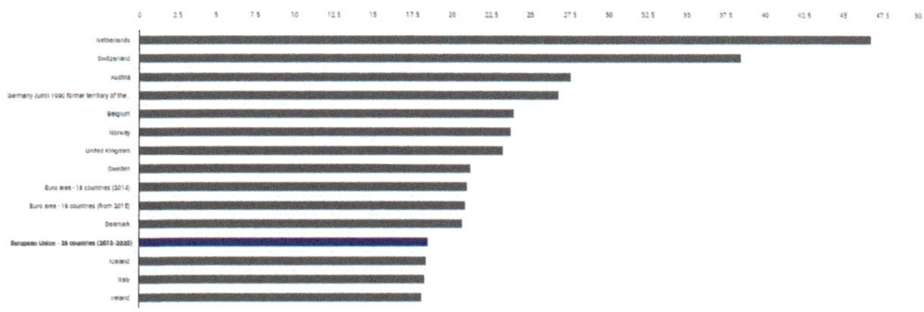

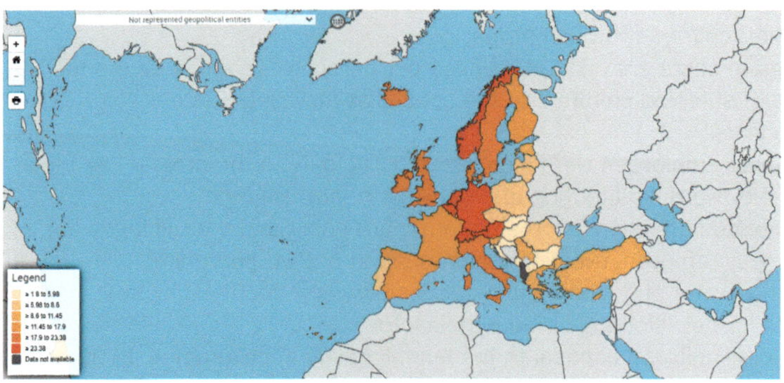

Share of part-time work in Europe in 2020

2.5.1 The 24/7 economy

The factors of time and place are less and less applicable to work these days. Technological developments have made it possible to be productive anytime, anywhere. This trend will continue, if only because of globalisation and the many multinationals that are active in different time zones. The internet and social media also change the work pattern of office workers. Technological developments have completely changed the concepts of accessibility and presence. Working outside the traditional office is called remote working. Remote work comes in different forms. Some people work remotely a few days a week and others whose companies are hundreds of miles away, so their entire job is remote. In the USA, for example, 66% of companies allow remote work and 16% are fully remote. Benefits of remote working include lower costs, greater autonomy, improved work-life balance and less time spent commuting. Social trends involving Millenials and Generation Z workers push businesses towards adopting more flexible working policies and allowing at least partial telecommutes.

Remote working

A less radical form of flexibility is the adoption of flexible working hours to personal preferences (limited by the buildings' opening hours). Flexible working arrangements are beneficial to a healthy work-life balance but can also contribute to solving social issues. For example, the Dutch government believes flexible working hours are a solution to traffic congestion problems and an opportunity to get (and keep) more women with children in work. The right to flexible working is even legislated for in the Netherlands. Flexible working arrangements require flexibly organised facility services. Here too, technology is an enabler. With facility apps and online self-service, customers can easily arrange things anywhere, anytime, without a 24/7 staffed service desk.

Facility apps

Now that employees are increasingly able to decide for themselves when and where they work, we are leaving behind two centuries of industrialisation. Many countries face a trend of increasingly shorter employment contracts, more flexibility, ip'ers (independent professionals), part-time workers and slash workers (people with multiple jobs) - and increasing spare time. However, flexibility does not only have positive effects. The 24/7 economy has drastically shifted work patterns. In many industries, the evening and weekend surcharge on hourly wages has been abolished because these are no longer exceptional situations. Work and private life are increasingly intertwined. For some people, this results in a better balance between work and private life, for others, being always accessible is a new source of stress. It is one of the explanations given for the vast amount of people who suffer from symptoms of burn-out or excessive strain. New occupational hazards require new measures. In Germany, a law has been passed stating that employers are not allowed to email or otherwise contact employees in the evening or weekend.

Independent professionals

Slash workers

The next illustration shows the evolution of the working person beautifully. A long time ago, people lived in tribes. Besides daily routines, such as breeding chickens or growing fruit, some activities required collaboration. When there was a need for mammoth meat, the men gathered and went hunting. After the mammoth was killed and slaughtered, the men resumed their daily activities - keeping chickens and growing fruit. Much later, people started specialising and distributing work. Specific crafts such as toolmakers, blacksmiths, skippers, carpenters and bricklayers emerged. People (men) learned the trade as a companion in guilds, and once skilled enough, most of them were in charge of their own business. The Industrial Revolution has drastically changed this. The conveyor belt work and the time clock forced working people into a straitjacket that did not fit in well with the nature of humankind. The three men on the far right of the picture depict the beginning of the industrial era (around 1850) to the present day. They are more and more burdened by life and do not seem very happy. Working in projects, in alternating teams, in communities with an entrepreneurial spirit, seems to suit people more. We will discuss this in more detail in Chapter 6.

The evolution of mankind

2.5.2 Growing pluralism

Social generations

Formative experiences

We have already indicated that society is always subject to change. This change is also visible in how social generations differ from each other in their views on what is essential and how you should live. The core idea behind the social generations' theory is that the social circumstances and historical events have shaped people during their youth. These 'formative experiences' would determine how they view life and the world for the rest

Chart 1: An overview of the working generations

Characteristics	Maturists (pre-1945)	Baby Boomers (1945-1960)	Generation X (1961-1980)	Generation Y (1981-1995)	Generation Z (Born after 1995)
Formative experiences	Second World War Rationing Fixed-gender roles Rock 'n' Roll Nuclear families Defined gender roles — particularly for women	Cold War Post-War boom "Swinging Sixties" Apollo Moon landings Youth culture Woodstock Family-orientated Rise of the teenager	End of Cold War Fall of Berlin Wall Reagan / Gorbachev Thatcherism Live Aid Introduction of first PC Early mobile technology Latch-key kids; rising levels of divorce	9/11 terrorist attacks PlayStation Social media Invasion of Iraq Reality TV Google Earth Glastonbury	Economic downturn Global warming Global focus Mobile devices Energy crisis Arab Spring Produce own media Cloud computing Wiki-leaks
Percentage in U.K. workforce*	3%	33%	35%	29%	Currently employed in either part-time jobs or new apprenticeships
Aspiration	Home ownership	Job security	Work-life balance	Freedom and flexibility	Security and stability
Attitude toward technology	Largely disengaged	Early information technology (IT) adaptors	Digital Immigrants	Digital Natives	"Technoholics" – entirely dependent on IT; limited grasp of alternatives
Attitude toward career	Jobs are for life	Organisational — careers are defined by employers	Early "portfolio" careers — loyal to profession, not necessarily to employer	Digital entrepreneurs — work "with" organisations not "for"	Career multitaskers — will move seamlessly between organisations and "pop-up" businesses
Signature product	Automobile	Television	Personal Computer	Tablet/Smart Phone	Google glass, graphene, nano-computing, 3-D printing, driverless cars
Communication media	Formal letter	Telephone	E-mail and text message	Text or social media	Hand-held (or integrated into clothing) communication devices
Communication preference	Face-to-face	Face-to-face ideally, but telephone or e-mail if required	Text messaging or e-mail	Online and mobile (text messaging)	Facetime
Preference when making financial decisions	Face-to-face meetings	Face-to-face ideally, but increasingly will go online	Online — would prefer face-to-face if time permitting	Face-to-face	Solutions will be digitally crowd-sourced

*Percentages are approximate at the time of publication.

Multiple generations in the workforce

of their lives. In other words, the generational idea is not about how old a person is, but about the period in which that person was young. Roughly every fifteen years, a new generation of people emerges who have had the same social experiences at an early age, and thus show similarities in attitude and behaviour. The overview of working generations below depicts the different generations and their most prominent characteristics. It is good to realise that there are many names for different generations in use and that the demarcation in time differs per source.

The differences between growing up in the time just after the Second World War and growing up at the turn of the last century are immense. However, one might wonder whether dividing people into generational categories will still make sense in the future. Generations, such as those described in this context, will follow each other faster and faster. They will minimise the differences between those social generations through the increasing ease of use of the technology they use. For example, the differences between generations Y and Z are not so substantial, while those between the baby boomers and generation X are distinctive. It is also important to realise that the characteristics attributed to social generations are strongly generalised images; the reality is often more nuanced.

An understanding of the needs of different social generations is essential for the facility manager. At present, facility managers' employees and customers can belong to as many as five different social generations. The oldest employees are part of the baby boom generation, and the youngest can belong to generation Z. Their working styles and workplace preferences are likely different. A customer-focused facility manager will have to respond to this variety of needs. For example, concepts such as 'bring your own device (BYOD)'. Herewith, people have the freedom to make their own choices regarding the tools they use, instead of a standard solution prescribed by the company. More freedom goes hand in hand with less standardisation and less control; this entails security risks, which the facility manager will have to monitor and cover carefully.

Bring your own device (BYOD)

Diversity in the workplace

Globalisation

The diversity of customer needs is increasing under the influence of globalisation. The economies of most countries have a strong international character through their trade with other countries. Internationalisation is noticeable in procurement in general and in European tendering in particular, but also in concept development and the management of employees. Due to the increasing number of multinationals (both clients and facility providers), facility professionals deal with employees and clients from a wide range of cultural backgrounds. It is up to the facility professional to translate this international context into individual and local circumstances. Facility management designs and delivers services and facilities in consultation with the end-user; this requires a flexible organisation and flexibility of its staff. This brings us to the experience economy.

2.5.3 Experience economy

Experience economies

Many societies can be characterised as experience economies. More and more office-based companies, such as banks and insurance companies, attach importance to hospitality concepts for their employees and visitors. Also, the healthcare industry increasingly realises the benefits of hospitality, as it reduces stress in patients and enhances well-being.
As a result, the providers of facilities services are specialising in this theme. Hospitality and related functions bring diversity to companies or organisations through the implementation of new combinations of services and products. This is called blurring. For example, in the concept of coworking hotels. These are hotels where facilities are also accessible to visitors so that they can meet and collaborate with hotel guests. In this type of hotels, the hotel room is no longer 'the primary product'. For modern hotel guests, the workplace is almost more important than the bed. The most successful coworking hotels create a community and offer events that facilitate connections between guests. By integrating the combination of work, residential and leisure facilities and offering hospitality services, coworking hotels respond to the needs of the digital nomads. Another example is working in theatre foyers. During office hours, theatres transfer their lobbies and catering facilities to flexible workplaces and meeting areas. Although this is a current trend, there is still potential for growth. Given the increasing number of independent professionals, this trend will continue to be decisive in the coming years.

Blurring

THE EUROPEAN COWORKING HOTSPOT INDEX

The European Coworking Hotspot Index, published by Cushman & Wakefield, looks at 'Where Next' for coworking in Europe. By understanding where coworking is likely to develop, owners and users can quickly anticipate the redevelopment of traditional office spaces. The European Coworking Hotspot Index is simple: a ranking of key European cities based on a weighting of elements including scale, business environment, people and catalyst factors. The index covers 40 key European cities, but the illustration shows the top 10 cities. London ranks first, with around 10 - 20 % of rental activity in flexible workplaces - including coworking. But the question is, where will the model move towards, and what does this mean for traditional office spaces?

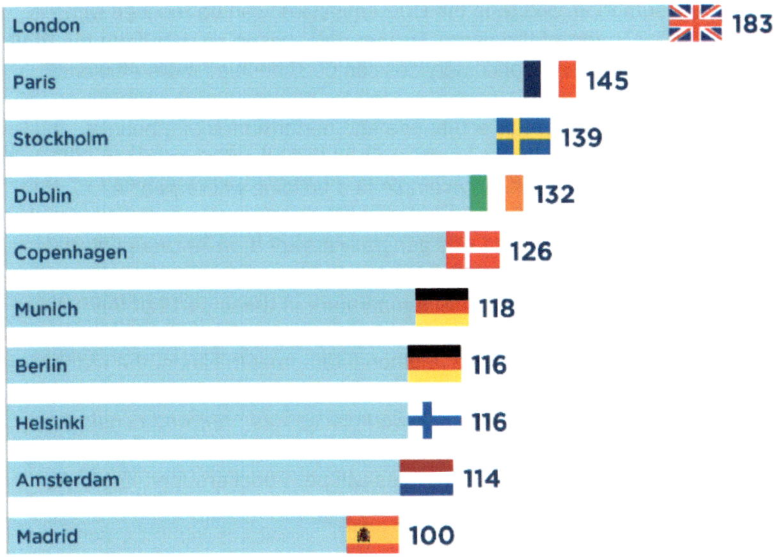

European Coworking Hotspot Index, Cushman & Wakefield (2019)

2.6 Technological developments

Our great-grandchildren will see us as the technology pioneers: the generation that experienced the advantages but also many disadvantages of the technical applications. An example of such weakness is the hardware: keyboard, mouse, beamer and cabling. The keyboard is a very clumsy, unhygienic, mechanical interface between man and computer. Besides, computers themselves cause annoyance: crashes, viruses, lagging, memory shortages and version updates are drawbacks and associated irritations during use. However, developments are moving at a rapid pace.

2.6.1 Technology, management and maintenance

In the built environment, smart buildings are a technological development that affects facility management. A smart building is a building that does something for the user. It goes beyond the automation of buildings; there are many installations in buildings like this, but in most cases, they do not or hardly communicate with each other. In a smart building, systems are connected by a building management system (BMS). The BMS allows them to communicate with each other over the internet. Connected systems can ensure not only a more efficient use of a building but also a more comfortable use of the building. Lighting and ventilation adjust automatically to the intensity of use. No users in a room? Then the system switches off. If the sensor detects people, the ventilation speed is adjusted to maintain indoor air quality levels, and the lights are switched on automatically. This way, the

Smart building

Connected systems

'Technology is never an end in itself. It is an enabler.'

— *Jan Hein Tiedema, Executive Managing Director Edge Technologies, in Smart Workplace, October 2018, p. 33.*

energy consumption (and thus the CO_2 emissions) can be reduced without affecting the comfort of the users. Furthermore, based on data from the BMS, the cleaning and maintenance services can be scheduled more efficiently.

Patterns

Apart from insights into the actual use and performance of a building, (big) data can be used to recognise patterns and to find explanations for all kinds of problems or complaints. An example is a building where people complained about the lack of workplaces. BMS-data analyses showed that the north side of the building was generally crowded, while there was enough space at the south side. From the BMS-data, the underlying cause became apparent. After noon, the temperature in these parts of the building rose to above 23C°. As most people are not comfortable with this, they seek a cooler spot. By installing additional (automatic) blinds, the root of the problem was removed. From big data, the effect could be measured. Indeed, the distribution across the available workplaces was improved considerably. Finally, business models - and thus the way of charging - are also changing thanks to digitisation. Soon, companies will pay suppliers 'per click', 'per lumen' or 'per use', instead of 'hour's bill', as also described in the following example.

In a smart building, it is possible to link occupancy data with other data sources so we can better understand the use of buildings. For example, occupancy data allows the cleaning provider to adjust its activities to the use and intensity of use of specific rooms. And sensors in soap, towel and toilet paper dispensers - no matter how high the intensity of use - ensure that they are timely refilled. Additional capacity can be planned on busy days and scaled-down during holiday periods. And if these data are linked to the weather forecast, one could even anticipate a greater or lesser degree of pollution of the common areas, for example.
The catering offer will also be better attuned to the number of occupations, which will reduce 'waste'. Security is positioned in places where there is an increasing amount of traffic. Equally, the occupancy of the reception area is always geared to the number of visitors. This can mean that suppliers no longer invoice hours in the long run, but base their billing on a price per user, visitor or supply and demand. This would cause a revolution in the facilities market.

Predictive maintenance

For the coming years, the challenge is to use the possibilities of big data generated by smart buildings. Such as for predictive maintenance. Installations send out a signal when parts are approaching the point at which malfunctions can occur due. Instead of waiting for these to take place, maintenance is scheduled in advance. It is even possible that an installation gives a direct signal to the maintenance company. This type of maintenance is more efficient than the preventive maintenance that normally takes place at regular intervals (e.g. every six months). Periodic maintenance is not wrong but cannot guarantee that faults will be prevented. While a timely, automatic notification of an expected malfunction prevents many problems. Predictive maintenance is expected to increase the lifespan of installations; with optimised the total costs of over the entire lifecycle. By the prevention of malfunctions, the continuity of services will also improve. However, periodic preventative maintenance will not be fully

replaced by preventative maintance. In some facilities, like operating centres, you will need to do both.
However, it is not enough to have smart systems. The organisation as a whole must have reached a certain level of digital maturity before the full potential of big data can be exploited. The next illustration shows the digital maturity model developed by TIMAF. At the top, you can see the development phases that organisations can go through in their growth towards digital maturity. The classification used is standard within the information management domain. The starting point is that an organisation develops its maturity step by step. Only at the highest level, will the predictive use of data be possible. There is no point in buying an expensive system if the organisation is not ready for its proper use.
The left side of the Digital Maturity Model depicts the preconditions in the field of organisation, information and systems. With the help of this matrix, organisations can assess themselves to determine their current position. Once the baseline is known, it is clear what steps can be taken for growth.

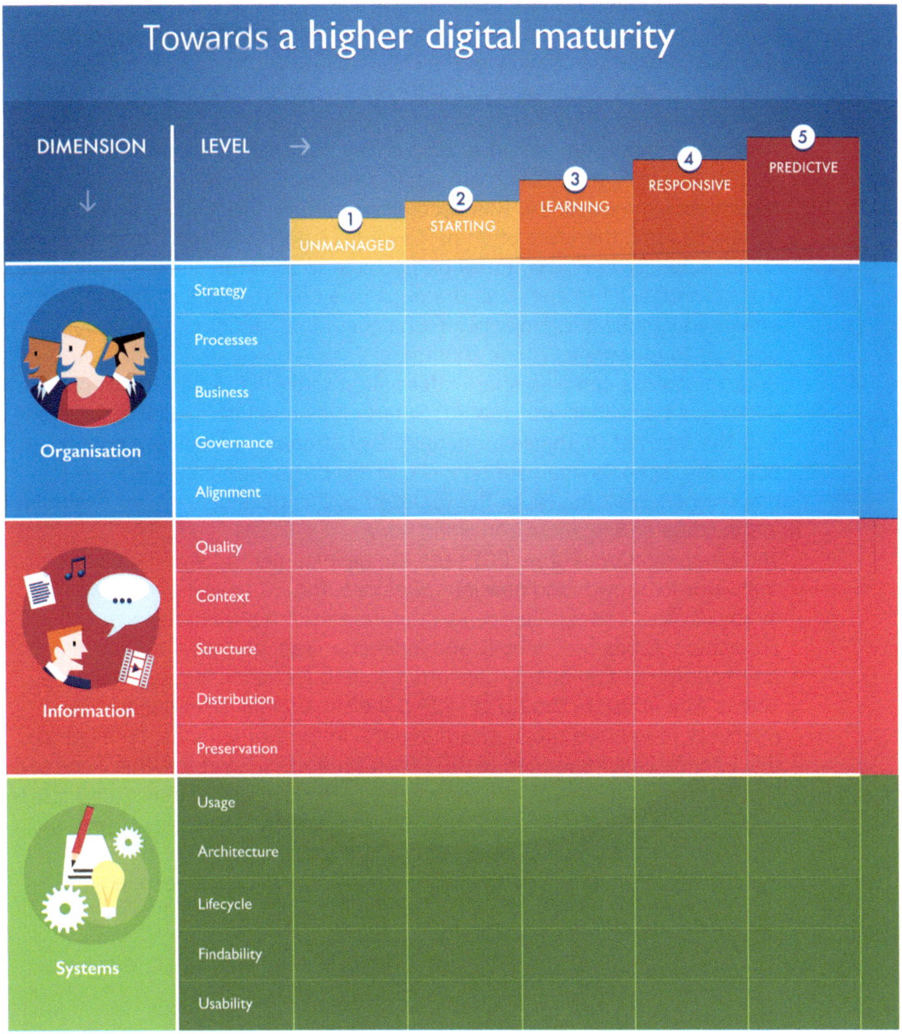

Digital Maturity Model (TIMAF)

Another form of technology that makes management and maintenance easier is the use of innovative materials. In an ideal situation, maintenance is unnecessary, especially in places that are difficult to reach, inspect or repair. Innovative materials may bring us a step closer to this ideal situation. Self-cleaning materials use a coating of nanoparticles. A nanometer is one-billionth of a metre, which is slightly larger than atoms or simple molecules. Nanoparticles come in all kinds of shapes, sizes and attributes. For example: by applying a special nano-coating, a surface is easier to clean, and therefore the maintenance frequency can be reduced. Aggressive cleaning agents are often no longer necessary. A dust or moisture repellent Nano application can achieve a self-cleaning effect, taking away the need for cleaning.

Innovative materials

Self-cleaning materials

An exciting new development is the self-healing materials. This development is also possible thanks to nanotechnology. Polymers, for example, can be provided with micro or Nanocapsules filled with a substance that is capable of repairing the damage. The content reacts with a catalyst contained in the material, and that substance fills a tear.

Self-healing materials

Nanotechnology

To take it one step further: in the future, it may be possible to create an adaptable finish and colour scheme through nanotechnology. As mentioned before, branding the identity of the organisation becomes increasingly important. Standards, values and the desired appearance of the company are at the basis of the brand. A good, well-considered design that reflects the organisation's brand makes employees and visitors aware of these values. However, more and more organisations opt for temporary and flexible accommodations. This accommodation is also available to other companies and therefore often has a neutral, general layout, finish and colour scheme. In future, it may be possible to alter the colours of floor and wall coverings, so to speak, at the push of a button. And thanks to advanced lighting and projection techniques, the interior can be flexibly adapted to the wishes of the customer.

Smart buildings can also share information with other smart buildings. In this way, smart cities are created. Smart cities use information technology and The Internet Of Things to manage and streamline utilities, transport, public life, schools and care on a day-to-day basis (see image with smart city components). Asia is at the forefront of this development, for example, in the cities of Singapore and Hong Kong.

Smart cities

The European Committee (EC) sees the development of smart cities as a motivator for growth and new jobs. Besides, the EC believes that investing in smart cities can put Europe at the forefront of renewable energy production. Cities are, directly and indirectly, responsible for approximately three quarters of all energy consumption and greenhouse gas emissions. This is why, in 2016, the EU launched the Urban Agenda to support EU-members to improve urban environments. In the Netherlands, for example, this agenda is incorporated in the Dutch Smart City Strategy. This Smart City Strategy aims to improve the quality of life of its inhabitants, strengthen international competitiveness, and support welfare, local employment and local economic activity. All in all, smart cities may be proof to a driver for sustainable, low carbon and environmentally friendly (local) economies.

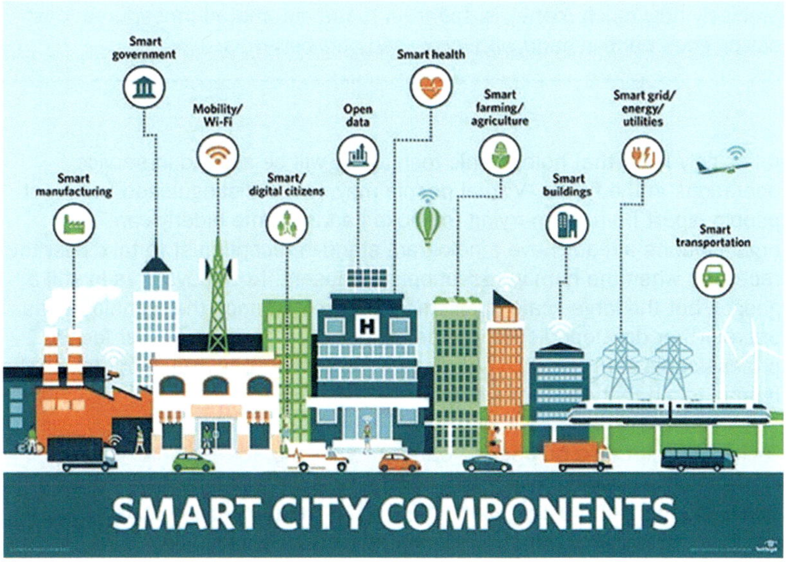

Smart city components

The use of technology also poses challenges for the operation of smart buildings. Think of issues such as user privacy and security. Since the attacks on the Twin Towers in New York, security has become a prominent value in societies. Buildings connected to the internet of things (smart buildings) can be hacked. Risk assessment, therefore, not only looks at the prevention of failure due to malfunctions but also at the risk of sabotage. This adds a whole new dimension to building security. Facility management is increasingly involved in safety issues relating to business continuity, governance and risk management. **Privacy**

2.6.2 Technology and services

Technological developments can have a positive impact on the experience of facility services. An example is the use of sensors, which makes it possible to measure real-time occupancy and thus helps users to find a suitable and available workplace or consultation space. And to find a workplace close to colleagues.

Through data analysis, customer needs and even customer satisfaction can be predicted. This knowledge enables facility professionals to forecast and act proactively. The following is an example of how cleaning company Hago Next is developing this feature in practice. **Proactively**

DATA ANALYSIS

The weather sensor indicates that it will rain tomorrow. The smart building suggests that there will probably be 1,000 people in the building tomorrow. The system proposes to do the following: perform additional cleaning rounds, install additional umbrella stands, lay extra cleaning mats and hand out ponchos. For each box ticked, the system automatically places orders with suppliers which results in a 75% time-saving in logistics. The costs are directly debited from a prepaid service charge, allowing the client to track precisely how much money is spent. In these automated procedures, cost-saving goes hand-in-hand with increased service levels.

Holographic technology

Hologram

Human touch

It is highly likely that holographic technology will be applied in service operations in the future. Virtual people may not be distinguished from real people, apart from when trying to shake hands. Some elderly care organisations already have a hologram stand-in receptionist to take over the reception when the human receptionist is absent. Nowadays, this is still a gadget, but the large-scale application is expected once the technique has been further developed. The interesting question here is whether facility professionals can use such solutions to deliver hospitality without losing the human touch that drives the customer's experience.

Technology can also be used to optimise services; for example, a smart cleaning trolley (the so-called smart cart), which measures the air quality during cleaning. If the air quality in a room is critically low, the cleaning employee receives a signal to intervene by opening a window.

Robotisation

The downside of technological development is the loss of jobs due to automation. According to OECD research on the future of work, 66 million jobs will eventually be lost due to automation and robotization, however, differences between the 32 countries included in the survey are substantial. An average of 32% of the jobs will change in such a way that it may be difficult for employees to keep up. Currently, mainly activities with a high degree of structure and routine are at risk of being lost. These include, for example, chefs, cleaners and physical work in construction and transport. Thus, robotisation will also affect the operational facility services.

As systems become more and more self-learning, less structured work can also be taken over by computers or robots. However, tasks that require perception and social intelligence remain difficult for robots to take over. While routine jobs are at risk to disappearring new posts will be created in the industrial and information technology industries. In total, almost half of all jobs will be more or less affected by robotization.

Gamification

Another development is gamification. This term stands for the use of a form of a game, for example, on a smartphone, for the playful stimulation of desired behaviour. Facility management can use gamification principles to influence the behaviour of customers and guests. The underlying idea is that a reward can stimulate guests to do certain things, for example, to report that a lamp needs to be replaced or to report malfunctioning. The desired behaviour is rewarded, for instance, with points that add up to a free lunch.
Another example of gamification is the use of virtual reality (VR) in training employees. With VR, new employees can look around 360 degrees in, for example, a hotel room or an operating room, as if they were there in person. Via interactive info-points, they can unlock information within the VR environment. VR not only makes training more realistic but also gives it a 'fun element'. And it allows (new) employees to gain skills wherever and whenever it suits them. Research by cleaning company Gom shows that participants who follow the basic cleaning training via VR generally achieve

better results in the exam than participants who do not use VR. Here the use of technology results in a higher level of quality.

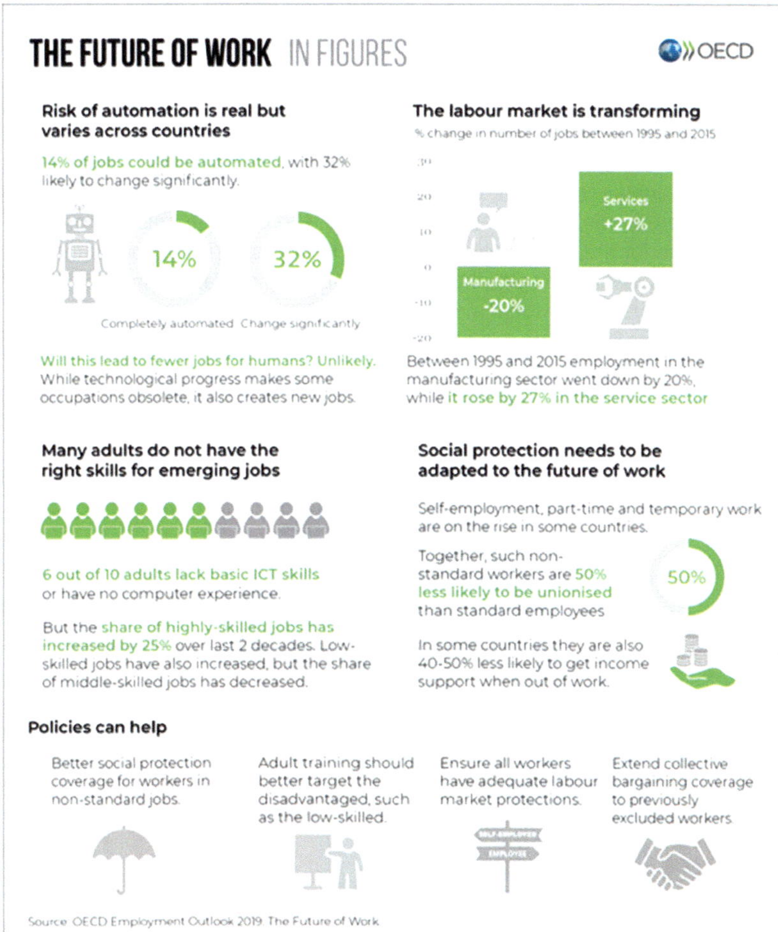

2.7 Ecological factors

Ecology refers to sustainability. Sustainability is based on the needs of humankind in the present without compromising the needs of future generations. Sustainability, therefore, looks at the current needs and how they can develop into the distant future, taking into account people, the environment, society and business operations. **Sustainability**

Currently, the world population counts seven billion people, and the number continues to grow. The United Nations forecast that the world's population will increase from 7.2 billion today to 8.1 billion in 2025. More people means more consumption of resources. But many of the resources we use today are scarce, such as oil and minerals, with rising prices as an annoying side effect. Moreover, the excessive consumption of fossil fuels is the main cause of climate change the world is facing. Sustainability is no hype because of the simple fact that there is *no planet B*.

2.7.1 The Triple P

Sustainability discussions tend to focus mainly on the environment, but - as the meaning of the word suggests - it is about general continuity, on a large and a small scale. Sustainability is about the survival of life on the planet, and the continuity of business and organisations. The scope of sustainability often referred to as the triple P; written as PPP, or People, Planet, Prosperity. The latter is also commonly named Profit. In other words: sustainability is about the survival of people, preservation of the earth, and the health of companies, now and in the future.

Triple P

A facility manager is the one who, in terms of sustainability within the organisation, is jointly responsible for people, the environment and the added value for the organisation. As far as the physical work environment is concerned, s/he is even ultimately responsible for sustainability. Finding a balance between the three P's requires careful consideration of various, sometimes conflicting interests.

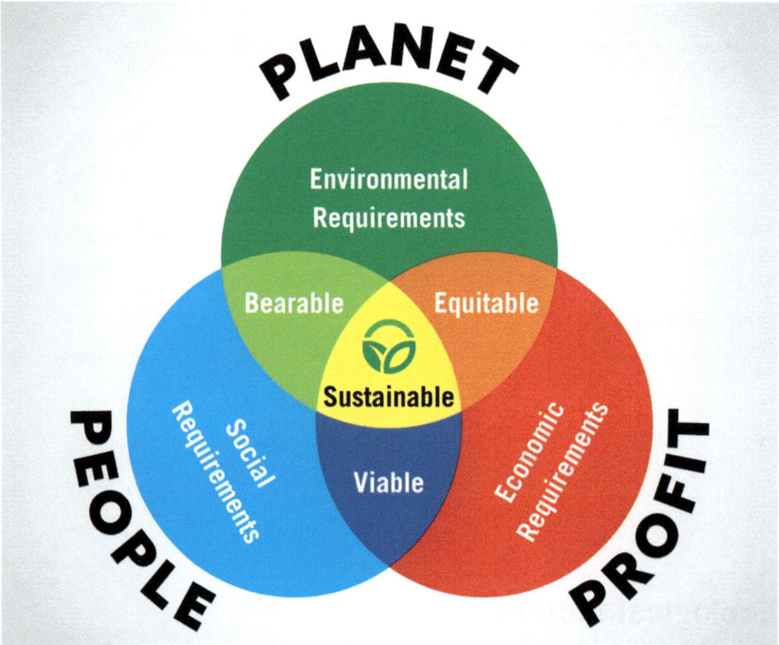

The Triple P-model

Footprint

Another widely-used term in the context of sustainability is footprint; a measuring instrument developed by William Rees and Mathis Wackernagel of the Canadian University of British Columbia. The guiding principle for their research was the question of how much fertile land would be needed to sustain the population indefinitely, no matter where in the world it is located. By this method, it is possible to compare the environmental impact of different types of consumption, behaviour (lifestyles), or of different population groups (countries).
The footprint reflects the extent to which the earth's capacity is affected. Interestingly, companies and even households are now starting to think in

terms of ecological footprints. The footprint can be useful as an indicator of the degree to which specific consumption patterns are sustainable.

2.7.2 People

Concerning the virtual and physical work environment, the facility manager is responsible for the welfare for the people in the organisation. One goal is to provide employees, visitors and suppliers with an optimal environment, services and facilities to do their work effectively. However, the facility manager is also responsible for stimulating the sustainable behaviour of employees and visitors, as far as - again - the work environment is concerned. Construction- and installation-wise a building may belong to the best of green buildings. Still, if users are not aware of the importance of sustainable behaviour, the building and business operations may be far from green. For example, a facility manager contracting for separate waste processing depends on the behaviour of users to make this a success. If users do not take the effort to put their waste in the dedicated bin, recycling will be of little use. Even worse, the facility manager runs the risk of being charged with extra costs by the vendor.

Welfare

2

Therefore, awareness is an absolute prerequisite for overall sustainable behavioural change. Much can be achieved with visual education and feedback on consequences of behaviour; for example, through dashboard panels that indicate energy consumption. Furthermore, facility professionals can use gamification to reward sustainable behaviour (see section 2.6.2).

But there are more perspectives on the P of people. For example, that of the 'inclusive society'. In July 2016, the UN-Convention on the rights of persons with disabilities became in force. This convention establishes the rights of people with disabilities. The ambition of the treaty is to ensure that people with disabilities can participate in society on an equal basis with others, despite clear differences.

Inclusive society

People with a disability have the right, just like everyone else, to earn their own money with work that suits them. We distinguish between physical and mental disabilities, and some people have both. Physically disabled people often need adaptations to their workplace and sometimes also adjustments to the accessibility of the building. People with a mental disability in particular often also need extra guidance in their work. In practice, it appears to be difficult for many employers to find employment for people with a work-limiting disability. Facilities services, but also facilities providers can help clients to organise suitable work. An excellent example of good practice is the project 'Inclusive-FM'.

Suitable work

Schiphol Airport has ensured that four competing cleaning companies work together to employ people with a distance to the labour market. To find out what works and what barriers companies experience in practice, Inclusive-FM was launched as a pilot project. Inclusive-FM is a collaboration between Schiphol Group, Aviation Community Schiphol, four cleaning companies, a mid-level vocational training institute in Amsterdam, and the unemployment benefits services. Together, they developed a one-year entrance course that combines practical and theoretical training. Participants go to school one day a week and work for one of the cleaning companies to practice their skills, build their CV, and earn a salary. For many participants, this is their very first diploma. Something to be proud of!"

Corporate social responsibility (CSR)

A final perspective is that of corporate social responsibility (CSR). Facilities clients and buyers have a major influence on the quality of the working conditions of the employees of suppliers through the procurement strategies they apply. It is important to be aware that the lowest price strategy could be at the expense of the quality delivered or unreasonable remuneration of those involved. These are often the employees of the contractor or the subcontractor.

Responsible procurement

To prevent unrealistically low quotations and associated substandard work conditions, more and more (professional) associations embrace a code for responsible procurement. Responsible procurement is defined as procurement activity that:

- seeks to build on and expand the scope of 'ethical' and 'sustainable' procurement to include environmental, economic, and social issues and their risks and benefits;
- includes a range of considerations of equal or greater importance than profitability and the generation of value;
- includes purchasing that minimises carbon emissions causing climate change, challenging the systems and industries driving it;
- identifies, prevents and mitigates the risks of human rights abuses in the supply chain, including forced labour, child labour and human trafficking;
- recognises that these considerations are interconnected and require a holistic approach.

An example is the Dutch Code for Responsible Market Conduct. This Code relates to the cleaning, catering, security and moving industries. In these branches, focus on lowest price and cost reduction is common. Other important aspects, such as the quality of the work and the working conditions, are under pressure. The Code aims to prevent excessive market forces; a situation in which the continuity of the business operations of suppliers is at risk because they compete too vigorously on price. The signatories of the Code undertake to take responsibility for realistic quotations and the responsible awarding of work. At the time of publication of this book, 1,400 Dutch organisations have subscribed to the Code.

2.7.3 Planet

CO_2 emissions

Climate agreements to limit global warming aim to reduce CO_2 emissions significantly. This is why companies and governments are committed to promote possible alternative production methods and resources and to generating ever available wind, hydropower and solar energy. As a result, the share of renewable energy is increasing but remains relatively low compared to the total demand for electricity. As such, we will remain dependent on fossil fuels for the time being. That is why it is so crucial that efforts are made to increase energy efficiency.

The built environment is responsible for almost 40% of CO_2 emissions. On the one hand, we need to take measures to reduce energy consumption; for example, by insulation and more efficient installations. On the other hand, we need to use renewable energy sources as much as possible (solar panels, wind turbines, solar boilers, etcetera). Energy-neutral buildings (also known as passive buildings) will become the standard. The reason for this is simple: what you do not use does not have to be generated or stored. Passive constructions have five core principles. These are optimal insulation; triple glazing with south-facing windows; balanced ventilation with heat recovery; optimal crack sealing; and facilities that prevent thermal bridges.

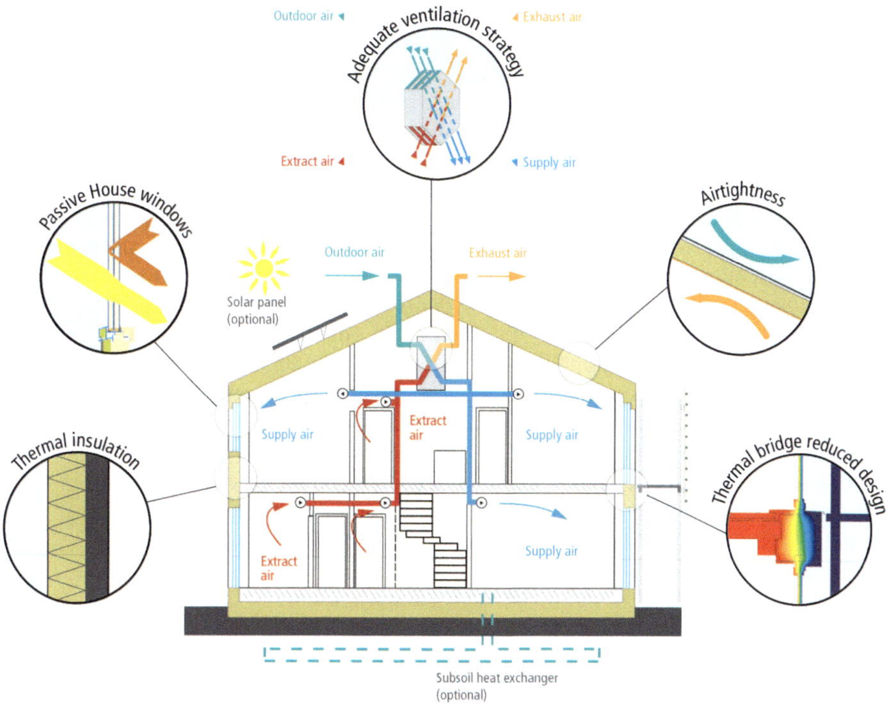

The five core principles of passive buildings

Facility management plays an essential role in stimulating sustainable choices in renovation and (re)development processes. If facility management has sustainability high on its agenda, it will reflect in discussions with its clients and suppliers. Instead of waiting for clients' demands to become more sustainable, facility professionals can take the lead and promote sustainable choices in procurement and projects.

Another way in which companies will limit their CO_2 emissions is by minimising the commute of their employees and the number of business travels, and by promoting more sustainable means of transportation (see section 2.4.4). Facility professionals influence the logistic kilometres by their supply chain management. From the perspective of the carbon footprint, fewer orders of larger batches from regional suppliers have a less environmental impact than daily deliveries from whatever supplier offers the lowest price. However, from a cost-only perspective, the latter is the sensible choice to make. Storage space and keeping stock is expensive, and regional suppliers may not offer the best price available.

FM TREND REPORT BELGIUM, 2019, P. 118

The facilities division of bpost (Belgium postal services) plays a crucial role in bpost's CSR strategy, says Koen van Gerven, bpost's CEO.

One reason for this is its responsibility for the management of physical assets. First and foremost, the buildings. With more than 800 buildings in Belgium - consisting of a mix of 5 gigantic sorting centres, offices and post offices in each of the 598 Belgian municipalities.
Furthermore, facility management is responsible for furniture, equipment, technical installations and specific assets such as sorting machines for letters and parcels and even robots. It is also responsible for fleet management. With 3,500 electric bicycles, 7,500 delivery vans and 450 large trucks, bpost operates one of the largest fleets in Belgium. All these resources require investment and maintenance to guarantee continuity of service. Facility management initiates projects such as the acquisition of electric vehicles, and sustainable energy generation, to contribute to the CSR-strategy. For example, bpost has approximately 20,000 m2 of solar panels and is working on its first wind turbine. By 2030, half of the small vehicles will be electric or equipped with a fuel cell, depending on how the technology develops in the coming years.

Circular economy

As we have briefly described in section 2.4.5, the economies of many countries will undergo a transition from a linear economy to a circular economy. A circular economy is a restorative and innovative system, that aims to maintain products, components and materials at the highest possible value and to create as little waste as possible. The goal is to give products and materials the longest possible lifespan and to facilitate optimal re-use and repair of products and materials. The circular system has two cycles of materials:

Cycles

1 a biological cycle, in which residues are safely returned to nature after use;
2 a technical cycle, in which product(s) are designed for re-use, in other words: in such a way that materials and components can be easily disassembled and re-used at a high-quality level. This ensures that the economic value is preserved as much as possible.

The circular economy is a complete transition that is not limited to circular construction. Facility management can contribute to the development of a circular economy through circular exploitation of buildings. In this, procurement officers play a crucial role. They stimulate and create a demand for circularly produced goods. However, closing the loops also involves enabling re-use of materials, which must be organised by the facility manager. Waste management is, therefore, an essential part of the sustainability strategy.

Waste management

- https://ecopreneur.eu/circularity-check-landing-page/

To help procurement officers, Ecopreneur offers a free online scan with a questionnaire of approximately 60 points, which determines the 'circularity score' of a specific product or service. Is the product circular and sustainable, and if so, to what extent? Each question receives a score and the higher the score, the better. The result of the scan is a percentage that indicates how circular the product or service is. The check gives partial scores for design, purchasing, production, delivery, use, re-use and sustainability.

2.7.4 Prosperity

The third P, Prosperity, refers to the financial health of people and of organisations. Financial health is a general condition for survival, as oxygen is for humans. It is not always about profit. Unlike the business community, government organisations (non-profit organisations) do not think in terms of financial gain. In the case of the government, it is about the budget.
This P is reflected in the payback period of investment in sustainability: how long does it take before an investment is recouped through lower annual consumption costs? At first sight, it often seems that investing in sustainability is a costly affair. But if the costs and revenues during the entire life cycle of a building or its installations are taken into account, the bottom-line may be quite different. This holistic assessment is referred to as the total cost of ownership (TCO). By including the cost structure during the entire life cycle (and any residual value) the business case for investments in sustainability becomes much more viable. The returns can also be seen in a broader context than the financial benefits. For example:

Payback period

Total cost of ownership (TCO)

- the positive effect on a company's image;
- gaining experience with new technology;
- or being eligible for government orders because the sustainability requirements that are imposed on suppliers can be met.

Sustainability is about the present, but especially about the future. It is, therefore, fair to consider the current and future social costs in the decision-making process. By broadening-up the assessment framework in terms of revenues and costs, you gain a better insight into the direct and indirect effects of policy. TCO-thinking may be somewhat new to facility management but is common practice in asset management. By definition, this domain focuses on the development of the value of buildings throughout their entire life cycle.

Social costs

2.8 Political-legal factors

Political factors determine the playing field for the business community and for (semi)public organisations. Some examples have already been mentioned, such as the raising of the state pension age, the transition to a circular economy and the agreement to help people alienated from the labour force find paid employment. Other ways in which politics influences business are tax measures, a licensing system and labour law regulations.
Also, geopolitical developments impact the economy, such as trade wars, political unrest and refugees, the agreement on the free movement of people

and goods within Europe, and European procurement rules. The latter especially affects soft service providers. Legislation and regulations are also subject to change and can have a significant impact on the work of facility professionals. Think, for example, of the introduction of the General Data Protection Regulation (GDPR), which standardises the rules for the processing of personal data by private companies and government bodies throughout the European Union. It is, therefore, important that facility professionals are well-informed about political developments and decisions.

General Data Protection Regulation (GDPR)

In countries with a multi-party system (such as Germany, the Netherlands, France, Canada and Italy), the government will always consist of several parties. The coalition agreement is therefore by nature a compromise of the positions of each governing party. As a result of this, the political course in this type of country is fundamentally moderate, with alternately a more socialist and a more liberal slant. Moreover, European countries must comply with European laws and regulations. National governments and the European Union steer not only through legislation and regulations but also by granting subsidies to initiatives that are in line with government policy. As a facilities entrepreneur, it is interesting to find out whether an innovative idea or project can be eligible for a grant to compensate for start-up costs partially. All European funding programmes can be found on www.europa.eu.

Compromise

European laws and regulations

2.9 From trends and developments to scenarios

Section 2.2.2 lists the four steps that lead to scenarios. Before giving examples of these steps, we describe how we think scenario planning sessions should take place. Useful scenarios are not created in isolation. The ideal is to put together a team with representatives from all parties involved: the facilities department, its clients (both on a strategic and tactical level), the supplier(s) and (external) service providers. The advantage of such a team is that ideas are brainstormed from all kinds of perspectives. Tunnel visions, which of course are undesirable, can thus be avoided. The advantage of making scenarios in a team is that, by definition, a collective opinion and vision emerge.

For a maximum result from scenario-developing workshops, there are some general 'rules of the game':

- The atmosphere must be informal for a productive brainstorm. This means that those present must dare to be vulnerable and creative: as a rule, avoid criticism of ideas. Nothing is stupid.
- All team members are entitled to each other's presence at all brainstorming sessions. After all, they have committed themselves beforehand to participating throughout the entire process.
- During the first session, it must be made clear to everyone that scenarios are not predictions of the future, but that they explore possible futures.
- Scenarios must have a certain degree of probability and reliability. Unbridled fantasy and fairy tales are of no use.
- The advantage of scenarios is that the organisation learns to deal with uncertainties. They are only really relevant when their impact on the organisation is clear.
- Support for scenarios, outside the team itself, arises especially when they capture the imagination of the audience. Think of a story, film, act or sketch.

Scenario planning is a continuous process. Through brainstorming and discussion, those involved gain progressive insight into the influencing factors of their business. With more in-depth insight, the need arises to adjust the scenarios; the four steps are repeated as often as needed. We call this an iterative process.

Next, we give an example of steps in which scenario planning can take place. We start from an insurance company where the facilities department wants to draw up scenarios as a basis for strategic policy:

Example of step 1 - Identifying trends and developments.
The first workshop(s) outlined the situation in which the insurance company finds itself and - as a derivative thereof - what the particular circumstances of the facilities company are. Next, the DESTEP factors will be discussed, i.e. what is going on in society at large. The result is shown (in staccato) in Table 2.1.

TABLE 2.1 Example of DESTEP factors for an insurance company scenario planning

Factor	Developments
Demographics	The team does see the impact of ageing on claims and population growth or decline on market size, but do not consider this to be a driving force for the insurer's future.
Economy	The question that immediately arises within the team in this context is: to what extent is the customer willing to pay increasing premiums?
Social	Although people's values, opinions and lifestyles will change, the team assumes that they will continue to need to be insured against the most common risks.
Technology	Technological developments have a significant impact on the insurance industry. The team is unanimous in its opinion.
Ecology	In itself, insurance does not have much to do with sustainability. Therefore, the team decides not to include ecology as such in the scenarios.
Politics	Regardless of which way politics goes, people want to insure themselves, is the proposition of the team.

Example of step 2 - Determining the driving forces.
driving forces
Based on the considerations in Table 2.1, the team chooses the following two driving forces: the economic and the technological, which are respectively the expected load-bearing capacity and the self-activity of the customer. On the one hand, the insurer's success depends on its competitive market position. The premiums must be acceptable and affordable for the customer, and the operating result must, therefore, remain positive. For this reason, the future development of financial strength is considered essential. On the other hand, self-service forces the client to be actively involved in the service process. He can compile the insurance policy to his needs and calculate the associated premium via the insurer's website. In the event of a claim, he is required to provide the evidence and necessary paperwork via the online portal.

Example of step 3 - Creating a scenario matrix.

Scenario matrix

The vertical axis of the scenario matrix in Figure 2.4 represents the financial capacity of the customer; from minimal willingness to pay for insurance premiums (bottom of the axis) to high willingness to pay (top of the axis). On the horizontal axis comes the future self-activity of the customer. On the far left-hand side of the axis, technological standstill is indicated (which is, of course, unlikely) and on the far right an enormous 'disruptive' (disruptive) technological development.

2

The four quadrants in the matrix are the scenarios. Lower left (LL) concerns low carrying capacity and little change in self-activity; lower right (LR) low carrying capacity and a strong increase in self-activity; upper left (UL) more carrying capacity and little change in self-activity and upper right (UR) both carrying capacity and self-activity.

FIGURE 2.4 Example of the construction of a scenario matrix

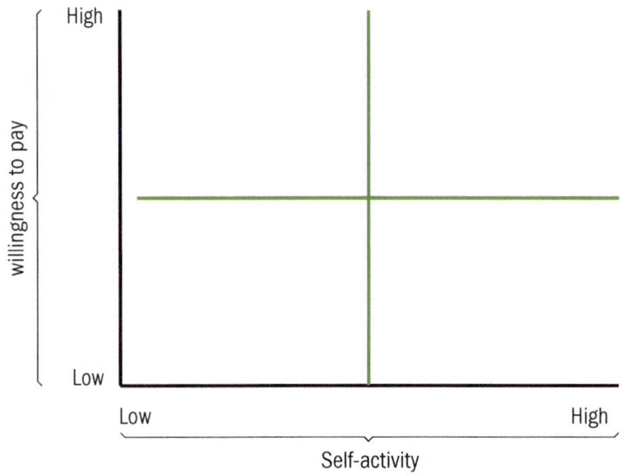

Example of step 4 - Translating scenarios.

These four scenarios are given a name (see Figure 2.5):

1 LL is called 'Price war'. Turnover is under pressure because the premiums will be cheap and competitive. At the same time, at least due to the absence of technological developments, no special savings can be expected.
2 LR is called 'Cutbacks' because technological developments can create more self-activity on the part of the customer, which makes it possible to reduce premiums.
3 UL is called 'Market conquest', assuming that the customer can afford a lot, because the carrying capacity is relatively high. We should not count on cutbacks due to automation here.
4 UR is called 'Innovation Battle', given the fact that extra money can be earned through higher premiums and automation, which can, for example, be spent on innovation.

FIGURE 2.5 Example of a completed scenario matrix

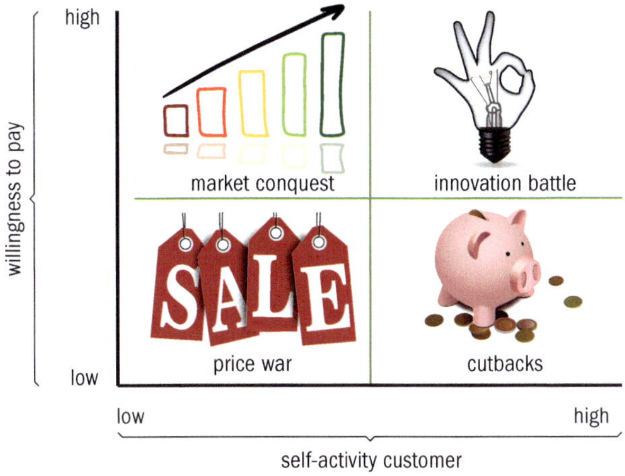

Impact analysis

Next, a so-called impact analysis is carried out. For each scenario, the probability and the impact on the organisation are determined. The scenarios with the highest scores in this analysis must be taken into account most. Table 2.2 shows the summary of the analysis.

TABLE 2.2 Impact analysis

	Probability	Impact	Total score
Price war	2	4	8
Cutbacks	3	3	9
Market conquest	3	2	6
Innovation battle	4	5	20

The analysis shows that the 'Innovation battle' is the most likely scenario and at the same time has the greatest (and most positive) impact on the organisation. This scenario must, therefore, be taken into account. For example, the facilities company will have to take into account staff reduction and innovative challenges. Ergo, the facility manager must be prepared for a reduced need for office space and even the more advanced infrastructure for the new technologies. These insights are the starting points for drawing up the strategic facilities policy.

Please note that this is just one example to provide insight into how scenario planning can proceed. In Chapter 6 we give an example of scenario planning for the future of the discipline itself.

Young Professionals in Facility Management

2

Joost Andela, Senior Global Project Manager at Sodexo global headquarters in Issy Les Moulineaux (Paris), France – graduated bachelor International Real Estate and Facility Management at BUAS.

From Breda to Issy Les Moulineaux, the Story of an international Career at Sodexo

In 2015 I graduated with a degree in International Real Estate & Facility Management from the Breda University of Applied Sciences. In this bachelor-programme, I developed my skills, knowledge and attitude by the support I received of the (guest) lecturers, coaches, and fellow students. This input helped to elevate my mindset to a strategic and autonomous level and has been highly valuable in building my career.

My career at Sodexo HQ started during my final year at BUAS when I was offered a 9-month internship as operational excellence manager at Sodexo. My responsibility was creating, presenting, and the following-up of advisory reports based on lean management, benchmarking and operational excellence to Sodexo colleagues and our clients.

Following on from this success, I became Transition Manager for the first vested contract Sodexo signed with a strategic client. This contract encompassed 200+ sites in 42 countries in the EMEA region. Aiming to ensure efficient transition processes from Kick-Off to Go Live and beyond, I created and delivered new tools, processes and procedures. Moreover, I challenged the proposed solutions and reviewed local contracts for sign off.

In my current role, I lead global strategic accounts from the transition phase to business as usual. Additionally, I support three regions in the world in the implementation and continual improvement of two global frameworks. The overall focus is to improve selling and mobilizing of contracts.

The biggest challenges I have experienced in these roles are influencing, challenging, and supporting multi-cultural international teams to gain a positive outcome. I found that being receptive, respectful, and clear in one's intentions when interacting with others, is a prerequisite for success in an international work environment. Therefore, attending cultural awareness courses is a good starting point!

Summary

- Companies face ever-increasing changes, and those changes follow each other ever faster (exponential times).

- Social developments affect how people work and the work environment and therefore, the work of facility professionals.

- Predicting the future is not possible, but developing scenarios is. Because facility managers take decisions that have a long-term effect, it is important that they are aware of social trends and developments. Also, it is essential that they can translate these into scenarios. Industries differ in the way in which social developments affect them. Although scenarios are a useful tool, it is important to note that, under the influence of social trends, reality will always deviate from what was expected.

- Scenario planning consists of several steps: (1) Identifying trends and developments (DESTEP); (2) identifying the driving forces; (3) drawing up a scenario matrix; (4) making an impact analysis and translating the chosen scenarios into the organisation itself.

- An analysis of the DESTEP factors provides insight into relevant social developments.

- Driving forces are megatrends. They determine a large proportion of the other trends and can have a significant impact on society.

- A scenario matrix is a tool for creating four images of society based on developments on two driving forces.

- An impact analysis determines the probability and impact of each scenario.

- Important demographic developments are the ageing population, densification of urban and metropolitan areas and population decline in non-urban regions, older employees (attention to vitality) and scarcity in the labour market (war for talent).

- The economic climate always influences the work of facility professionals: in times of economic contraction, the emphasis is on efficiency; in times of economic growth, there is more room for innovation. It is up to the facility professional to balance effectiveness and efficiency. A far-reaching development is the transition to a circular economy. The rise of the sharing-economy and networked economy also affects the work of facility professionals.

- Social developments have everything to do with increased flexibility and individualisation: the 24/7 economy, the growing number of independent professionals, the increasing diversity of client needs and types of employees (multiple generations, globalisation) and the importance of paying attention to experience. National and organisational cultures have a significant impact on if and how these trends emerge and develop.

- Technological developments can be disruptive because they affect other trends and thus society at large, and because they can take place at breakneck speed. Some developments that facility professionals have to deal with are smart buildings, smart cities, data-driven facility management, smart materials, predictive maintenance, robotisation, gamification and virtual reality. Technological developments offer opportunities but can also have a downside, e.g. loss of jobs.

- The facility manager is partly responsible for people (People), the environment (Planet) and the added value for the organisation (Profit) in the field of sustainability within the organisation. Through facility purchasing, the facility manager also has an indirect influence on the working conditions of employees of suppliers (People), the use of virgin materials by suppliers (Planet) and the continuity of the business operations of suppliers (Profit) by awarding them a healthy margin.

Questions and Assignments

2.1 Why is it essential that a facility manager should study social developments and not only the events within his or her field of expertise?

2.2 What do you think a working day at an office in 2030 will look like? Create a word cloud with at least ten keywords that describe your idea of the future of work.

2.3 It is impossible to imagine our society without social media. However, many facility managers are still reluctant to use Twitter, Instagram or WhatsApp in their services. Give two reasons for this.

2.4 Municipalities and other government bodies are increasingly investing in projects to improve safety, connectivity and quality of life within neighbourhoods and regions. Search the internet for job offers for the following positions: area manager, business improvement district manager and community manager. What is the relationship with facility management?

2.5 Looking at the developments in the labour market ('war on talent'), organisations make every effort to present themselves as an attractive employer. Can facility management also contribute to this? Can you name some of the facilities that can help you stand out as an employer in the labour market?

2.6 Explain what contribution the facility manager can make to 'the inclusive workplace'.

2.7 In recent years, robotization has also made its appearance in the facilities industry. Mention at least five facilities activities that you believe will be taken over by robots in the foreseeable future. Support your answer with media reports.

2.8 From Baby Boomers to Generation Z; what does the multi-generational workforce mean for an organisation's facilities services?

2.9 'The facility manager of the future is a data analyst', according to a speaker at a conference. Explain what message the speaker is trying to communicate with this statement.

2.10 An organisation has to deal with vacancies in its 8-storey office building. After mediation by a real estate agent, two parties are found that would each like to rent two floors for the coming three years. The facility manager is not yet able to assess what this new multi-tenant environment will bring. Can you name two opportunities and threats of a multi-tenant building?

3 Experiencing the work environment

In this chapter, we will discuss the following questions:
- What does the working environment consist of?
- How do facility professionals influence the experience of customers?
- What is the difference between real estate management and facility management?
- What are the facility manager's responsibilities with regard to accommodation and space management?
- How can service processes be described?
- What are the characteristics and critical success factors of the most common facility services?
- What is the importance of service providers' behaviour in relation to customer satisfaction?
- Are all customers and employees the same?

◄ Through the eyes of the facility manager:
'In this hospital many healing environmental aspects have been applied. Natural daylight, planting, and the many coloured and natural materials used contribute to a pleasant stay and a hospitable atmosphere.'

Expert Column

We need to build offices for everyone, not for some

A tide is changing in workplaces and attention is being put on the workplace as an experience. An experience that attracts top talent, an environment that promotes individual health, cognitive performance and the best collaboration in/between teams. What is not often said is that the typical working environment is poor, promotes sickness and lacklustre if not lousy collaboration.

But as we just said, a tide is changing. Interestingly, one of the primary motivators in the industry is not money, opportunity or praise and honours. It's young people who are refusing to work in 'bad environments'. While Millennials and Generation Z are often criticised for their impatience and stubbornness, it seems like these two factors may end up improving workplaces for us all. Research has shown that the younger generations will compromise pay and benefits to work with a company that fosters an environment that promotes health, balance, purpose, flexibility and creative collaboration.

Simultaneously, researchers are seeing more and more opportunity for improving the health of workplaces and the way organisational systems, culture and buildings can foster a better way of working. For instance, a recent study out of Harvard Business School suggests we should not sit with our team but instead meet with them for specific reasons. Space should be used purposefully for value-added initiatives, rather than the sake of showing up.

Additional research is starting to acknowledge the differences in people's biology and the opportunity to capitalise on those differences. It is time now to combine these learnings to build new work experiences that meet the natural needs of the people using them. At LAB we are doing just that, working on new research to discover who the employees are in a building and changing the environment for them. For instance, is your team mostly introverts and need quieter areas for focused work? Are your teams a later chronotype (evening people) who love to collaborate and socialise? We need to build offices for everyone and not only some. So, we say to you: *stay impatient and push the change in the way we all work!*

Elizabeth C. Nelson
Head of Research and Innovation at LAB Learn Adapt Build
Author of The Healthy Office Revolution.

Nicholas C. White
Managing Director, LAB Learn Adapt Build
Publisher of The Healthy Office Revolution

3.1 Introduction

By now, you will recognise that increasing attention is being paid to the perception of the work environment. We define the work environment as any environment used for the professional delivery of services. Thus, the work environment covers any business setting, like an office, a hospital, a hotel, a recreation park or a factory. The private or residential environment is excluded unless professional services are provided there (like in residential supported living). The term 'work' environment may give rise to the suggestion that facility management only focusses on supporting workers. However, as we discussed in Chapter 1, the nature and scope of facility services depend on the context. In some industries a facility manager mainly supports external clients, such as patients, hotel guests, visitors to a cinema or residents of a nursing home. Of course, these customers do not work in the environment, yet they are very important users. For the sake of readability, we generally use the term' work environment', unless it also refers to the experience of external customers, in which case in this book we will refer to 'environment'.

Broadest sense

This chapter discusses the work environment in the broadest sense of the word. The structure follows the common distinction made between the hard services and soft services. In general, building-related activities fall under hard services, such as the maintenance of buildings and installations. By soft services, we mean all user-related services, such as catering. Table 3.1 presents an overview of 49 areas of facility services expertise. The ranking indicates the relative importance of services, as shown by a study by Hospitality Group in 2012. However, in this book, the table is used to illustrate the richness of facility services. The order of importance is different for each organisation and in each country, and therefore, cannot be used as a general criterion. Also, the ranking will have changed under the influence of other ways of working. Office relocation service, for example, has a top-5 position. However, in organisations that adopt flexible working few people have a fixed workplace; relocations are much less necessary and therefore a less dominant area of attention.

TABLE 3.1 Facility focus areas in order of importance (De Bekker et al. 2012)

Focus area	Ranking	Focus area	Ranking
Cleaning services	1	Facilities Service Desk	26
Waste management services	2	Archive management	27
Foodservices/catering services	3	Real estate management	28
Building operations management	4	Rooms and spaces for rent for meetings and events	29
Office relocation services	5	Art	30
Security	6	Fleet management	31
Contract management	7	Company clothing	32
FM Procurement	8	General procurement management	33
Building renovation, painting & decorating	9	Taxes and duties	34
Mailroom services	10	Risk management	35
Indoor greenery	11	Quality management	36
Office supplies	12	Insurance	37
Inventory management	13	Workwear and linen services	38
Energy and water management	14	External workspace management	39
Wayfinding and routing	15	Administration	40

TABLE 3.1 Facility focus areas in order of importance (De Bekker et al. 2012) (continued)

Focus area	Ranking	Focus area	Ranking
Space management	16	Information management	41
Grounds maintenance & green services	17	Telecommunications	42
Reception services	18	Hardware	43
Parking management	19	Internal logistics	44
Print room	20	Library services	45
Janitorial services	21	Software	46
Technical service	22	Network management	47
In-house emergency response	23	External logistics	48
Occupational Health & Safety (OHS)	24	Travel desk	49
Distribution & transport	25		

There are more factors with which facility professionals can shape the experience of customers or users than hard services and soft services. After all, the hospitality of the facility staff has a great deal of influence on the perception of the customer or user. The third building block, therefore, is the user experience. Although user-perception of the environment is subjective and personal, facility professionals can influence or even stage a user's experience. The image below shows how the building blocks of facility management: people, hard services, and soft services, are (inter)related.

Perception

Three building blocks of facility management

In the centre of the image, all circles overlap in the 'FM sweet spot' Here FM can have a direct impact on both the employee experience and the performance of the company. In other words, this is where added value of FM is created.

FM sweet spot

But what is experience? What are you supposed to do with it? How can you influence or create the desired user experience? To make this rather abstract concept somewhat more tangible, we use the diagram in Figure 3.1. This diagram is based on the principle of supply and demand. The customer, employee or visitor (number 1) wants an optimal (work) environment which is provided for by the facilities company (number 2). The customer (or user) has only five means to perceive the experience, namely his senses (number 3). The facilities provider has the following means at its disposal to influence this experience (number 4): the soft services, the environment and the behaviour of the facilities employees. While he is in the environment, the user experiences the ambience by - in order of importance - *seeing, hearing, smelling, touching and tasting* (number 5).

FIGURE 3.1 Creating experiences

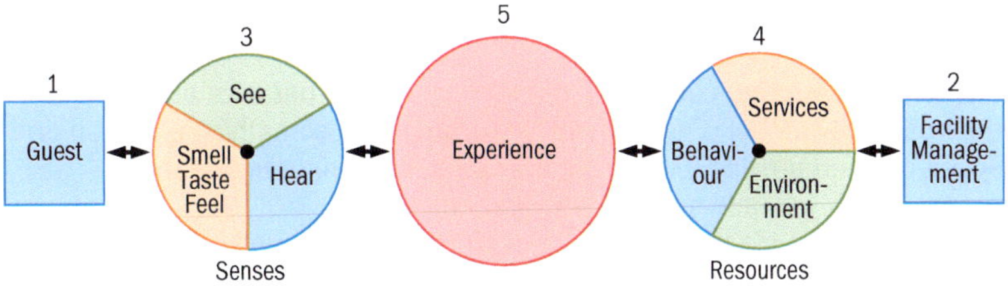

It may sound clinical, but simply put, the facility manager has hospitality (behaviour), services and resources at his disposal to create an experience with the customer through his senses.

The hospitality model of Daane Bolier (1988) is a model that clearly shows the relationship between services, environment and behaviour, and the customer experience. Figure 3.2 shows that each customer has needs (N) and objectives (O). The service provider can meet these needs and objectives by using a combination of products and services (P in the model) in a specific environment (E). Because service provision always involves a form of interaction with the customer, the behaviour of service providers (B) is an inseparable part of the service offered. The combination of the behaviour of the employee, the service or the product and the environment in which it is delivered, provides an experience for the customer. The customer's assessment of the service is determined by what extent the service provider succeeds in meeting the needs and objectives of the customer.

FIGURE 3.2 Hospitality model

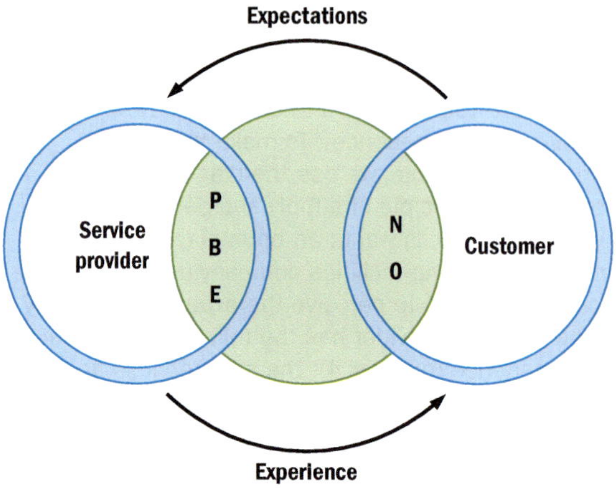

P = Products (services)
B = Behaviour
E = Environment

N = Needs
O = Objectives

Research shows that customers do not consciously perceive all elements of the hospitality model. Environment, for example, is something a person is aware of when arriving, but which, over time, if the situation meets expectations, will fade into the background.
A product or service that meets a perceived need is usually seen more consciously, for example, a delicious cup of coffee or clear signage, when someone first arrives somewhere. Other services in the environment are hardly noticed, like cleaning or security. Because service provision always involves a form of interaction between a service provider and customer, the behaviour is very noticeable. When people discuss an experience, they often mention what was striking about the service provider's response, for better or for worse. Therefore, the service provider's behaviour is a vital prerequisite for customer satisfaction. As a result, hospitality plays an increasingly important role in the work environment. It is the kind of hospitality we know from tourism, retail and the hospitality industry. Professional, effective hospitality is achieved when behaviour, services and environment form a well-considered combination.

Figure 3.3 shows the focus of this chapter: the experience of the work environment. This chapter builds on the social trends and developments discussed in Chapter 2.

The chapter starts with an exploration of hard services in section 3.2, followed by a discussion of the most common soft services in section 3.3. Since experience and behaviour are closely related, we conclude this chapter with a section on behaviour (3.4). With this, all elements of the PBE model (Figure 3.3) will be discussed.

FIGURE 3.3 The focus of this chapter

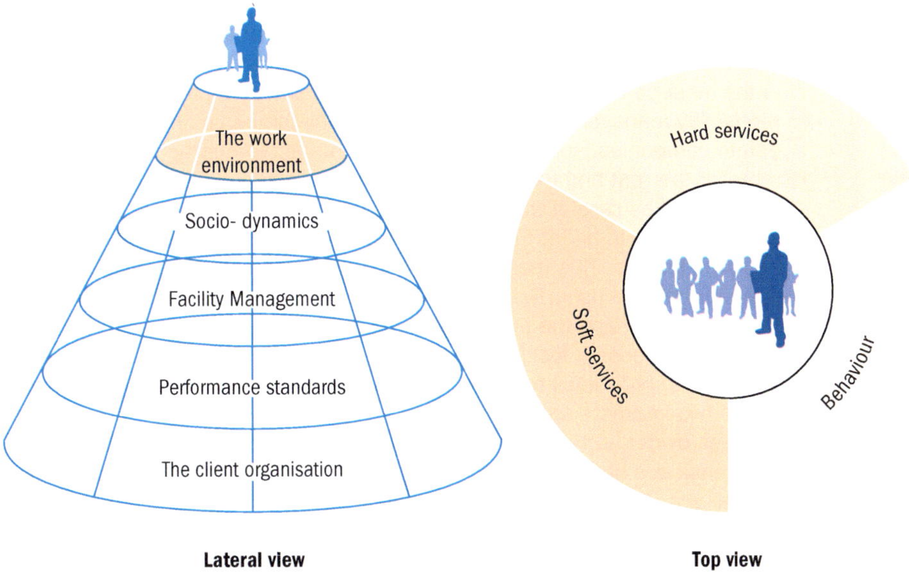

3.2 Hard services

The most tangible aspect of the experience is that of accommodation. Almost everything that has to do with buildings falls under this heading. Accommodating means offering shelter. The activities associated with accommodation fall under hard services. As far as buildings are concerned, the facility manager always has to deal with real estate management in one way or another. Real estate managers develop, manage and dispose of buildings and land, usually with the aim of achieving financial returns for the long term. In general, real estate managers see a building as an investment property and a business resource. In many cases, they work for investment companies, such as pension funds, banks and insurers. You could say that the real estate manager represents the interests of the building owner.

Hard services

The difference between facility management and real estate management is that facility management focuses on the internal support of the primary business process, while strategic real estate management deals with real estate as an investment object.

Real estate can be an important strategic asset. The strategic issues involved in the evaluation, design and realisation of buildings are therefore diverse in nature:

- In which market does the company operate?
- What are the mission and values of the company?
- What is the image, brand and identity of the company?
- How can the organisations' culture be characterised?
- What is the strategic policy, and what are the organisation's culture?
- Which scenarios of growth and decline should be taken into account?
- Are changes needed and, if so, what changes?
- Is there a preference for ownership or lease?

- Is there a preference for new construction or renovation?
- Which workplace concept is considered?

3.2.1 Real estate management versus facility management

Property managers

Broker

Owner

3

Not all users of a building are owners of the building; renting office buildings is the most common accommodation strategy. In such a rental situation, the facility manager inevitably has to deal with companies that work for real estate companies, such as real estate agents and property managers. With a broker the rent and the rental period are agreed on, and the property manager takes care of the maintenance of the building. The owner desires his real estate to be let for as many years as possible at the highest achievable rents. This means that the owner has an entirely different interest than the user. The owner will weigh all the wishes and requests of the user concerning the building against the short- and long-term yield. However, the user is looking for an optimal amount of office space with an attractive service level to serve the organisation as effectively as possible. These two interests are often contradictory, which creates a field of tension between owner and user.

Owner's interest

This tension is healthy and should apply not only if a property investor owns the building, but also if the organisation itself is the owner of the building. In that case, the owner's interest and the user's interest must also be balanced. The owner's interest is represented by corporate real estate management, which has the institution that all available square meters should yield money. The facility management function represents the demand side. This contradiction results in a supply and demand game, and some field of tension between 'rival' real estate management and facility management. The result is the highest possible occupancy of an optimal amount of office space, with an optimal service level at the lowest possible cost. Too many square metres result in too high accommodation costs, and too little in sub-optimal working conditions.

Accommodation management

Financial position

Accommodation management, or hard services, is the main focus for facility managers in many countries. In other countries, like the Netherlands, the focus is primarily on user services, at least at an operational and tactical level. The more strategic the position of the facility manager, the higher the chance that he is also involved in real estate. Real estate always requires a strategic approach, as many of the strategic real estate choices have a significant impact on the financial position of the organisation. A few examples:

- A lease of an office usually has a term of three to five years (although contracts with a much shorter lease are now also possible). All this time, the organisation has committed itself to pay the agreed rent. For a building of 12,500 square metres, this can quickly amount to more than 3 million euros per year.
- The construction of a standard office building with five hundred workplaces (12,500 square metres) requires an investment of between 18 million and 25 million euros.
- The construction of a university hospital requires an investment of 500 million euros.
- The entire construction process of a hospital takes an average of eight to nine years.

To get an idea of the demand for space, here are some examples of surface sizes:

- Retail area Ikea Concept Centre Delft, The Netherlands: 39,000 m2 (excluding offices);
- Floor space skyscraper 'the Gherkin' in London city, United Kingdom: 47,950 m2;
- Retail space Les Halles in Paris, France: 75,000 m2 (excluding offices);
- The Ministry of Security and Justice and the Ministry of Home Affairs in The Hague, The Netherlands: 132,000 m2;
- De Rotterdam (shops, restaurants, homes, offices) in Rotterdam, The Netherlands: 160,000 m2;
- The seat of the European Central Bank in Frankfurt am Main, Germany: 185,000 m2;
- Erasmus University Medical Centre in Rotterdam, The Netherlands: 210,000 m2;
- The Pentagon in Virginia, United States of America: 620,000 m2, the largest office building in the world with 26,000 users;
- Beijing Capital International Airport, Terminal 3, in Beijing, China: 986,000 m^2
- Aalsmeer Flower Auction: 990,000 m2, the world's largest auction building according to the Guinness Book of Records.

3.2.2 Site selection

After the decision that new premises are needed, the search for a suitable location follows. The location affects the visibility and image of a company, on the accessibility for customers and employees, and their safety. And in terms of costs; the city centre of a metropolis has a different price tag than a business park in rural areas. As the choice of location affects both the property portfolio and the use of the site, the facility manager and the property manager work closely together on the location choice. **Location choice**
Considerations are:

- financial feasibility in respect of the budget;
- market conformity of the building, in price but also the type of building;
- the sustainability of the building or the possibility of sustainable construction;
- futureproofing (flexibility);
- image and appearance of the location;
- safety and security of the environment;
- the quality of the site and the greenery;
- area-development in the vicinity;
- accessibility of the building; think of the disabled people but also suppliers;
- accessibility by public transport;
- available (or to be created) parking facilities and bicycle storage.

3.2.3 Building types

Many types of buildings can be distinguished in the work environment. Table 3.2 shows examples of building types divided into industries. **Building types**

TABLE 3.2 Examples of industry-specific building types

Industry	Industry-specific building types
Business	Office buildings Call centres
Healthcare	Hospital Nursing home Rehabilitation centre
Education	School Laboratory Auditorium
Industry	Factory, plant Storage, shipyard
National administration	Government building
Local Government	County hall
Municipality	Townhall
Logistics	Distribution centre Railway station Airport terminal
Retail	Shop
Social functions	Sports hall Prison Community centre
Public utility	Power plant
Housing associations	Social housing Offices Social real estate

3

Office-based organisations

Besides, there are special categories of buildings, such as museums, event halls and crematoria. Many companies use office buildings in addition to industry-specific structures; we call these office-based organisations. Section 3.4.4 is dedicated to office buildings, as this is the most common building type. In addition to workplaces, in enclosed or open spaces, an office building consists of areas with dedicated functions. Think of the company restaurant, the mailroom, reception, coffee corner(s), meeting rooms, and (sometimes) in-house fitness centre. Moreover, with an increased diversity of cultures, more and more companies are providing a quiet space where one can pray or meditate. Offices thus have a high intensity of facility management, meaning that the guest or user is explicitly in touch with facility management. With a low intensity, facility management plays a role 'behind the scenes', with less direct contact between facility staff and guests or users.

3.2.4 Office buildings

The facility manager is responsible for providing an optimal work environment. In the case of office-based organisations, the availability of an office building is the first point of focus. In the case of a new company (or business unit) that is to be established, a real estate agent will be consulted. Such a person knows the (near future) availability of offices in the area the company wants to establish itself in. If a building plot is available, consideration can be given to building something new, even when the company prefers to rent its office space. In this case, it is necessary to find a real estate investor that is willing to include (take ownership of) the building in his real estate portfolio. A broker needs the following information to preselect suitable premises:

- the required floor area (FA) or the number of square metres;
- specification of specific rooms and functions;
- the budget for purchase or rental;

- wishes concerning the location;
- requirements on sustainability (certification class);
- requirements in terms of accessibility and parking capacity;
- requirements in terms of visibility and appearance.

Surface measures

Traditionally, the required floor space is calculated by multiplying the number of employees by the gross area needed for one workplace (in the range of 18 to 25 square metres). This results in the GFO, or gross floor area (GFA). This is the aggregate surface area of all floors, measured from the outside of the facade. External walls, corridors, stairwells and lift shafts are also included. The gross internal area (GIA) is the area of the building measured to the internal face of the perimeter walls at each floor level. This is not the same as usable space; this is measured by the NIA. NIA is the abbreviation of net internal area: the usable area within a building measured to the inner face of the perimeter walls at each floor level. It does not include walls, staircases, loading bays, access ways or car parking areas or any area occupied by building installations. The NIA is also known as usable floor area (UFA).

Gross floor area (GFA)

Gross internal area (GIA)

Net internal area (NIA)

For benchmarking and estimating and cost planning purposes, the measurement of the floor area must be consistent across all building types. However, there are various standards in use. Relevant standards are the ISO 9836 - the definition and calculation of surface area and volume indicators, and the Royal Institute of Chartered Surveyors' Property Measurement standard. It is important to note that floor measurement standards will change over time to reflect the realities of building design, as well as changes in the real estate market.

The ratio between UFA and GFA reveals much about the efficiency of the building design. Even though only the useful surface is of interest for the organisation, the traffic and installation space, walls and layout-loss, also need to be paid per square metre. On the other hand, higher rents will be charged for efficient buildings compared to inefficient buildings.

Efficiency of the building

By no means do all employees need a permanent workplace. Many people work part-time, which means that if they had a fixed desk, their workplace would be unused part of the week. But even people who work full-time do not sit at their desks all day. They may be in meetings, work in a project team, follow a course, or visit external business relations. And, of course, full-time workers also sometimes take a day off. As a result, it is not uncommon that the average workplace occupancy rate is less than 40%. An occupancy rate of 40% means that on average, workstations are empty for 60% of the time during office hours! If you multiply this by the costs per square meter, you immediately see the potential cost savings. However, in the aftermath of the corona virus, the way people work may change dramatically and this could change the way we think about and the need for workspace.

Occupancy rate

● www.measuremen.io/cases/netflix

Netflix

Netflix is the world's leading provider of streaming media and is growing at a tremendous pace ever since 1997. Netflix' global real estate and workplace strategy is aimed at accommodating growth, and to attract and retain talented international oriented, highly IT educated knowledge workers. A complex task in a very turbulent environment. This task can be broken down into more operational workplace related questions to check and balance the workplace strategy. How efficiently and effectively are my current offices used? Is there any space for growth and if yes; how much and where exactly? Netflix EMEA hired Measuremen to conduct a Workplace Occupancy Study. The outcome of this study showed that there was ample room to accommodate more people. The flexible office space has an average occupancy rate of 36.2%, with peaks at Tuesdays and Thursdays and Fridays as the calmest day at the office. This pattern is equal to most organisations in Western Europe. The use of the workplaces shows interesting facts: it's mainly computer work (41,3%) or 'in use, but unoccupied' (46,4%). The last percentage shows that staff are often engaged in short meetings, leaving their belongings at their desk if not working at their laptop. The meeting rooms, in all types and sizes, have an overall occupancy rate of about 30%. The most popular meeting room types are the regular ones and the meeting tables in the restaurant.

Flexible working

It is possible to improve the occupancy rate of workplaces by having more people use the same number of workplaces. This saves the organisation valuable square metres and a considerable amount of money. The principle whereby several employees use the same workplace is called flexible working. Storage space is often centrally arranged (and more limited); working with digital documents is especially encouraged. However, the Netflix example shows that even in one of the most flexible organisations, flexible working doesn't necessarily lead to significant changes in occupancy rates. Also, non-financial arguments may be more critical than the cost of the inefficient use of space. For example, national and organisational culture are of strong influence. In hierarchical cultures, a workplace is used to communicate the status of a person. Even if a manager is in meetings most of the day, s/he will never agree to desk-sharing. In cultures with a low power-distance, desk-sharing may be used as a positive affirmation of the equality of managers and staff. And if leadership manages on employee-presence instead of output, flexible working arrangements will not easily be embraced, resulting in higher average occupancy rates.

Since the introduction of flexible working calculating the floor space required for an office building based on the organisation's workforce in full-time equivalents (FTEs) no longer applies. To compensate for expected low occupancy rates, a reduction factor (or flex factor) of 0.7 or less is now commonly applied. Seven workplaces for ten employees was revolutionary about ten years ago. Yet, now there are companies that assume six or even five workplaces for ten employees. A difficult point for facility professionals is that the number of employees fluctuates during the week and that attendance patterns are not always predictable. Every organisation has attendance peaks, for example, on the fixed meeting day, and times of day, or days, with very low attendance. In Western Europe, this is generally the Fridays (like in the Netflix example), and many part-time workers have one or more free afternoons when their children are in primary school. A possible solution is to calculate with a ratio of the number of employees per workplace, instead of with FTEs. Of course, this has an effect on the reduction factor used.

Attendance peaks

Matching the number of workplaces to the peak days is unwise because it leads to a too low occupancy rate in the rest of the week. To smooth out the peaks, organisations are experimenting with spreading meeting times and steering employees on working days. Workplace shortage in peak times can be (partly) solved by a broader definition of the workplace. Bar tables, open seats and the company restaurant can double as a work and consultation area.

Rijnstraat 8 in The Hague: winner of the ARC17 Architecture Award

Vacancy

In some countries or regions, the office market is characterised by an extensive supply and a high vacancy rate. In the Netherlands, there are forty to sixty million square metres of office space, of which some has been vacant for a very long time. This is partly due to the rise of remote working, in which offices are increasingly used only as a meeting place, and there is less need for square metres. And some of it is due to outdated buildings that require far-reaching renovations to match modern standards.

But moreover, too many new office buildings have been built in recent years. Although there is slightly more demand (take-up volume) for office space than in previous years, the vacancy rate is still substantial. Many organisations are now taking fewer metres in the new situation than they used to use in the past, see Figure 3.4 (Cushman & Wakefield, 2018). This 'buyers-market' has a positive side-effect: an increasing awareness of the needs of end-users among real estate professionals.

FIGURE 3.4 Developments in the Dutch office market mid-2018

OFFICE JOBS ▲ 2,445,350 m² 1.4% 2017 as compared to 2016	**AVAILABILITY** ▼ 6,240,000 m² 3.3% mid-2018 as compared to mid-2017	**STOCK** ▼ 47,491,000 m² 2.3% mid-2018 as compared to mid-2017
TAKE-UP VOLUME ▼ 558,000 m² 3.9% mid-2018 as compared to mid-2017	**VACANCY** ▼ 5,488,000 m² (11,6%) 3.7% mid-2018 as compared to mid-2017	**RENT** ▲ € 137 per m²/jaar 1.9% mid-2018 as compared to mid-2017

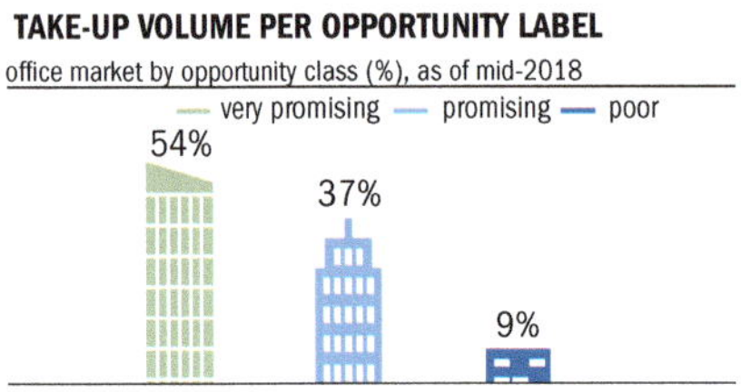

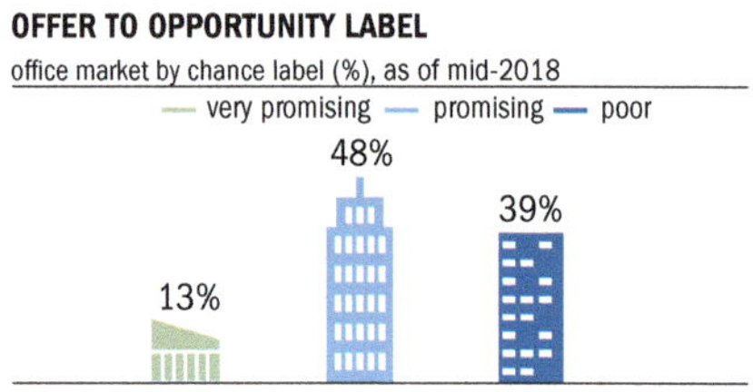

In 2018, the proportion of vacant office space fell to almost 5.5 million square metres. By mid-2018, the vacancy rate amounts 11.6% of the total stock of office space in the Netherlands. 39% of the office space on offer belongs to the category 'poor'. As Figure 3.5 illustrates, the risk of structural vacancy for these properties is high.

3.2.5 Space management and relocation management

Space management

An essential activity of the housing department is space management. Space management is a collective term for various activities:

- workplace management;
- inventory and allocation of space;
- management of common areas;
- design of spaces;
- signage;
- landscaping and art management.

Central and decentralised facilities

An important consideration in space management concerns the balance between central and decentralised facilities. Centralised facilities include meeting venues, catering facilities and flexible workstations that transcend departments. Examples of decentralised facilities are 'private' areas for departmental meetings, departmental workstations and pantries in the departments. A pantry is a service spot that combines vending machines, copying facilities, pigeonholes (if still in use) and lockers, often for a department or building floor.

The advantages of centrally offering facilities are:

- The management is effective and efficient: sharing means more capacity and better utilisation.
- Shared spaces can emphasise the culture and identity of the organisation.
- Shared spaces can support knowledge sharing through unplanned encounters.

The risks of central facilities are:

Central facilities

- During peak hours, demand can exceed supply. The question then is which reservation is given priority and especially what the alternative options are.
- Users feel (and take) less responsibility. For instance, regarding the cleanliness of the rooms after use and for reporting malfunctions. This places higher demands on the janitor function.

Brief

The allocation of space to organisational units is about serving the customer as well as possible and also about making optimal use of (expensive) square metres. In some cases, it's about allocating existing spaces; in others, it's about redesigning a building. In the latter case, from a brief (also referred to as Programme of Requirements, or specifications) to a spatial design (see Figure 3.5), three intermittent steps need to be taken.

FIGURE 3.5 From brief to furnishing plan

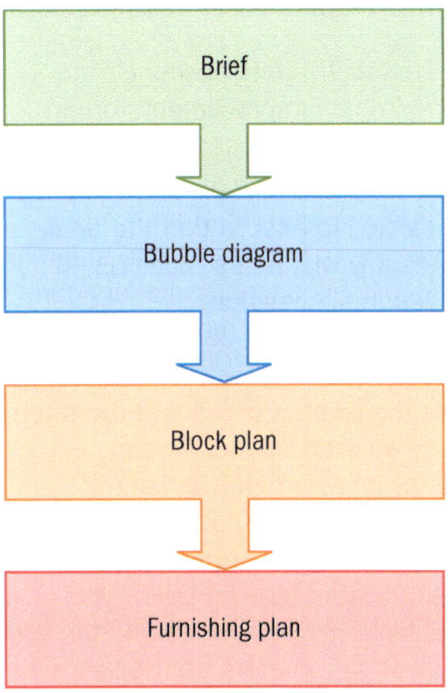

Brief

Relationship diagram

Later in this section, these steps are explained. Part of the brief is a relationship diagram, which indicates which functions can be clustered and which should not. Figure 3.6 is an example of a relationship diagram. On the interface of two functions, the nature of the relationship is indicated by the letters p, s and a. The letter p indicates that there is a primary relationship between functions; for example, between a manager and his secretariat. In primary relationships, it is important that such functions are adjacent. Secondary relationships (s) are relationships that have a close interconnection and therefore need to be placed in each other's vicinity, but not necessarily next to each other. Anti-relationships (a) are those functions that should not be placed next to each other, such as a meeting room and the reception desk in Figure 3.6. The relationship diagram will be different for each organisation, depending on functions, processes, culture and working style.

FIGURE 3.6 Example of a relationship diagram

	Groups	Manager	Secretariat	Policy officers	Function A	Function B	Function C	etcetera	Storage	Meeting room	Informal meeting area	Reception desk	Flexible workstations
1	Manager												
2	Secretariat	p											
3	Policy officers	s											
4	Function A												
5	Function B	s	a	p									
6	Function C		s	s									
7	etcetera												
8	Storage	a	a	a	s	s	s						
9	Meeting room	p	p	s					a				
10	Informal meeting area	a	a	a	s	s	s		a	s			
11	Reception desk	a	a	a	a	a	a		s	a	s		
12	Flexible workstations	s	s	s	s	s	s			s	s		

p = primary relationship (adjacent)
s = secondary relationship (in each other's vicinity, same floor)
a = anti-relationship (cannot be placed next to each other)

Based on the brief, various designs are possible. To ensure that the final result meets all requirements and and meets any wishes as much as possible, the following design steps are taken in succession.

Step 1: Bubble diagram
As the design process advances, the higher the level of detail and the more costly it is to make adjustments. That is why we start with a bubble diagram. This sketch visualises alternative layouts without the risk of spending too much time working out the details of plans that will change.

Bubble diagram

A bubble diagram consists of one or more floor plan drawings in which the use of space per department or function is indicated by a 'bubble'. The level of detail is relatively low. Walls, doors, furnishing elements and the like are not indicated. Thus, the exact number of square meters per department or function cannot yet be calculated. Still, it does show an indication of relative size, the location, and the mutual relationship between the bubbles. A bubble is created on the basis of:

- the required floor space, as stated in the spatial brief;
- a relationship diagram;
- and the available space in the building plan.

It is not often that the required square metres can be fit-in exactly; as you have to deal with the space between the façades, traffic space, constructive walls, columns and the like. Fixed elements and zones like these are part of a so-called construction drawing, in which the bubble diagram is made (see Figure 3.7).

Design

Usually, the design of the bubble diagram starts with the general facilities. After that, based on the requested and available square meters, and the desired mutual relationships between these departments, a rough arrangement can be made for the departments.

FIGURE 3.7 Example of a bubble diagram

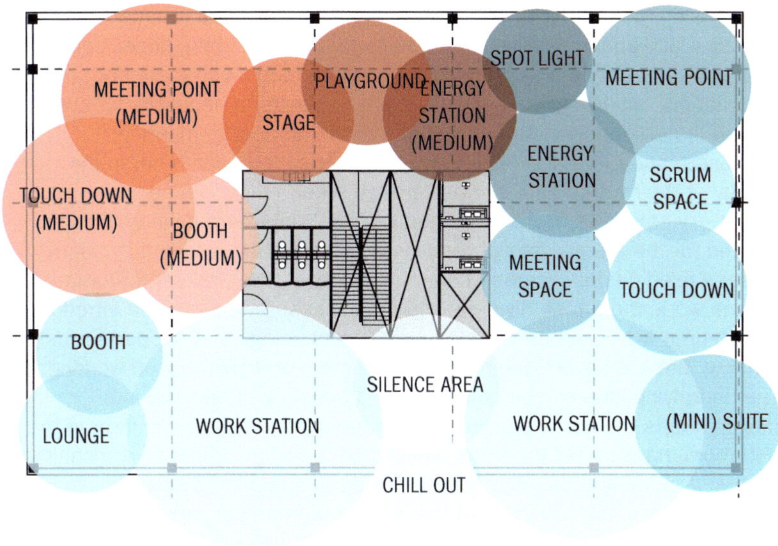

The following information is required to determine the floor space required per department:

- staff headcount in full-time equivalents (FTEs) for workplaces;
- expected presence and the associated occupancy rate of the workstations (flex ratio);
- workplaces and specific areas (in capacity and size);
- any decentralised facilities, such as dedicated meeting and project rooms, with associated capacity calculation;
- the surface area of installations, machinery and equipment and space needed to operate them;
- relationship diagram, in which the relations between departments are laid down.

The surfaces of traffic areas and areas for general facilities (such as sanitary facilities, storage areas) need to be added, e.g. based on a percentage of the total area required or by applying standards, such as the number of users per restroom.

Step 2: Block plan

After completion of the bubble diagram, it is time to determine the position of the partition walls. This results in a block plan, another important document for the design process (see Figure 3.8). **Block plan**

An organisation's culture is strongly related to the way people work together and therefore influences the workspace design. In traditional, hierarchical organisational cultures, managers will have their own office, often on a higher floor than the employees they supervise. In less hierarchical organisations, employees, regardless of their function, share the same working environment without walls that separate the managers from the (subordinate) employees.

When office buildings came into being, almost all employees had their workspace. There were no computers yet - a considerable amount of intensive and administrative work was carried out with pen, paper, calculator and typewriter until the seventies of the last century. With the introduction of automation, especially the network, it was no longer a necessity for everyone to have a personal workstation. Moreover, the number of meetings increased, and more and more employees consulted each other in the vicinity of their workstations in designated areas - the meeting rooms. But still, every employee had a dedicated workstation, where s/he could be located for the greater part of the day. Over time, the average occupancy rates of fixed desks decreased dramatically. Job development (automation of routine work, more task diversity and project-based working), the introduction of part-time work, and the development of technology that supports remote working, are the drivers behind the introduction of activity-based working. The underlying idea can be compared to a house, because the interior design meets the same principle. You sleep in the bedroom, shower in the bathroom, have breakfast at the dining table, read a book in an armchair in the living room, and so on. In a house, a person relocates according to what he or she does. It is very logical that what has been customary in the house for a long time, is also applied in the work environment. After all, people don't do the same thing all day long. Thus, in an activity-based workplace concept, workers can choose different types of workplaces, according to their activity. Regardless of the function, most divisions of the organisation will be able to use the work environment without having to move a single table, desk or cupboard. Only secretarial and purely administrative functions still have a fixed place. The main advantages of such a workspace concept lie in a potentially higher occupancy rate. However, the example of Netflix shows that this effect is not guaranteed.

Activity-based working

Many organisations are dynamic in nature. Reorganisations and restructurings often result in downsizing or growth and the splitting or merging of departments, resulting in changes in the need for floor space overall and on department-level. To adapt to changing needs, many internal relocations take place. Less frequent but impactful are external removals: from one building to another. It is recommended to standardise the furniture so that as little inventory as possible needs to be moved when relocating. This is another reason for the popularity of the activity-based working concept. Examples of such workplaces include:

Relocations

- private places for work that requires concentration and confidential conversations
- project rooms for collaborative work;
- workplaces in an open plan setting for routine work and ad hoc consultation with colleagues;
- library-like spaces for working in silence;
- relaxation areas for meeting and informal consultation;
- meeting rooms for formal consultation.

Another important determining factor for the layout is the structure of the building, as the design of a building imposes restrictions on the layout. Whether a building is suitable for a particular user cannot be determined solely based on a brief, relationship diagram and available surface area. In basis, the structure or form and the building grid determine the suitability. One of the methods used to assess this suitability is to determine the ratio of circulation space to the total surface area. In most cases, the circulation space cannot be changed. The layout of the circulation zones is dependent on the location of stairs, lifts, entrances and escape routes. Also, sanitary areas, installations, pipe shafts, meter and work cabinets limit the layout possibilities of a building. Moreover, fire safety regulations apply to circulation zones. In addition, a building can contain empty spaces, such as sharp corners in the façades, or voids (that is, spaces over several floors, without mezzanine floors) and excessively narrow residual spaces.

Building grid

In the construction industry, standardisation is often applied to the dimensions of building elements, such as concrete slabs and prefabricated concrete beams. Usually, the construction of a building is dimensioned at a multiple of 30 centimetres. Standard distances between walls and columns are 1.80, 2.70, 3.60, 5.40 and 7.20 metres.

FIGURE 3.8 Example of a block plan

In office buildings, for example, the columns are very often spaced 5.40 or 7.20 metres apart. This is referred to as a heart-to-heart distance of 5.40 or 7.20 metres.
Between these structural elements, internal walls can be placed that connect to, for example, a frame and 'the heart' of a column. Many buildings have a grid for placing the inner walls.

Step 3: Furnishing plan
Furnishing plan
After completion of the block plan, the drawing becomes detailed by placing furnishing elements: the furnishing plan (see Figure 3.10). This plan provides information about the location of the furniture in the various rooms. Examples of standard workstation elements are desk, chair, cabinet or locker, meeting facility. The following must be considered:
- adjustable or non-adjustable furniture;
- flexible use or permanent user;

FIGURE 3.9 Workstation ergonomics

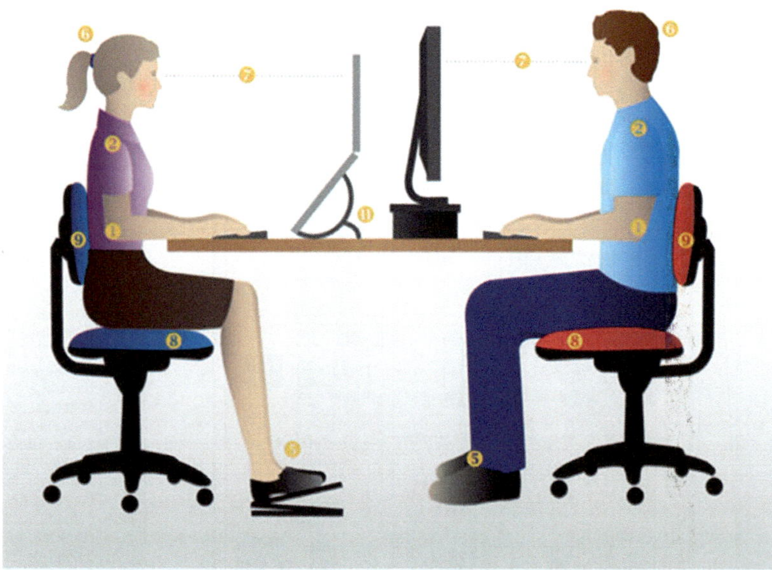

1 **Elbows**
Above the desk, at 90-110 degrees
2 **Shoulders**
Relaxed as opposed to hunched
3 **Wrists**
In line with forearms
4 **Hips, Knees, Ankles**
At 90 degrees whilst seated
5 **Feet**
Flat on the ground or footrest
For prolonged standing, consider a mat
6 **Head**
Upright with ears aligned with shoulders
7 **Eyes**
Looking at the top third of the screen. Consider the use of a laptop raiser with your laptop
8 **Seat length**
Should be long enough to provide support beneath thighs
9 **Backrest**
Angled at 90-110 degrees with adequate lumbar support in line with lower back
10 **Keyboard and Mouse**
G and H of keyboard aligned with your nose. Mouse gripped loosely
11 **Laptop**
Used with a riser, external keyboard and external mouse

- short- or long-term use of the workplace;
- desired atmosphere;
- storage capacity and if applicable: physical archive space.

The furnishings must comply with the applicable guidelines for working conditions (Working Conditions Act). These guidelines are detailed and depend on the duration of use. Figure 3.9 shows an example of the Occupational Health and Safety Directives concerning computer workstations.

The furnishing plan is essential for many reasons:
- to know what inventory is needed (purchasing);
- to determine the required cabling;
- to know what furniture should go where (relocation);
- to be able to charge the costs of the use of space to the budget holders;
- for the management of the inventory after delivery (identification, stock, and replacement after depreciation).

FIGURE 3.10 Sample of a 2D and 3D representation of a furnishing plan

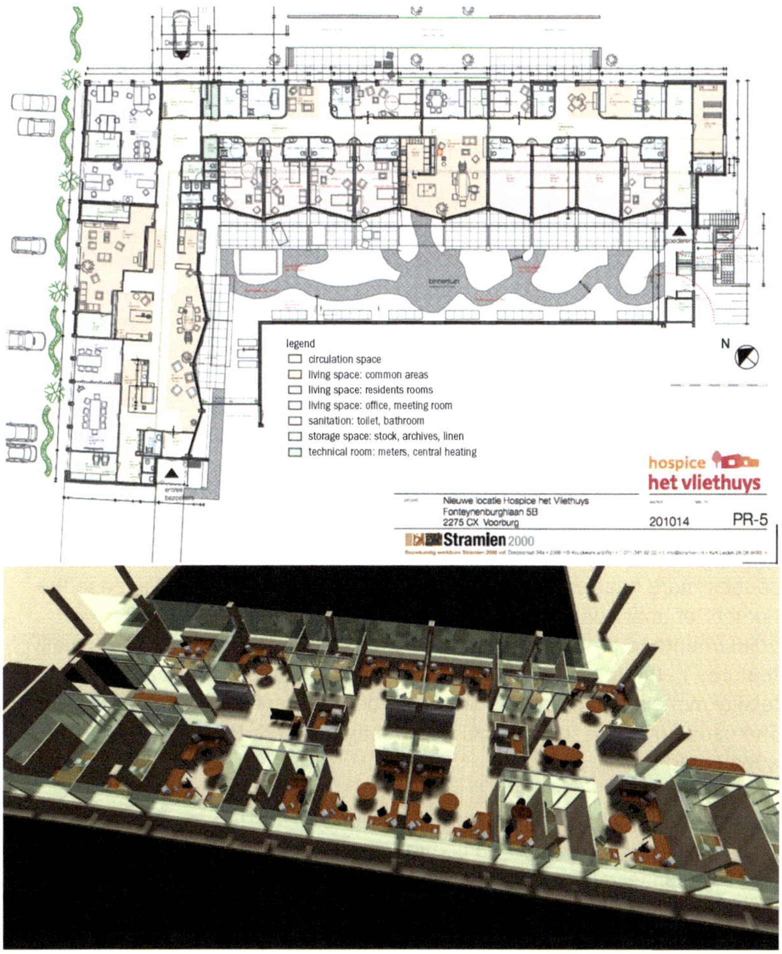

Future users are involved in the first two steps of the design phase through user consultation. However, for many users, it only becomes clear what their future work environment will look like once the layout plan is completed. Still, to many users interpreting floor plans is such a challenge, that defects or errors are too often only detected after commissioning. Adjustments afterwards come at a cost. Fortunately, there is a solution: 3-dimensional drawings and virtual reality simulations to help users interpret the plans (see Figure 3.10). 3D images can be displayed on paper or screen or via projection and simulations. In virtual reality (VR), the observer uses special glasses that completely enclose the eyes. Looking left or right, the image of the future working environment changes along with it: in 3D, interactively and 360 degrees. This technique allows everyone to experience design as if it were already there, thus making it much easier to identify shortcomings in the planning phase.

3.2.6 Building management

So far, we have discussed the steps in the creation or renovation of facilities. The next phase is the maintenance and optimisation of the environment. If the building is owned, first and foremost, the construction should be highly resistant and insured against fire and storms. In all situations, the facility manager is responsible for hedging the risks related to the use of a building, as regards: furnishing, furniture and user-equipment.

Maintenance
Building maintenance

The maintenance of the building and the building installations also belongs to building management. Building maintenance comprises all building-related activities intending to maintain and optimise the functional value of the building.
Regular maintenance is essential for aesthetic and safety reasons. Many malfunctions can be prevented by effective preventive maintenance. To avoid unpleasant financial surprises, a so-called long-term maintenance plan (LTMP) must be compiled for each building. In this plan, the necessary maintenance and the required budget are distributed over a more extended period (five to ten years). A new development is the use of predictive maintenance. By means of the internet, building installations can communicate with a server. Self-learning algorithms can be used to convert the acquired data into information, to predict when an installation will fail. In this way, maintenance can be scheduled timely, and malfunctions can be prevented.

Long-term maintenance plan (LTMP)

Also, changes in the use of buildings may require interim adjustments, for example, to the climate control system. An increasing number of computers causes more heat production, and flexible working concepts increase the number of users per room. Moving internal walls may also require adjustments to the climate control installation. The indoor climate usually features in the top 5 of complaints. It is a continuous point of attention for facility managers. New technology may offer a breakthrough in this area. The cause of users' claims can be investigated with the help of virtual and augmented reality.

Indoor climate

Three parties play a role in maintenance: the client (the owner of the building or facility manager), a contractor that carries out the maintenance and the planner who schedules the maintenance. This triangular relationship creates a game of interests: if a too-tight schedule with a

budget is submitted, the contractor will object to it. The planner will then present arguments as to why he thinks the plan is realistic. This division of roles will keep the parties on their toes.
The immediate surroundings of buildings also fall under building management, such as gardens and grounds, and their facilities. Think of outdoor lighting, sewage and other piping, access control and surveillance systems (cameras, barriers). Here too, maintenance plays a role, such as paving and landscaping. Parking space also is a part of building management, both in a parking garage and outside on the grounds. In addition to climate control, this is the most sensitive issue for the organisation's employees within building management. There seems to be no subject that can harm the image of a facilities company more than a lack of parking spaces.

Surroundings

Parking

Moreover, energy and water management are fundamental aspects of building management. World-wide our energy consumption is increasing, Sources of fossil energy are depleting, and the earth is warming due to the emission of CO2. As buildings account for about 40% of energy consumption and subsequent CO2 emissions, the built environment is an essential focal point for fighting climate change. Approximately 40% of the built environments' CO2-emissions is related to energy consumption. According to European agreements, the built environment must be ultimately energy-neutral by 2050. This means that the building's energy consumption (heating, hot water, installations) comes from CO2-neutral, renewable sources (solar, wind, hydro). A vast challenge every facility professional is facing!

Energy-neutral by 2050

Challenge

First and foremost, it is essential to use as little energy as possible. For the latter, an energy consumption monitor is visibly present in many buildings. This monitor shows the current consumption (also of water) and the deviations from, for example, the previous year. Such a monitor can create a positive atmosphere of competition: employees turn it into a 'sport' to consume as little as possible - in other words, to achieve declining trends.
Dwindling supplies of safe drinking water is a major problem impacting every continent. By 2050, it is projected that at least one in four people is likely to be affected by recurring water shortages. Water scarcity affects more than 40 per cent of people around the world, an alarming figure that is expected to increase with the rise of global temperatures as a consequence of climate change. Almost three-quarters of the earth consists of water. Of all that water, 98% is in the seas. Only two per cent of the world's water is fresh, but half of it is frozen on mountain tops or at the North or South Pole. Of all the water in the world, we can only use one per cent for drinking water. Most water consumption is indirect and used for the production of food, clothing and industrial products. In fact, 2% is in our daily water consumption and 98% in our production processes. To save water, besides direct consumption of the fresh water for building operations and consumption (e.g. coffee and tea), facility managers need to consider the freshwater needed for the production of goods they purchase.

Water scarcity

To encourage the business community to take the necessary sustainability measures, the government imposes standards on the energy performance of buildings. It is needed to know precisely how sustainable a building is, and which improvement opportunities make the most significant positive contribution. Specific assessment methods have been developed to measure the level of sustainability of properties, of which BREEAM and LEED are the best known in the Netherlands (see also section 4.8.1). Both methods also offer the possibility to certify buildings by an independent assessor. Real estate investors and parties that rent out real estate commercially make use of it in particular, as the sustainability of a building can be an attractive feature for a potential tenant. For example, because the building has lower energy costs or because it fits in with the sustainable image of the company. Certification can also be attractive for building owners who use a building themselves. A building that proves to be sustainable makes an organisation's sustainability policy visible to stakeholders and contributes to the desired image.

Certify buildings

● https://www.worldgbc.org

Net Zero Carbon Buildings Commitment

In September 2018, the World Green Building Council (WorldGBC) officially launched its Net Zero Carbon Buildings Commitment as part of the momentous Global Climate Action Summit. Businesses across the world (representing US$ 22.95 billion in revenue throughout the building and construction supply chain), have set ambitious targets to eliminate operational carbon emissions from their building portfolios. Collectively, these organisations are committed to reducing a cumulative total of 209 million tonnes of carbon emissions equivalent (CO2e) from their buildings by 2050. That is the equivalent of 44.7 million cars off the road for one year. Leaders from some of the world's biggest cities plus two major regions have committed to enacting regulations and/or planning policy, that will require all new buildings within their jurisdiction to operate at net zero carbon from 2030; and all buildings, including existing, to operate at net zero carbon by 2050. Some cities, states and regional governments, have additionally committed to ensure the municipal assets they own, operate and develop are net zero carbon by 2030.

To enable parties to act on the Commitment, WorldGBC's network of almost 70 national Green Building Councils (GBCs) will continue to provide a range of transformational and capacity building support including advocacy, policy development, education and certification.
WorldGBC's next priority is to address embodied carbon, which is the carbon dioxide emitted during the manufacturing, transport and construction of building materials, and the end of life emissions. Embodied carbon emissions must reach net zero by 2050 to achieve a below 2-degree scenario. Besides a contribution towards slowing global warming, green buildings contribute to the realisation of the Global Sustainable Development Goals (SDG).

3.3 Soft services

A soft service is a process or a collection of operations to achieve an agreed facility result. **Soft service**

3.3.1 Service, process, activity and experience

A process consists of a series of interrelated activities which are carried out in a certain order, to transform inputs to outputs that create value for the customer. As this is rather abstract, the following is a practical example of the company catering service. For foodservice, several processes need to be carried out to achieve a certain result. This result can be: **Result**

- the possibility to have lunch for all employees and visitors who wish to do so;
- a supply of coffee, tea and soft drinks with sufficient capacity and within a reasonable distance from the employees' workstations.

The result consists of physical, tangible food and beverages. In this example, lunch in the company restaurant and drinks from vending machines on every floor of the building. Various activities are required, such as taking orders, preparing lunch, preparing the lunch assortment at the distribution desk, collecting used crockery and washing dishes. In general, the result of service is tangible. Still, there are also intangible results, such as a professional reception at the entrance to the company restaurant, 'delivered' by the catering employee. **Tangible** **Intangible**

Figure 3.11 shows the relationship between the concepts of service, process, activity, result and experience.

FIGURE 3.11 Relation of service, process, activity, product and experience

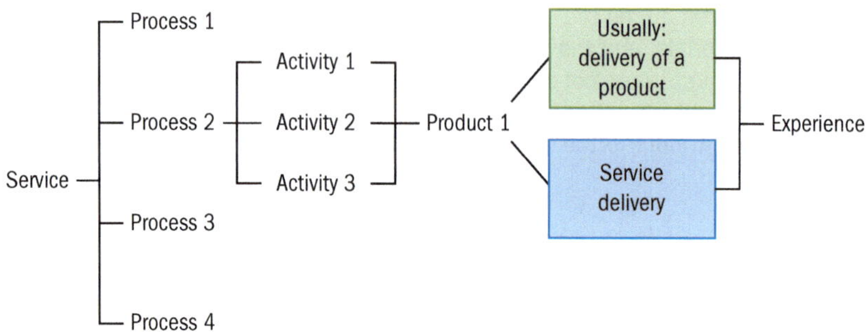

The delivery of a good, like a meal or coffee, is a tangible result. Non-tangible results are, for example, the interaction between people, like help or assistance, and the ambience in which the service is delivered. Delivery, ambience and support together create an experience, both for the facility employee and the customer. Figure 3.12 shows examples of processes, activities and results that belong to the company catering service.

FIGURE 3.12 Matrix of examples

Service	Process	Activity	Product/result
Food services	Lunch	Unpack delivered food and ingredients	Meal delivery Assistance: distribution and checkout Experience: pleasant relaxation
		Food preparation	
		Collect dirty cutlery and waste	
		Dishwashing	
		Disposal of waste	
	Beverages	Supply vending machines	Provision of beverages Assistance: none Experience: prompt service at any time
		Vending maintenance	
		Emptying vending machines	

Table 3.3 describes the characteristics of these three main types of processes: primary, support and management processes.

TABLE 3.3 Characteristics of different types of processes

Type of process	Description
Primary processes	The core activities of the organisation. For example, the production of shoes, curing people, or the processing of a mortgage application at a bank.
Support processes	All processes that support the primary processes; these processes occur in departments such as facility management, real estate management, IT, financial administration and human resources management.
Management processes	The management processes, the strategic activities to lead the organisation or business unit; examples are policy preparation, planning and control and strategic reporting.

The most striking differences between organisations are evident in the primary processes. Primary processes, also called core processes, include the activities to produce the product or service for the external customer. For a production company, these are the planning, purchasing, production and sales. For a healthcare institution, these are the intake, treatment and aftercare. This classification shows how (throughput), the input (means of production) is converted into output (products or services).

Primary processes

For primary processes to take place, supporting processes are required. All supporting processes that are related to the physical work environment fall under facility management. In a hospital, building maintainance must be scheduled, patients must be registered at reception, and security is needed to prevent expensive medical equipment from being stolen. For patients' recovery, it is crucial that the hospital is clean and that meals match the patient's diet. These processes are necessary to provide medical care to patients. Still, they are not a direct part of the care process. In an office-based organisation, support processes offer a healthy, safe work environment, payment of salaries, IT infrastructure, etcetera.

Supporting processes

Management processes are related to the planning, evaluation and adjustment of the organisation. The management processes take place throughout the entire organisation, because each department, including facility management, has its management processes. These processes almost always consist of steps from the plan-do-check-act cycle (see Figure 4.13).

Management processes

3.3.2 Describing and depicting processes

The main facilities process outlines how the facilities organisation is managed. From there, you can derive the work processes (service processes) with further detailing of the input - throughput - output. Work processes are more detailed but still encompass multiple employees or departments. Subsequently work instructions are derived from the work processes. A work instruction describes successive activities and actions to be taken. Figure 3.13 shows the relation between the main process, the work process and the work instructions.

Facilities process

FIGURE 3.13 Example of three-tier process description in a restaurant

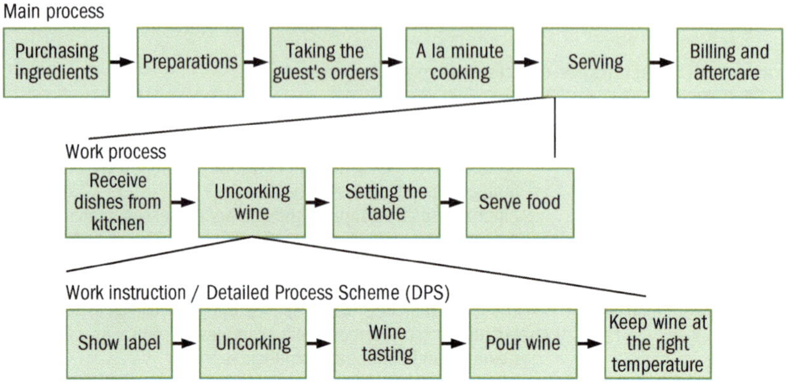

Process descriptions

Process descriptions form the basis for standardised and measurable services. A process description is a schematic representation of a process. By visualising the connection between processes, it is possible to identify opportunities for improvement. Sometimes processes do not fit well together, or work is duplicated without people knowing it. Optimising processes by removing unnecessary actions is in line with a 'lean' way of working (see section 5.4.4).

Role obscurity

For service processes to run smoothly, every employee must know precisely what is expected of him, and what the role of the employee is in the overall process. In reality, this is not always the case. Employees without sufficient training or insufficient knowledge, information or skills to do the job properly, will not be able to carry out the processes as optimally as possible. In these situations, we speak of role obscurity. Role obscurity can be prevented by providing work instructions. However, clear roles can only be described if the service process has been mapped out based on a flow chart or blueprint. Naturally, role obscurity is more common in services that fall outside the standard package.

Flowcharts

Flowcharts can be used for a graphical representation of the dynamics within a process. They represent the flows of successive tasks within an organisation in chronological order. Individual steps in the process, with associated information flows, are connected using lines. The level of detail in which a flowchart must be drawn up depends on the type of service and its complexity. In the event of a multitude of processes in an organisation, it is advisable to divide the main process, work processes and work instructions. A series of flowcharts then provides an overview of the interrelationships between processes in the organisation.

Figure 3.14 shows the sequence of activities for dealing with a malfunction reported by a guest of a bungalow park. The diamond shapes indicate a decision in the process; they always contain a question to which several answers can be designated by arrows that follow the decision diamond. In this flowchart, the process is provided with two questions or options: whether the mechanic is present at the location where the malfunction is reported and whether the failure can be resolved by him/herself. The possible answers are yes or no. But there may also be three, for example for a traffic light: is it red, orange or green? The answer to the question determines the further course of the process: whether the process can be continued, whether intermediate steps must be taken, or whether feedback must take place. Appendix 2 contains the symbols used in flow charts and their meaning.

FIGURE 3.14 Example flowchart for handling a malfunction report

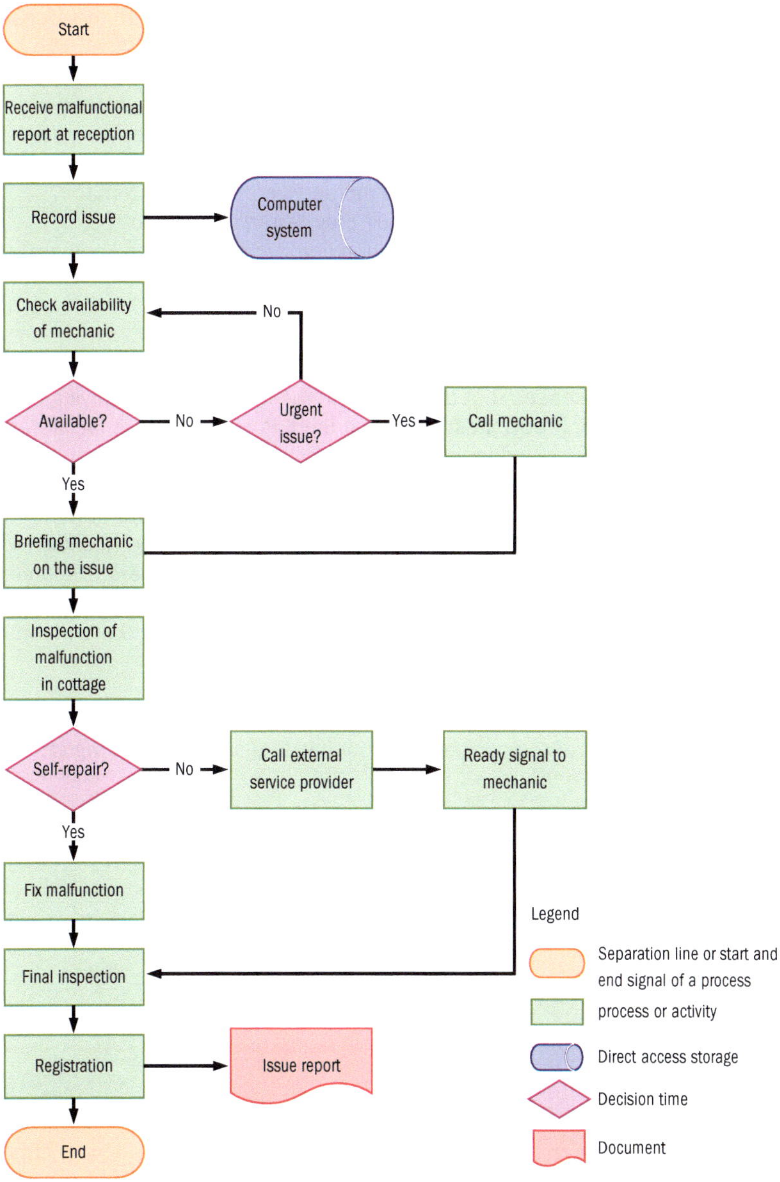

Shostack developed a technique based on flow charts but refined for the service process: the service blueprint. It is a visual representation of a service process from the user's point of view.

Service blueprint A service blueprint offers the possibility to design, implement and monitor the what (final result) and how-quality (how the result is achieved) (see also section 5.4). Reducing the number of flaws within a process is one of the most important goals of the blueprint. Also, measures can be built into the design of the service to reduce the consequences of any errors.
This method uses two principles for improving quality: the lines of visibility and the fail points. The lines of visibility divides the blueprint into the steps visible to the user (frontoffice) and the non-visible steps (backoffice). The facility manager should be aware that these invisible steps have a major impact on the steps that are directly perceptible by the user. By identifying the fail points, the points that are most sensitive to irregularities become visible. These are, among other things, the moments when the interaction takes place between the customer and the service provider (external interaction line) and between the service provider and its supporting department or supplier (internal interaction line). The interaction lines are depicted as dotted lines in Figure 3.15.

Lines of visibility

Fail points

For a reliable service blueprint, four steps must be taken:
1 Identify the processes. The first step in creating a blueprint consists of mapping the processes that together form the service. By identifying the various activities, it becomes clear what input is required. This facilitates the control, analysis and improvement of the entire process.
2 Isolate any vulnerabilities. Once the relevant processes have been mapped out, it is possible to identify the potential flaws in the system.
3 Establish a time frame. Because all services depend on time (a critical cost driver), the designer will calculate and determine standard execution time.
4 Analyse the costs. Based on the standard execution time and rates, the cost price is determined.

FIGURE 3.15 Example of a service blueprint for handling a malfunction report

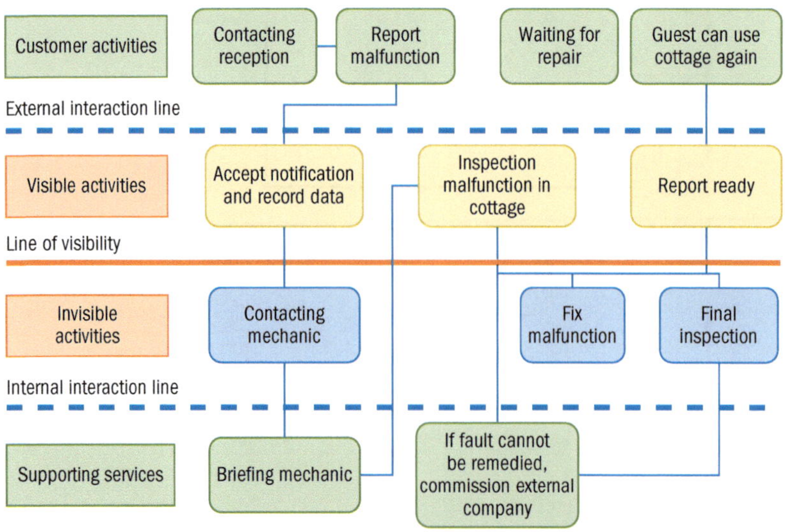

To be able to control service processes, indicators of the quality of the process are needed: the so-called performance indicators (PI's). These are quantifiable metrics that can be used to determine whether the service was provided following the agreements (see Chapter 5). Examples are waiting time, handling speed, customer satisfaction and the number of complaints. These correspond to the meters on a dashboard of a car. In a service blueprint, performance indicators are translated into tolerance margins. Tolerance margins can be set regarding time (for example, a waiting time of three minutes for a meal is acceptable), but also in the form of orders or output. The tolerance margins may differ per customer type. This way, any variation in the performance of the service can be minimised, while still fitting the interactive and thus somewhat unpredictable course of services processes.

Tolerance margins

A service blueprint can also be part of a customer journey map (see the image in section 3.4.3). In such a case, the external interaction points are viewed from the perspective of the different personas - fictional characters created to represent a customer group. Creating a blueprint is a laborious and meticulous process. However, the method offers so many advantages in terms of design, management and control of the service process that they certainly outweigh the disadvantages.

To effectively manage processes, the facility manager must know what the critical success factors of a product or service are. These are the conditions that always require the organisation's attention in order to be successful. The performance indicators that measure whether the critical success factors are achieved are called the key performance indicators (KPI's). These are the essential indicators used to manage and steer the organisation. (see also section 5.4.3).

Critical success factors

The following is a description of the services that occur in almost every facility management organisation. For each of these services, the characteristics are described, as well as the critical success factors.

3.3.3 Reception services

Reception

First impression

The reception is an essential point of contact for visitors and customers of the organisation, both physically (at the counter) and by telephone (via the organisation's general telephone number). Based on this contact, the first impression is given to visitors or customers. The reception is therefore considered to be the business card of the organisation. Professional, representative reception staff that create a sense of welcome for visitors or customers have a positive influence on the image of the organisation as a whole. The following aspects determine satisfaction with the reception:

- The appearance of the reception. This is largely determined by the friendliness and representativeness of the receptionist (i.e. behaviour), but also by the decoration and ambience of the reception area.
- Accessibility of the receptionist. In addition to the physical desk, the reception desk can also be reached by telephone, e-mail and sometimes by WhatsApp. An easily accessible reception staff member responds quickly and adequately to questions via all channels.

- The availability of the receptionist (desk). This aspect relates to the staffing of the desk. With correct manning, the waiting time is minimal and always acceptable for users. If the reception desk is also used by security staff, peaks can be dealt with flexibly by involving security staff in the handling of desk traffic.
- Quality of call forwarding. This is about finding the right person quickly and providing correct information for the transfer (name and reason for the call).

Critical success factors of reception services are:
- the appearance of the reception suits the identity and image of the organisation;
- accessibility and availability;
- speed and reliability of the provision of information;
- friendliness and service orientation of the reception staff.

3.3.4 Facilities service desk

Facilities service desk

The facilities service desk, or front office, can provide a user with information about facility services and handles service requests, bookings, reported malfunctions and complaints. The facilities service desk can be reached in various ways: physically (desk), but also by telephone and digitally (self-service). The service desk is the customer contact point of the facilities organisation. The image of the facilities organisation is therefore partly formed by the service quality provided by the service desk. In particular, user-friendly digital accessibility is critical in light of an increasing degree of self-service. Aspects that affect the quality experience of the user are as follows:

Image

- The convenience of the service desk. Users differ in their preference for making contact. The more user-friendly channels a user has at his disposal, the more satisfied customers are with the accessibility of the service desk.
- The user-friendliness of self-service. Increasing numbers of organisations use a facility management information system (FMIS) with a self-service portal or app for users.
- The transparency of follow-up processes.
- The service employees' expertise.
- The customer orientation of the employees of the service desk. The speed and timeliness of action.
- The turnaround times.

Service desk employees act on work instructions that describe how they should handle a request for information, a reported malfunction or a booking. However, there are situations where an exception to the standard is desirable or even necessary from the customer's point of view. A facility service employee who adheres to the standard procedures at all times may be perceived as non-customer-oriented and bureaucratic.

Critical success factors of a facilities service desk are:
- accessibility of the service desk;
- user-friendliness and efficiency of self-service;
- customer focus and expertise of service staff, such as proactive thinking and suggesting alternatives, a multidisciplinary approach;

- standardisation of work processes, so that every employee of the facilities reporting centre accepts, enters and finalises reports in the same way;
- coordination between the facilities service desk (front office) and back office (that act upon notifications);
- informing the customer with progress updates and the expected processing time;
- processing time from start to finish;
- meeting service level standards.

3.3.5 Safety & security

As we have already shown in Table 3.1, the facility manager is also responsible for the safety of users and the security of the physical environment. Safety and security are complementary and are, therefore, often mentioned together.

Safety concerns the safety of users in a building. Every country has regulations regarding fire safety and evacuation. EU legislature has established a system of basic principles of safety management, which must be transposed into national law by the Member States. General basic requirements for safety and health at workplaces are laid down in Workplace Directive 1989/654/EEC concerning the minimum safety and health requirements for the workplace (referred to as WPD). The WPD aims to introduce minimum measures designed to improve the working environment, to guarantee a better standard of safety and health protection. However, you can only implement the right measures if you are aware of where the risks are in your company and what their potential impact could be. A Risk Assessment & Evaluation, is a list of all health and safety risks involved in business operations and a plan for their solution. A RA&E is mandatory for all business owners with any employees under the Dutch Working Conditions Act. in-house emergency services play a vital role in safety at work. An emergency response officer is an employee who has followed a training programme allowing him or her to assist others in case of minor incidents. The primary duties of an emergency response officer are to:

Risk assessment & evaluation

In-house emergency services

- provide first aid in case of accidents,
- limit and fight a starting fire,
- alert and evacuate all persons present in the building in case of an emergency,
- assist other first-aid organisations, such as the fire department and ambulance service.

An emergency response officer plays the role of the outpost for the fire department and other first-aid organisations. Emergency response officers intervene after an accident or fire until the professional first-aid organisations take over this task.

Critical success factors of in-house emergency services are:
- the presence of an up-to-date in-house emergency response plan (First Aid Fire Safety plan) that meets the requirements of the Working Conditions Act;
- sufficient FAFS personnel;
- the expertise of the in-house emergency response workers (training, diplomas);

- familiarity with the in-house emergency response organisation among users;
- familiarity with alerting procedures;
- accessibility in the event of an alarm notification;
- response time in the follow-up of an alarm call;
- the method of evacuation;
- communication about the central assembly point in the event of an evacuation.

In 2015, the number of fatal accidents per 100,000 employed persons ranged from less than 1.00 in Germany, the United Kingdom, Sweden and the Netherlands, to more than 3.50 fatal accidents per 100,000 persons employed in Portugal, Bulgaria, Lithuania and Romania. Across the EU, there were, on average, 1 513 non-fatal accidents per 100,000 persons employed in 2015. The range for incidence rates among the EU Member States was from less than 100 accidents per 100,000 persons employed in Bulgaria and Romania to more than 2 750 per 100,000 persons employed in Spain, Portugal and France. The highest rate was recorded in France, at 3 160 non-fatal accidents per 100,000 persons employed. The occurrence of non-fatal accidents could be biased by differences in reporting, as the reporting of non-fatal accidents is not obliged. However, these numbers do illustrate the importance of Occupational Health and Safety standards. The graph shows the distribution of fatal and non-fatal accidents across industries.

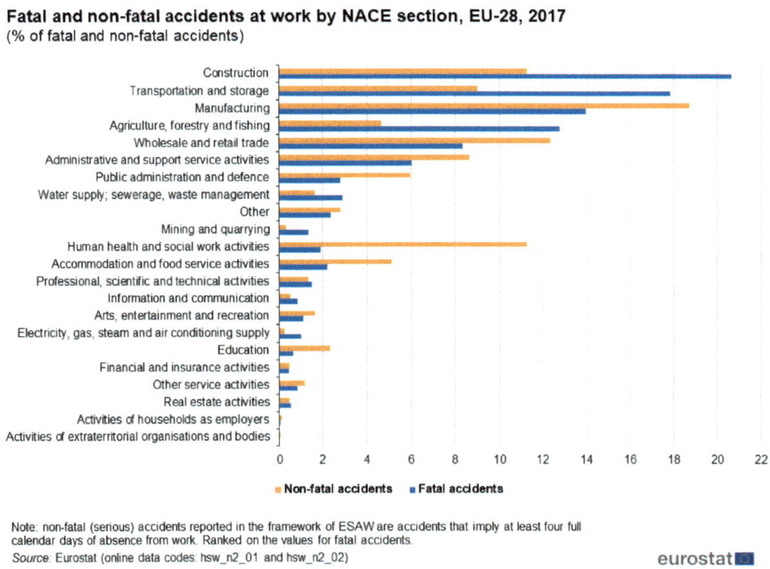

Distribution of % of fatal and non-fatal accidents 2017 (Eurostat)

Security

Security is about ensuring the optimal physical and social security of a location. Security is a service that is often outsourced to specialised companies, for reasons of expertise and the efficient provision of 24/7 monitoring and alarm response. Security is more than surveillance; it always involves a combination of organisational, constructional and technical security measures. Organisational measures include covering risks using procedures, increasing safety awareness among employees and choosing to deploy security staff. Constructional measures relate to zoning, routing through the building and compartmentalisation. Areas to be secured should preferably not be adjacent to freely accessible spaces. A safe, the storage of hazardous substances, or the storage of information that is sensitive to the company's activities, should not be situated next to the main entrance hall, for example. Technical measures are camera surveillance, detection gates, electronic locks on doors and lifts among other things.

Security staff is present in more than half of the organisations. Security staff is responsible for enforcing the access policy and for safety in the building and related areas. Approximately 10% of workers will ever experience a case of theft in the workplace. All employees can commit misconduct. Or unwelcome visitors, if access is easy and social control within the organisation is limited. Even though many premises have CCTV, the possibilities for recording the data of employees and visitors for security reasons are not unlimited. The General Data Protection Regulation (GDPR) safeguards the privacy of employees and visitors. This regulation has significant consequences for the facility manager as the responsible person for access control, camera surveillance, and as 'owner' of other facility data.

Access policy

General Data Protection Regulation (GDPR)

End-user satisfaction with security is determined by:

- The sense of protection within the building. The visibility of security guards influences this, and the familiarity of users with safety procedures, the accessibility security staff in alarm situations, and the appearance of the environment.
- The access policy. An explicit access policy that is visibly maintained gives users a sense of security.
- The building premises. This is mainly determined by the layout of the outdoor area. A well-organised, well-lit environment provides a sense of safety, especially if other people are present there as well. Abandoned, cluttered areas (e.g. due to the presence of high bushes or obstacles in the sightlines) and inadequate street lighting create a feeling of insecurity, as do vandalism and litter.

Critical success factors of security are:

- prevention of unsafe situations through an optimal mix of organisational, constructional and technical measures;
- adequate alarm response;
- the expertise and customer focus of the security staff;
- clear procedures and protocols;
- compliance with the General Data Protection Regulation (GDPR).

3.3.6 Foodservices

Foodservices, or catering services, encompass all activities related to the consumption of food, refreshments and beverages. It is in the top 3 of facility focus areas in Table 3.1. Catering mainly concerns four categories: lunch facilities (in the company restaurant), workplace and meeting services, drinks and refreshment facilities (usually with vending machines) and event catering (such as receptions). This also includes the

Foodservices

management of the company restaurant and the kitchen. In most organisations, employees have to pay for lunch and sometimes even for coffee and tea. The organisation often subsidises this by bearing part of the costs, so that employees pay attractive and acceptable prices for food and beverages. What is customary differs per industry and per country. The level of service, and thus price, is determined by many factors, including opening hours, the product mix, self-service or serving, and the quality and diversity of the coffee and banqueting (providing lunches and meetings).

Level of service

In many cases, catering is outsourced, with increasing use of so-called commercial catering. In these situations, it is not the organisation (the facility manager), but the caterer who comes up with a catering concept and revenue model that fits in with the culture and frameworks of the organisation.

Commercial catering

The guest judges the quality of catering services by tangible aspects and experience:
- The general hygiene of employees and products. If a company restaurant is visibly dirty or the catering employees look untended, the guest will assume that the kitchen's cleanliness is not good either.
- The freshness of the products. A salad bar with fresh salad is a prime example of quality, while sandwiches made in advance with soft lettuce are not.
- The price level. The price level should correspond to the quality, service and ambience delivered. A guest accepts that a lunch in a modern company restaurant with a wide choice of fresh products costs more than a minced meatloaf in an atmosphere-free canteen.
- Friendliness of the catering staff. Friendly staff ensure that guests feel welcome and that they are happy to come back.
- Choice of products. The employees and guests love variety in the offer, especially if they regularly use the company catering. A broad and varying range of products continues to surprise customers.
- Opening hours and waiting times. A company that uses standard working hours for everyone also uses fixed breaks. A company with flexible working arrangements has to deal with more variation in the break times of its employees. And increasingly, the company restaurant can also be used as a work and consultation environment. Too limited opening hours cause a great deal of traffic at drop-off points and cash registers in a short time, resulting in queues and a lack of seats. On the other hand, excessively long distribution times cause high personnel costs.
- Atmosphere. Of course, it is better to stay in an attractive than in an uncomfortable environment.
- In vending machine supply, the quality of the coffee, the availability of vending machines and a variety of options are also important.

Critical success factors for food services are:
- hospitable, a customer-oriented attitude of catering staff;
- a vision of what standard product range, luxury additions and reasonable prices are, with the possibility of a partial compensation by the organisation;
- multi-functionality of the company restaurant, enabling it to be used as a flexible working environment, meeting room and events such as presentations and receptions;

- alignment with new developments, such as commercial catering;
- responding to the external environment and the retail offer in the immediate vicinity to distinguish itself from the competition and to attract external customers.
- segmentation of target groups; no one-size-fits-all for all guests but something for everyone (for example through a food court).

3.3.7 Waste management

Waste management

Waste management is ranked second in the table of facility focus areas (Table 3.1). This is not surprising, since every company has waste streams, and every facility manager has to deal with them. Waste or garbage are substances, materials and products that the organisation wants to dispose of. An accredited waste collector must collect the trash. This waste collector chooses a processing method depending on the characteristics and origin of the waste. But waste management is more than just disposing of residual flows, the collecting waste and safely storing it until the waste processor comes to collect it. It is also about preventing waste and enabling recycling through smarter purchasing. And about increasing end-user awareness of the need for reducing and separating waste.

Waste separation

Hazardous substances must always be collected separately and disposed of safely. With waste separation of non-hazardous waste, there are two possibilities: separating waste at source (by users) and sorting waste by the waste processor after collection. Separation at source contributes to the organisation's environmentally conscious image and helps users' environmental awareness. Moreover, post-separation is not possible for every type of waste. Paper, organic waste, textiles and glass can only be recycled if they are collected separately. For example, paper and textiles that are part of the residual waste become wet and dirty: the recycling company can do absolutely nothing with them. Vegetables, fruit and food residues in the residual waste stick to other debris. They are therefore difficult to separate, and there is also a lot of 'contamination' between them (e.g. small pieces of plastic). Iron (cans, screws) can be easily extracted from residual waste using magnets, but this is much more difficult with aluminium cans. Only for plastic packaging and drinking cartons it does not matter whether it is collected separately or together with the residual waste. Post-separation is a solution for situations in which separate collection is difficult. What also plays a role in the choice is the available capacity of sorting machines.

Although it is cheaper to offer waste separately than mixed, it does present a financial risk. If not correctly sorted, the facility manager will receive an additional invoice from the waste processing company. At the same time, the facility manager has limited influence on the quality of waste separation by users. Above all, the possibilities lie in the establishment of the preconditions. These are sufficient (separate) waste collection points so that the user does not have to make an extra effort, and a simple system to prevent mistakes. Above all, success depends on the end-users' motivation and behaviour. However, it is not easy to convince people. Anyone who uses a school canteen knows how much waste remains on the tables (or the ground). Adequate waste disposal, therefore, has a significant effect on the quality of the environment. A clean environment more pleasant and is also easier to keep clean because it implicitly encourages people to clean up their rubbish.

Waste management is becoming increasingly important, because in a circular economy (see Chapter 2) the aim is to re-use raw materials fully. It is said that 'Waste is food'; meaning that the value of raw materials in residues is recognised, and collected in such a way that they can be re-used. Waste management is not an end in itself; it is part of the environmental care system (see section 4.8).

Critical success factors of waste management are:

- the choice of a waste management system is in line with the environmental care system and legal requirements;
- adequate collection, in line with the agreements made in the contract with the waste processing company;
- sufficient waste collection points;
- safe storage;
- clear communication about what is expected of users, and about possible risks (in the case of hazardous substances);
- addressing users about undesirable behaviour, motivating the desired behaviour;
- exemplary behaviour of facility staff.

3.3.8 Document management and information management

Document management

Document management is the part of facility management that deals with information processing, i.e., archiving, copying, formatting, printing, scanning, publishing, mailing and distributing. A limited number of specialised suppliers are active in this market. Due to the ongoing digitisation, the scope of this service is extended to the field of information management. Information management is a process that ensures that the information needs of organisations are translated into information provision. It, therefore, goes much further than taking care of document flows; it concerns the accessibility and quality of information for work processes. As a result, many companies are busy converting physical information flows into digital information flows, such as scanning documents and records.

Information management

Archive management

Archive management is important because the legal retention period must be met, and old records destroyed correctly. Paper archives take up much space and are difficult to consult. To save space, and thereby, costs, many organisations adopt paperless operations. Remaining paper documents are digitised as much as possible and stored on servers. A point of concern here is information security. The media regularly report on confidential information that unintentionally ends up out on the street, whether or not due to crime. Nowadays, a data breach is a major risk of image damage or financial damage to the organisation. However, it is not enough for organisations to protect their computers and other devices against attacks from outside; a large proportion of the incidents are the result of negligence on the part of their own employees. Just think of the many confidential documents stored on laptops that people take with them wherever they go. Or to the well-intentioned donation of depreciated computers to primary schools, which still contain large amounts of confidential business information. It is often customary for employees to take company information home or to another external workstation. This poses a risk as information and information carriers are not always handled with the highest care. One solution is to work in the cloud, but then it is vital that people work via secure networks and not, for example, via free Wi-Fi connections on the train or social hotspots. This will remain a massive challenge for the years to come.

As indicated, document management also includes printing and copying. Although almost all information is available digitally, there is still a lot of printing and copying in organisations. Many users prefer to read longer texts on paper rather than on a screen, and the meeting documents are often printed out for meetings as well. To curb the printing and copying behaviour of users, many organisations limit the number of printing and copying devices to a few central facilities per department. Many of these devices only print after log-in on to the device (with user card or tag), to stop employees from printing unnecessarily.

Critical success factors of document management are:
- enabling (primary) processes to run smoothly and more efficiently;
- indexation and authorisation (tracing and access to documents and digital information);
- insight into opportunities of new developments, e.g. in the field of sustainability;
- information security; think of secure physical storage, efficient backup and destruction of digital archives plus the effective authorisation;
- compliance with privacy requirements in the General Data Protection Regulation (GDPR).

3.3.9 Cleaning services

Cleaning service

Cleaning service is probably the oldest and most stereotypical service within facility management, which explains the first ranking position in Table 3.1. Cleaning services are outsourced almost everywhere. Professional associations help to develop the industry, amongst others by training programmes and drafting standards. ISSA is a worldwide cleaning industry association (founded in 1966 in the USA). Similarly, the European Facilities and Cleaning Industry (EFCI) is the voice of the European cleaning and facility services industry in Europe since its founding in 1988. In the Netherlands, the Association for Cleaning Research (VSR; founded in 1979) develops all kinds of techniques and measurement methods.

Quality of cleaning

Nowadays, cleaning companies and customers alike are well-informed of how important it is to maintain a healthy work environment. Amongst others, cleaning services are essential to control dust and the prevention of spreading infection and disease. Moreover, the cleanliness of an environment profoundly impacts the user experience, and thus customer satisfaction. Hence, the importance of quality control in cleaning. There are different angles to assess the quality of cleaning:
- the extent to which the execution of activities and the deployment of personnel are in line with the commitments made;
- the perception of the quality of the cleaning work by the end-users;
- the technical quality of cleaning.

Not so long ago, cleaning companies were judged on - let's say - hygiene. But what is clean? Experience has shown that 'dirty' and 'clean' are subjective concepts to which different meanings are assigned in different situations. The need arose for an objective measurement based on uniform, clearly defined principles. In Europe, the EN 13549 standard was created following legal acts, hygiene standards, best common practice, and cleaning industry experience. The standard is intended for organisations' quality control systems, as well as for visually assessing the quality of cleaning services, using selected models and objective criteria.

Input specifications

Output specifications

Historically, cleaning services have been based on input specifications. Under such contracts, the service provider essentially delivers some predetermined resources within a specific time frame at an agreed price. In most markets cleaning services provisions are now based on output specifications in which the focus is on driving cost-efficiency via service analysis, knowledge sharing and benchmarking, a fundamentally different form of contracting.
An input-specified contract describes specific cleaning operations and frequencies. When the contract is settled, it is checked that the work has been carried out properly in the correct frequencies. A characteristic of quality control is that inspectors do not question whether an element is clean, but if the element has been cleaned following the work programme.
In an output-based service contract the provider designs the best-fitted service solution based on customer needs and expected outcome in terms of service levels, quality and frequency. In other words, this type of contract shifts the focus from the number of people and hours (the input) to the result achieved (the output). Working with result instead of input-based service contracts allows providers to 'think out of the box' and create a much more flexible, cost-efficient and value-added service solution for the benefit of the customer.

European Result Oriented Quality Measurement System (ER-QMS)

The quality of cleaning in an output specified contracts can be assessed with the European Result Oriented Quality Measurement System (ER-QMS). The ER-QMS is based on the EN 13549 standard. The basis of the ER-KMS is:

- the maximum number of permitted contaminations per type of room is set in advance;
- determine assessment categories per room: main inventory, residual inventory, walls and ceilings, floors and parts that are difficult to assess;
- determine the quality level per room type: 0 = no cleaning, 5 = highest cleaning level;
- record measurement results in types of dirt: waste, loose dirt, adhesive dirt and miscellaneous.

Although the technical quality is essential, the quality of cleaning is nowadays also assessed in terms of customer satisfaction and experience. Research has shown that technically clean toilets that smell fresh score a full point higher than without the use of toilet freshener. Also, some colours are associated with cleanliness (white, blue), whereas others are not (brown, orange). We also know that even non-clean toilet areas can be highly appreciated just because of the friendly attitude of the cleaning employee. The cleaner who cleans the cups in the meeting room. The cleaner who not only removes the dust from the windowsill but who also cleans the inside of the fridge.
Thus, a clean experience takes more than a clean building: side issues appear to have a much more significant impact on the experience than purely technical 'clean'. The experience of users can be measured using direct feedback with so-called happy-meters (see image) that register perception in real-time. A big advantage, because cleaners can immediately check the cause of the problem if customers indicate that they are dissatisfied.

Happy meter for instant feedback

Aspects that influence the experience of cleaning services are:
- Absence of visible dirt. For example, shreds of paper, dust and coffee stains.
- Absence of debris - a tidy room seems clean.
- Fragrance. Citrus fruit fragrances and the smell of bleach are associated with cleanliness by end-users.
- Visible cleaning. When people see that cleaning is taking place, they also feel that it is cleaner than when cleaning takes place outside working hours. Also, users pollute the environment less if they are acquainted with the cleaning employee;
- Customer-friendly and representative cleaning staff. A cleaner who doesn't smell fresh himself can do his job so well, yet he won't be highly appreciated by the end-user.

Critical success factors concerning cleaning services are:
- users are aware of the terms and conditions regarding cleaning services. It is often stipulated that cleaning staff are not allowed to move papers on the desk. A full desk is, therefore, not cleaned. If users do not know this, this leads to an unjustified negative opinion about the cleaning services;
- clear, objectively measurable contractual agreements between the client and the cleaning company regarding the technical quality and the quality of the experience;
- appropriate cleaning level per room type. An operating room has the highest level, a storage room the lowest;
- the expertise of the cleaning staff and supervisor;
- efficiency in the execution, tailored to the intensity of use of the rooms.

3.3.10 Workwear and linen services

Workwear and linen services

Workwear and linen services include the care of uniforms, linen and personal clothing.

The provision and supply of uniforms is an important area of attention for facility management, especially in public (enforcement) functions, the manufacturing industry, retail, hospitality and wellness industry and healthcare. The use of uniforms is job-related.
Clothing can have a protective or hygienic function, such as for operators in the industry, firefighters and surgical assistants. Other uniforms serve to indicate the role of employees or to radiate unity, such as among supermarket employees, judges and conductors. Sometimes workwear serves multiple purposes, such as police uniforms. Operational facility staff -such as technicians, cleaners, catering staff, security guards and receptionists- often use workwear.
Company clothing ordered to measure and employees receive several sets, often with a RIFD-tag or chip for traceability. Through RIFD or chip-identification, it is possible to automatically sort and issue uniforms by name, size and model. Large users, such as hospitals, make use of a Workwear Dispensing System that collects and distributes uniforms. Data is automatically imported into a Uniform Management System. By synchronising the data from the Uniform Management System and the laundry, the order is automatically prepared for delivery. In other situations, employees themselves are responsible for washing and maintaining their uniforms.

Linen includes all textiles that are not clothing, such as towels, tea towels, bed linen, napkins and tablecloths. Linen is widely used in the hospitality, wellness and care industries.

Many organisations nowadays lease their company clothing and linen. This can be done at laundries, but also at suppliers of workwear. The lease agreement covers the use of company clothing and linen, the cleaning thereof, replacement in the event of wear and tear, and the logistics of pick-up and delivery. The main advantage lies in pay per use or metered billing, instead of large periodic expenses for the purchase and replacement of garments and professional laundry and drying machines. The ease of handling in the process, the specialist cleaning options (in case of persistent stains) and the flexibility offered by a lease contract are also advantages. Workwear Dispensing Systems are also available for lease.

Critical success factors of workwear and linen services are:
- the quality of the fabric;
- timely availability of the right type and size of textiles on location;
- Simple dispensing and collection;
- Traceability to prevent mix-up and loss;
- quality of the cleaning;
- detection of quality problems (replacement/repair).

3.3.11 ICT

ICT is the abbreviation for information and communication technology. It includes everything to do with computers, tablets, smartphones, video conferencing, digital learning environment, servers, printers, etcetera. ICT does not always fall under facility management. Often it is a separate department, for example, because there is too much risk involved, it requires specialised knowledge or because of high investments. Think of a bank or insurance company, where money transactions and policy processing are part of the primary process, and the potential damage is high. You may reason that everything will fall under ICT as a result of digitisation. In practice; in many cases, the ICT responsibilities are divided. All workstation-related facilities and services fall under the responsibility of facility management; the rest is the responsibility of a dedicated ICT department (frequently outsourced to external parties). The development of smart buildings contributes to a blurring of the clear delineation of the focus areas of facility management and ICT, as well as working independently of location and time. In the case of internal removals, this means that an intensive collaboration between the two departments is required. The higher the (perceived) distance between the two departments, the more complicated such a relocation usually becomes.

Separate department

Intensive collaboration

In addition to hardware, software, cabling and wireless facilities, support for users is always very important. ICT-departments usually have a helpdesk, which can operate via chat functions, digital self-service and live at a desk. A relatively new development is that users bring privately-owned hardware to work and expect support from ICT. As mentioned before, this trend is known as bring-your-own-device; and mainly concerns smartphones, notebooks and tablets. This practice imposes challenges concerning cybersecurity.

Critical success factors of ICT are:
- user-friendliness and accessibility of the systems;
- keeping track of technical developments;
- effective authorisation;
- adequate security and protection of data to prevent external intrusion and data leaks;
- minimal downtime.

3.4 Attitude and behaviour

An important area of attention for facility managers is guiding the attitude and behaviour of employees. Whereas in the past, the focus was on buildings and later on services, it is now clear that the attitude and behaviour of services employees are drivers for users' experience and satisfaction. In practice, hospitality is often mentioned with hostmanship in the same breath. Hostmanship is the art of making people feel welcome. It first and foremost relates to welcoming one's employees. As Figure 3.16 illustrates: invest in satisfied employees, because it will result in happy customers, and satisfied customers are the precondition for successful business operations. If the company is doing well, this is good for the employees, which completes the circle.

Attitude and behaviour

Hospitality

Hostmanship

FIGURE 3.16 Satisfied employee, happy customer

3.4.1 Hospitality

From organisations in the tourism industry, retail and hospitality, we have learned that attitude and behaviour are among the most decisive factors in achieving success. These days hospitality principles are commonly applied in schools, in the office and care environments. For example: in the choice for further education, students visit open days. The impression the pupil and the parents get during this first visit is critical. Schools in which the pupil and the parents do not feel well-received can be rejected regardless of the quality of the education offered. For hospitals, hospitality is a means of increasing patient satisfaction and positively influencing the ranking of a healthcare institution. In the office environment, hospitality is a means of increasing attractiveness as an employer. For organisations with flexible working arrangements, it is also a means for tempting employees to come to the office to meet their colleagues. This may sound strange, but with the advent of time- and location-independent working, employees have a great deal of freedom as to where and when they work. Still, to foster team spirit and commitment, it is important that people regularly meet in person. Moreover, physical meetings are much more effective than digital communication in terms of the intensity of getting to know each other, understanding each other and exchanging knowledge. Facility professionals respond to this by developing concepts for meeting and collaboration that excel in comfort and hospitality.

Meeting and collaboration

Customer focus

Customer focus means that the facility organisation aims to fulfil the client's wishes and needs. The term hostmanship indicates that facility management staff offer sincere attention, safety and comfort to the client to the best of their ability. The result is perceived hospitality. It is essential to understand that hospitality is not synonymous with unlimited compliance with the customer's wishes. Hospitality, too, is about an appropriate level within the frameworks set by the paying customer (see Figure 3.17).

FIGURE 3.17 The link between hostmanship, customer orientation and hospitality

Attention to hospitality pays off, as demonstrated by Integron's annual study into the facility experience in the Netherlands. No fewer than 92% of the clients of the best-performing facilities companies state that employees of the facilities company are hospitable. By way of comparison, the national average is 62%. The group of customers that indicated that they highly enjoy their work scored an average of 1.1 points out of 10 higher than the average in the Netherlands when asked about specific products and services (Integron, 2018).

3

But what is sincere attention in a professional context? The story goes that a well-known fast-food chain trains its employees to thank each other every time they pass on a tray, for example. This way, thanking becomes natural behaviour, and the employees will also act sincerely towards the guests. Friendliness thus becomes a basic attitude. The employees are free to interact with customers in their way, but will always say 'please' and 'thank you'. This prevents people from acting, giving them an authentic and pleasant attitude.

Basic attitude

We will also give an example of ineffective posture. In a drugstore chain, staff are instructed on how to interact with customers. Standard questions should be asked, such as: 'Did you find everything you need?' and: 'Are you familiar with this drug, or should I explain it to you?' Customers will notice that in all stores they are being addressed in the same way by the staff. The opposite effect of hospitality is achieved when these phrases are voiced without real interest, and without looking at the customer. Customers become irritated rather than pleasantly surprised.

'Everyone who works here not only has a task but also a role as host or hostess.'

— Marijke Ris, Rotterdam Central Railway Station Manager in Facto

3.4.2 Types of customers and customer experience

It is the client who assesses whether the facility staff are hospitable and whether the service is customer-oriented. After all, it's all about his experience, and this experience depends on his needs. That is why facility professionals must first and foremost understand their clients' needs. In

Basic types of customers

Sweden, research was conducted into the predictability of these needs. Based on this study, three basic types of customers are distinguished, all of whom are given a common Swedish surname: *Andersson, Svensson* and *Hansson*. Of course, the reality is more refined, but this classification can be used as a starting point.

Andersson is the most common type (50% of the surveyed population). *Andersson* wants to know absolutely everything. Andersson has unlimited information need, and the more he knows, the better he feels. A customer of this type would like to be informed in great detail about what to expect.
Svensson also likes to prepare himself but bases himself on what you can logically expect. A customer of this type assumes that if the website or catalogue states that the meeting venue is outfitted with all the necessary facilities, the basic facilities will be available. A quarter of the research population can be considered the *Svensson* type.
Hansson is the impulsive type that organises matters at the very last minute. He does not need any additional information in advance, but he will make arrangements on the spot. A quarter of the population studied belongs to the *Hansson* type.

Information needs

The information needs of these three types of customers are quite different. Facility professionals, therefore, need to adapt how and how much information they provide about the products and services they offer to individual customer's need.

3.4.3 Service design thinking

To provide facilities services in a genuinely customer-oriented manner, we need more than just insight into the information needs of customers. What are their other needs, for example, and the importance they ascribe to them? What are the typical customer behaviours? What do customers need to be productive, to be able to learn, to heal or to feel entertained? Service design thinking is a useful tool in this respect.

Service design thinking

Principles and steps of service design thinking

The service design is built on the following principles:

1 Take customer experience as the starting point for designing services. To do this, you need to understand the behaviour, feelings and thoughts of your target group to gain an empathic insight into unmet needs.
2 The customer journey. In service design thinking, we consider service as a circuit of interrelated events. Customers do not distinguish between your services and those of others; on the contrary, they have a holistic experience.
3 Co-creation with all stakeholders. Both the customer and the employees (the service ecosystem) actively participate in the design process.
4 Physical elements. Services are intangible, but physical attributes can strengthen the experience and lasting memories.
5 Sensory approach. Involve as many senses as possible (see, hear, feel, smell and taste).

Service design thinking is a flexible, human-centred design method based on co-creation with all stakeholders. The design process is iterative; meaning that steps can be repeated in the interim if necessary. The process consists of the following five steps:
1 Empathise - gather as much information as possible to understand the customer experience.
2 Define the problem from a human perspective - what are unfulfilled needs?
3 Ideation - generate ideas for innovative ways to meet these needs.
4 Prototyping - make the idea tangible.
5 Test minimal viable products - don't wait until your idea is perfect, but repeatedly test during the process to improve the product.

The aim of empathising is to gain an in-depth understanding of the customer and other stakeholders through research. For example, by observing a service environment (like a fly on a wall), observing customers while undergoing the service (shadowing), conducting interviews, or by performing data analysis on customer data. The customer's experience from the first contact to the moment of departure can be mapped. Customer journey maps are based on the four stages of the guest cycle:

Guest cycle

1 *Pre-arrival*. This is the phase in which the customer is not yet at the location physically but orientates itself places a reservation or makes an appointment and is on his way. What the customer experiences in this phase partly determines the state of mind in which he will arrive. Unfortunately, the facility professional usually has no control over the factors that influence the customer's experience in this phase, such as traffic jams, car breakdowns or train delays. However, they can meet the customer's needs by offering an accessible website, various ways to make reservations and by providing information about accessibility (route description, accessibility by public transport and up-to-date traffic information).
2 *Arrival*. The customer arrives and has its first impressions at the parking lot and on entering the location. The availability of a parking space (preferably close to the entrance) and clear signposting influence the overall hospitality experience. Likewise, the welcome at the reception and speed of registration add to this or may compensate a negative start. A friendly question like: "How was your journey?" can make all the difference.
3 *Stay*. In this phase, hospitality relates to the friendliness of employees, the range of services, the level of service and the ambience.
4 *Departure*. The experience of hospitality at departure is influenced by how the client is shown out. Think of asking whether everything has been as desired and if the client still needs any help, the issuing of an exit ticket for the parking facilities and the provision of up-to-date travel information.

Customer journey mapping using personas

Personas

The insights gained into customer needs are translated into personas. Personas are detailed customer profiles to bring knowledge about target groups to life. Personas give a name and face to a particular type of customer, with their distinctive motives, desires and goals in life. It is a simple but useful tool to better tailor services to user groups. Personas ensure that the abstract concept of 'customer' becomes tangible. They also indicate quite specifically how the company should approach customers. To illustrate this, we provide two descriptions of the day in the life of personas who represent certain types of employees of an organisation: Persona Anka and Persona Paul.

Persona Anka is a management trainee in her late twenties. She is full of ambition, has a challenging job and a busy social life. She is living with her partner and has no children. Status is important to her, and she likes to show her success in life. She drives an expensive car and attaches importance to the quality of everything she uses. Because she has such a busy schedule, speed and flexibility of service are also important. Anka wants to be able to park her new car safely near the location where she works, except on Mondays, when she usually works from home. About twice a week, she has breakfast at work because she starts early to beat traffic. When working overtime, Anka likes to have a nutritious hot meal. Likewise, she wants to be able to have a good lunch with business relations. Anka likes to relax with a freshly ground cup of coffee and a healthy snack. She does not need a fixed workplace, but she does need a workplace to be available where she can concentrate when preparing for an important meeting or when she has to compose a report. Anka works from her laptop, bought from her personal IT budget. She needs good wifi to be able to safely connect her computer to the network anywhere in the building. She likes to consult in an informal, open setting with comfortable furniture where she readily meets other people. To stay healthy, she works out during the lunch break. She appreciates the gym with a locker room, a good shower, a hairdryer and a make-up mirror. Anka is used to organising her life digitally, which is what she expects at work. She always uses the self-service capabilities of the employee portal. She prefers to chat online with helpdesk staff rather than calling or queuing up for a service desk; she doesn't think that's efficient.

Persona Paul is an administrative employee of 54. He has been working for the organisation for twenty years. Paul is married and has two children in university. Paul is loyal and reliable, proud of the company he works for and expects to stay here until his retirement. Paul likes to help people. In his spare time, he is the chairman of a local football club. There, too, he has been actively involved since his children were small. Paul considers the price of goods more important than quality. He likes to buy things at a discount and is proud to announce that he has made an advantageous purchase. Every day Paul starts at 8:15 a.m. He comes to work by bike, which he wants to store securely and dry in the indoor bicycle shed. Every morning Paul greets the security guard and the ladies at the reception. If Paul needs to arrange something, he prefers to go to the service desk because he likes personal contact more than contact via the self-service portal. Paul has a permanent workstation in the finance department. Everyone he works with is on the same floor. Paul enjoys having lunch with a group of colleagues at a fixed time. He likes the fact that the employees of the company restaurant know him by name. Every day, Paul brings four whole-wheat sandwiches from home and buys a cup of soup in the company restaurant. On Friday he eats a fried snack as an appetiser for the weekend. At 5:30 p.m. Paul goes home by bike. He never works overtime.

In the description of the personas, you can see that the needs of Anka and Paul customers are different. To get ideas on how to improve the experience of these customer groups, we will also compose customer journey maps for the personas.

Customer journey maps

There are many different templates for creating customer journey maps. They differ in the number of dimensions and the level of detail of the information they contain. The employee experience map illustrates the first day at work from the perspective of a persona. In the upper right corner, you can see the persona to which this customer journey map relates (Newbie Ruby). Several dimensions have been worked out for this persona. You can see the steps in the onboarding process (in text and visually). The emotion curve illustrates how satisfied or dissatisfied she feels during these activities. The bottom of the map shows the pain points and opportunities to improve the experience of onboarding employees.

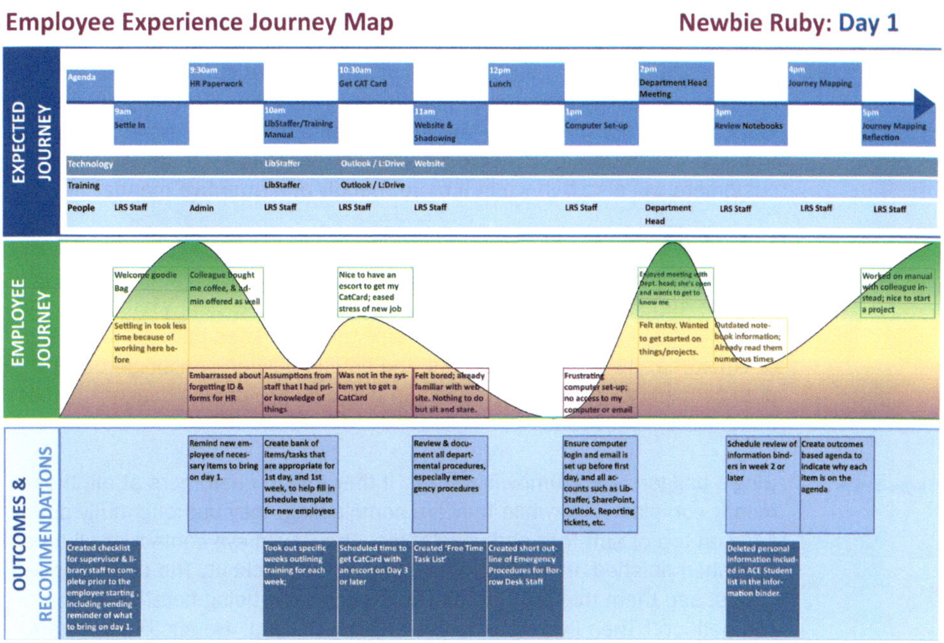

Example of an employee experience map for onboarding employees

A customer-focused (internal or external) facility service provider offers a service concept that fits the goals of the organisation, and that meets the needs of all types of users. We call it suitable for its intended use (fit-for-purpose). This is the only way to create added value. The personas Paul and Anka do not exist but are representatives of types of employees of an organisation. The use of these types of personas creates a similar understanding of the customer throughout the organisation. If the company knows the relative share of each persona in the organisation, it can better predict how much need there is for certain services. For example, let's assume that the organisation that Paul and Anka work for wants to implement a culture change to maintain its competitive position in the market. The HR-policy of this organisation will focus on attracting young,

Fit-for-purpose

ambitious employees (Anka-type). To be an attractive employer for this type of employee, a luxurious coffee bar, a gym, a flexible workplace concept, and healthy catering are amenities that work. However, a facilities consultant will not simply advise this. He or she will start by mapping out the customer profiles of *all* types of users in the organisation.

Co-creation

Personas also help to understand the wishes and problems of customers better. From this deep understanding of customer needs, ideas for service optimisation can arise in co-creation with the customer and all other stakeholders. The company chooses the most promising approach and works it out in a workable prototype for testing in practice, and alters the prototype based on the feedback that comes from experimentation.

Service design thinking is a fast and creative way to optimise customer experience and increase satisfaction, which is applied on a large scale by service companies.

3.4.4 Types of employees

The importance of enticement was discussed earlier in this chapter. The behaviour of facility staff plays a significant role in this. Not only because it is so pleasant for the people involved, but also because it serves as an example to others. Customer-focused behaviour is contagious. As a result, customers will also behave in a more friendly and attentive manner.

An example that illustrates that customer-oriented behaviour is contagious is the story of Paul Meert (infotainer in Belgium) about single and double baggers. In some supermarkets, the groceries are packed by a clerk while the customer pays at the checkout. The story distinguishes between two types of employee: the single bagger and the double bagger. The single bagger puts as many groceries as possible in one bag; the double bagger handles the groceries with care.

Single baggers

Single baggers are grumpy in nature. If they talk to customers at all, they mostly complain. And while they are complaining, they unthoughtfully put cans on top of soft fruit and vegetables. They avoid eye contact at all times, and when finished, they glace at the queue for checkout. The customer can almost see them thinking: 'What are these people doing here! Go to the competition!' Then I would not have to work so hard.' No customer likes to be helped by the single bagger.

Double baggers

Double baggers have a very different outlook on life. They are always in a good mood. They have a chat with the customer while they pack the groceries. They put cans in a separate bag because they would crush the soft products. They ask questions and occasionally make a joke.
What is the difference between the single and the double bagger? The single bagger is in a downward spiral and the double bagger in an upward spiral. Double baggers spend their days making pleasant conversations. Consciously or subconsciously, they are not only busy packing but also evoking a good mood in the people they are servicing. Single baggers, on the other hand, will only trigger a negative attitude on the part of the customers. In their opinion, the working day will take much longer than for the double bagger. The essence of this story is that an employee should be a double bagger, even if only for his own sake (and the employer!), as it makes work (and life) much more pleasant.

Organisations that excel in working in a customer-oriented manner are called customer-driven organisations. Figure 3.18 shows the difference between working in a customer-oriented way and working in a customer-driven way. The double bagger, who provides personal care and attention to the customer, is customer-driven. Organisations that want to guide their employees towards customer-driven behaviour need the management to set an example. If the manager of the supermarket is a double bagger, there is a higher chance that all employees will behave in the same way, as can be concluded from Figure 3.17.

Customer-driven organisations

Who is responsible for solving customer issues? The management or the employees? Delegation of power and responsibilities result in higher involvement of employees. And if employees themselves are responsible, do they also have the possibility to deliver customised solutions if necessary? Are they allowed to deviate from imposed protocols on their initiative? In other words: can the commitment to a customer and understanding of the circumstances also be translated into the most appropriate solution or should the employee at all times only follow the standard procedure?

Involvement

3

FIGURE 3.18 Customer-focused versus customer-driven

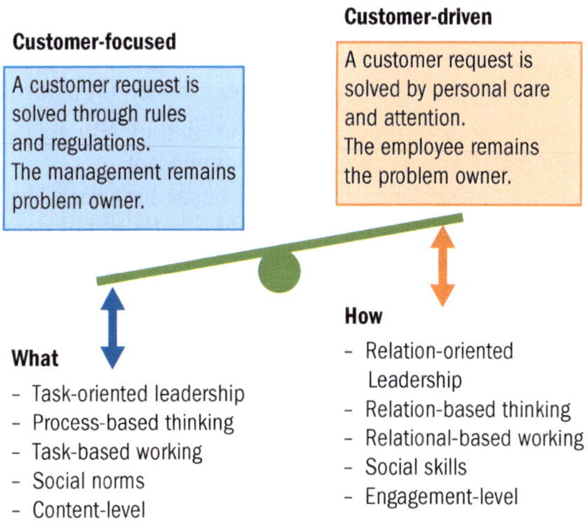

3.4.5 Hostmanship

It is unlikely that you will come across facility companies that do not pay any attention to hospitality in their policy plan. All commercial providers of facilities services have identified hospitality as an essential aspect of their services. Just take a look at the websites of for example Compass Group (Eurest), ISS Facility Services, Sodexo and Facilicom.

Specialist consultancy firms can help facilities organisations to improve their hospitality. Still, organisations also need to incorporate hospitality in their daily routines. The focus on hospitable behaviour can be maintained by setting goals and by including the subject as a standard item on the agenda. Besides, thematic meetings can be organised in which the value and effect of hospitable behaviour can be illustrated based on best

Buddy

practices. What also helps to establish and maintain cordial routine is the use of buddies. A buddy is a critical friend. In the facilities company, every employee could ask a colleague to be his buddy about hospitable behaviour, that gives feedback to improve personal performance. A buddy is not an auditor or an inspector and is therefore not threatening but assisting.

As is clear now, hospitality is a matter of behaviour. The essentials of the required behaviour must be defined, to make evident to employees what is expected of them. Even if the company outsources, it will somehow have to be established what kind of behaviour is expected from the provider and what contribution that behaviour makes to positive customer experience. In doing so, the company must be vigilant for the far-reaching regulation of action through internal procedures, routines and standard communication (see the blue box in Figure 3.18). For example, eye contact and undivided attention are decisive for the (subconscious) assessment of receptionists by the customer. Receptionists must ensure that they have eye contact as soon as possible - in an authentic manner - upon receipt of a customer. And when they talk to people, they should not be dealing with other things at the same time. A schedule of requirements could state, for example, that the receptionist should have eye contact with the customer within 15 seconds and should speak to the customer without any other action being taken. A mystery guest is used to check this. A mystery guest is someone who checks the contract compliance (called an auditor), by observing 'incognito' - just like an undercover agent.

Mystery guest

Clear procedures contribute to the uniformity, consistency and measurability of the service, but have the disadvantage that employees are not encouraged to think about the best solution for the individual customer. The human touch and sometimes a simple solution may cease to be self-evident if the emphasis is mostly on compliance with procedures. To be able to act in a welcoming manner honestly, the employee must have both the responsibility and the freedom to act on his or her understanding. This is the message of the orange box in Figure 3.18: 'employee remains problem owner'.

Responsibility and freedom

'People will forget what you said, forget what you did, but people will never forget how you made them feel.'

— Maya Angelou, the late American writer, poet, singer, dancer, civil rights activist and Professor of American Studies.

- www.facto.nl

5 ways to increase the responsiveness of your department

Rogier Jongejan

Five measures that you as a manager, can directly apply to make the service more responsive to the customer are:

1 *Gather your feedback from customers*
Listen to customer conversations regularly. Ask customers for feedback on your organisation's services. Be inspired and share this with your team. This will underline the importance of customer feedback for your organisation.

2 *Call your organisation*
Experience for yourself what it's like to be a customer. Call your organisation with a frequently asked customer question and share your experience with your team. This will encourage your employees to look positively and critically at their services and to incorporate the customer perspective in their daily work.

3 *Enable every employee to connect with customers*
Employees who do not have direct customer contact have a less clear picture of how customers experience the service. Have every employee collect immediate feedback from the customer on an annual basis. Linked to this is a personal action to improve their customer focus.

4 *Share successes*
Employees are not always aware of their behaviour and opinions about customer-focused differences. Share the positive examples you see among employees and collect them in a yearbook so that excellent customer-focused behaviour becomes visible. This document is also useful for training employees.

5 *Let others inspire you*
Employees are aware that the service can be improved, but they often do not know how to do it. Organise an inspiration tour with your team and be inspired by other organisations, especially outside your industry.

3

Young Professionals in Facility Management

Loesje Ophuis, Associate Project Manager Taurus Investment Holdings LLC - India, Boston, Massachusetts, USA, graduated bachelor's in International Facility Management at Hanze University of Applied Sciences in 2016

From Groningen to Boston to Trivandrum

Soon after I started studying International Facility Management at Hanze University of Applied Sciences in Groningen, I realized that I wanted to pursue a job abroad. Studying the International programme set me up to do just that. After I spent a semester each studying in Macau SAR and interning at a Brussels hotel, I chose to go West for my thesis project.

In February 2016, I started my internship at Taurus Investment Holdings LLC, a global real estate investment firm based in Boston, USA. My thesis project focused on the Downtown Trivandrum Project, a 5.5 Million square foot, mixed-use development in Trivandrum, India. I focused on how the design of the core retail component influences the user experience of its various stakeholders. This subject allowed me to leverage the knowledge that I had gained in studying IFM with a focus on real estate development and management.

The beauty of IFM is that it allows you to explore different aspects of how facility management can be applied. Having realized that real estate is my passion, upon graduation, I returned to Boston to work with Taurus. In my job, I am responsible for the property management of assets in Boston, New York City and Vermont. Moreover, I currently work as the project manager for the development of 2.5 Million SF of office, 1.4 Million SF of retail, a business hotel and a residential tower in India.

Just like in IFM, working with Taurus has allowed me to explore various roles within real estate, enabling my continued development and growth as a real estate professional.

Summary

- Facility management supports the primary process; demonstrating the strategic importance of facility management is a point of development in the field.
- The facility manager has three ways of influencing the perception of the working environment: behaviour, soft services and hard services.
- Almost everything relating to buildings is included in the area of accommodation.
- Real estate management is responsible for managing the real estate portfolio. Real estate is one of the assets of an organisation and is, by definition, strategic in nature. Increasingly, real estate is seen as a means of conveying a company's identity.
- Several essential considerations play a role in relocation: new construction or renovation, owner-occupied construction or rental, scenarios for growth and decline, choice of location, workplace concept and use of materials and space management in the light of sustainability and circularity.
- When choosing a location, the following factors are taken into account: image and image, market conformity, area development of the surrounding area, accessibility, parking facilities, safety, grounds and landscaping.
- Space management is a collective term for various activities: inventory and allocation of space; workplace management; management of communal areas; design of workplace facilities; signage; the provision of green space and art management.
- The steps to follow in buildings design are brief - bubble diagram - block plan - furnishing plan.
- A soft service is a process or a collection of processes to achieve an agreed result. Processes can be divided into primary processes (core processes); supporting processes and management processes.
- To be able to steer processes, performance indicators (PIs) are established (derived from the agreed result). Performance indicators that are related to the critical success factors of service are known as critical performance indicators (KPIs).

- Processes are often presented schematically to show how input is processed into output, for example, by flow charts and service blueprints. The graphical representation of processes makes it possible to identify overlaps or gaps in processes. The main objective is to reduce the number of vulnerabilities in the process and to implement measures that reduce the consequences of errors.

- Common facilities services are reception, facilities service desk, in-house emergency response and security, company catering, document management, waste management, workwear and linen services, cleaning services and ICT.

3

- The end-user assesses the quality of these various services based on his or her experience. Each service has specific characteristics and associated critical success factors.

- Hospitality plays an increasingly important role in the work environment. Professional, effective hospitality is created when attitude and behaviour, services and environment form a well-considered combination. The customer journey map can be used to map out the customer's experience, from the very first contact with an organisation up to and including departure. The basis for this is the guest cycle, which consists of four phases: pre-arrival, arrival, stay and departure.

- Not all customers are the same, and consequently, not all customer needs are the same. A well-known customer typology is that of 'Andersson, Svensson and Hansson', based on differences in information requirements.

- Service design thinking is an iterative, flexible and human-centred design method for improving the customer experience. This method uses personas to classify customers according to personality characteristics, customer needs and behaviour, as well as customer journey mapping to map out the customer journey. Characteristic is the involvement of all stakeholders in the design process (co-creation).

- Not all employees are the same. Employees naturally differ in the way in which they perceive the world and how they respond to the customer (single and double baggers).

Questions and Assignments

3.1 Daane Bolier's hospitality model illustrates that hospitality involves more than the attitude and behaviour of facility staff members. Give an example that demonstrates that hospitality is a combination of service, environment and attitude/behaviour.

3.2 Visit the websites of two commercial providers of facilities services. Do you recognise the difference between hard and soft services in their services? Motivate your answer.

3.3 Facility Management and real estate management have been increasingly converging in recent years. Give two reasons why both disciplines are increasingly joining forces.

3.4 The average occupancy rate of a workplace is around 40%. Why is this percentage so low?

3.5 At a Ministry, a flex factor of 0.7 applies, while at the adjacent town hall, the standard is 1.2 FTEs per desk. How can you explain this difference?

3.6 People of many different nationalities work and stay at Amsterdam Schiphol Airport. What does this mean for the product range of a commercial caterer based at Schiphol?

3.7 The bi-annual survey of developments in the Dutch facilities market (FMN/TwynstraGudde, 2019) states that 'waste management is the only facilities service that has been 100% outsourced'. What explanation can you give for this?

3.8 Organisations that have introduced flexible working, usually opt for an output-specified cleaning contract. What is the underlying reason for this choice?

3.9 Carefully read the following review of a visit to an amusement park:

'Upon arriving the barrier did not open. We tried the emergency button to talk to an employee. Through the intercom, a distant 'hello' reached us, and after that all we heard was silence. Frustrated, we parked the car outside the gates. My husband has difficulty walking; so annoyed and tired we entered the park, but we didn't feel like it anymore…'

What could you do in the next phases of the customer journey to change such a negative start to a positive overall experience?

3.10 The annual employee satisfaction survey at the head office of a major health insurer shows that employees are dissatisfied with the number of meeting rooms available. Because almost everyone meets on Tuesdays and Thursdays, meetings are increasingly booked at external locations. An undesirable (and above all expensive) situation. Facility staff members report that scheduled meeting rooms are often not used, although not cancelled. Despite the users' dissatisfaction, the facility manager does not intend to increase the meeting room capacity. What would your advice to the facility manager be?

DEFGH
Gates
Information
Transfer T4-9
Baggage hall
Arrivals
Exit
First aid
Eerste hulp
Airline lounges 25-52
By invitation only
BC
Gates
Transfer T2-3
SHOP & COLLECT
GRATIS SERVICE
Foodcourt
Mediterranean Sandwich
Grand Café Het Paleis
CENTRE

4
The functioning of the facilities organisation

In this chapter, we will address the following questions:
- What are the tasks of a facility manager?
- Who are the customers of a facility manager?
- How are supply and demand matched between client, contractor and supplier?
- What role does the identity of facility organisations play in the management?
- How do you establish a facilities policy?
- How do you manage customer expectations?
- What is information management?
- What is facilities procurement?
- What is quality management?
- What is environmental care?
- What is risk management?

◀ Through the eyes of the facility manager:
'It is a real art to manage the continuous flow of travellers in a pleasant, secure and well-signposted environment.'

Expert Column

Towards secondary revenue growth

Facility organisations are subject to a change in their form, characteristics and tasks within companies. In most companies, facility divisions are still seen as a cost factor to support the core business; as a secondary service. The traditional FM topics such as maintenance and repair, service control, cost-saving, and adequate space management and value retention are the main focus.

However, depending on the type of organisation, FM services support additional value and the development of the portfolio for secondary revenue growth. The upheaval in modern companies, the continuous scarcity of resources and the war for the best talents (especially in the tech industry) require additional FM services to broaden their field of expertise and support functions. They support new work environments and agile forms of work through a wide range of amenities.

In line with this development, corporate FM extends its service to an internal service provider for employees. Facility managers describe the employees of the core business as their customers and support the reputation of the company and their internal and external perception (Net Promotor Score). The FM organisation identifies Key Performance Indicators and develops Service Level Agreements. The overall aim is to focus on customer satisfaction and to develop a holistic approach to modern work environments in close cooperation with HR and Corporate IT and provide the connecting element of services (amenities), culture and the provision of an efficient workplace.

In this way, FM provides valuable support in recruiting and improving the work-life balance of existing employees, leading to a higher Net Promotor Score and increased attractiveness of the employer.

Jens Schlüter
Head of Corporate Real Estate Enabling Functions at Zalando SE, Germany

4.1 Introduction

In Chapter 3, we explored the activities of the facilities organisation. This chapter focuses on *how* facility managers organise the matching of supply and demand. In essence, facility management is the same in every organisation: creating and managing a clean, dry, comfortable and safe (work) environment. Accommodation and services are, therefore, in the focus of the majority of facility managers. Logistics, procurement, contract management and risk management are also typical responsibilities of the facility manager. However, as we have already discussed in Chapters 1 and 3, the range of tasks differs per function and industry in which the facility manager works. The work environment at an international bank differs from that at a large chemical plant, a university or a hospital. Moreover, a facility professional who is employed by an (global) provider of facilities services also engages in entrepreneurship.

Entrepreneurship

Facility managers spend most of their time managing people, activities and resources. The average facility manager is mainly a people manager with strong competencies in business administration. Because the quality, customer satisfaction and cost of services must be well balanced, a facility manager must be able to control these in an integrated manner.

Quality, customer satisfaction and cost

The following figures give an impression of the size of the budgets (according to the EN 15221 classification). In 2018, the cost per square metre of lettable floor space (LFS) for a workplace in the Netherlands was 490 euros, excluding VAT. Given the median value of 19.8 square meters per workplace, you arrive at about 9,700 euros per workplace. However, it should be noted that 'cost per workplace' is no longer a useful measure in a benchmark due to the great diversity of workplace concepts. Of the costs per square metre, 55% consists of costs for buildings and infrastructure (accommodation, grounds including parking, cleaning and installation packages); 15% of the expenses for ICT; 25% of the costs for people and organisation (hospitality, logistics including document management and security) and 5% for strategic facility management and overhead (NFC, 2019).

Figure 4.1 shows that many of the responsibilities of facility professionals are hidden from view. Examples of such invisible tasks are the development of facilities policy, (internal) marketing and expectation management, procurement and contract management, information management, environmental management, quality management and risk management. At the same time, one's own brand should be paid attention to. The brand identity determines the profile in the market (commercial providers) and the positioning in the organisation (internal facility management organisation).

Sections 4.3 and following explain all the terms in the section of the model. We will first describe the matching of supply and demand in section 4.2, as this is the underlying principle.

FIGURE 4.1 The focus of this chapter

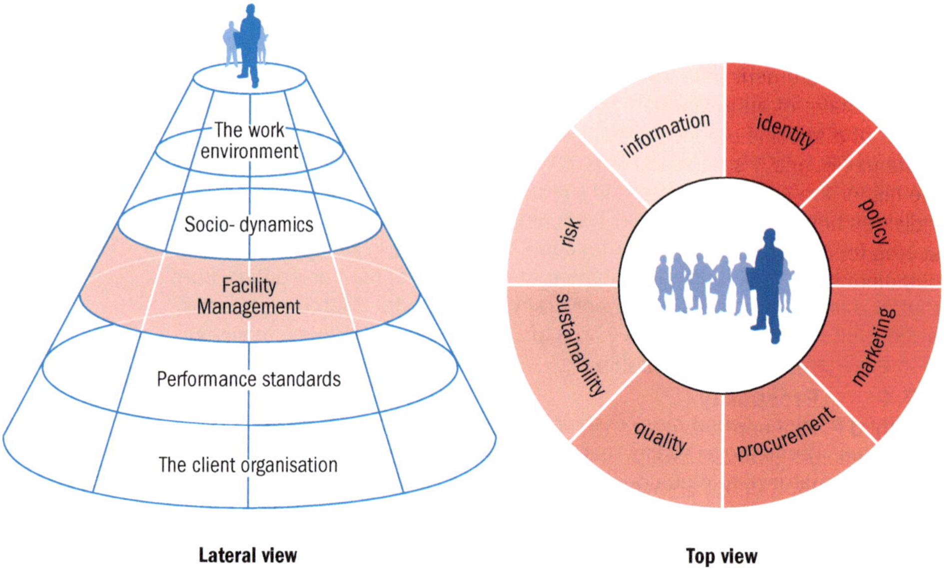

4

4.2 Matching demand and supply

The matching between demand and supply is based on knowledge of the customer and insight into the customer's requirements, as you have already read in Chapter 1. As we described in Chapter 1, there is no such thing as 'the' customer. Facility professionals have customers at all hierarchical levels of the organisation.

4.2.1 The stratified customer

Table 4.1 shows the so-called customer 'stratification'. First, we'll focus on the customer group that is targeted by customer satisfaction measurements: the users (benefitting customers). In an office environment, these are all employees working in the core process. In a hospital, patients are also beneficiary users, and in the recreation industry, the guest or visitor. In consultation with the budget owners (the paying customer), the facility manager determines the level of service and the range of products. Their agreements must comply with the frameworks defined by the company's top management (the defining customer). Other terms for the defining customer are a (principal) client or the demand organisation.

TABLE 4.1 The stratification of the FM-customer

Level	Focus of FM	Agreement
Strategical Defining customer: decision-maker and leading-edge	Translating client strategy into facilities policy	Policy framework (Master Service Agreement (MSA))
Tactical Paying customer: ordering party	Aligning services concept and product range with the policy framework	Service Level Agreement (SLA)
Operational Benefitting customer: (end) user (staff, visitors, students, external customers)	Placing orders and making deliveries following the service level agreement (SLA)	Product and Services Catalogue (PSC)

To illustrate the above, we use a bridge as a metaphor (Figure 4.2). This bridge has three parts: the bridge deck 1 in Figure 4.2) and two pillars (2 and 3). In an ideal situation, the coordination works as follows. At the highest level within the company - the bridge deck - the General Board, in consultation with the facility manager, determines the policy framework for the facility services (see Table 4.1). The facility manager translates the wishes of the board into a services concept and product range. The effects on service level, customer satisfaction, costs and added value are all taken into account, to ensure that the needs the paying and the benefitting customer, are met.

4

FIGURE 4.2 The bridge

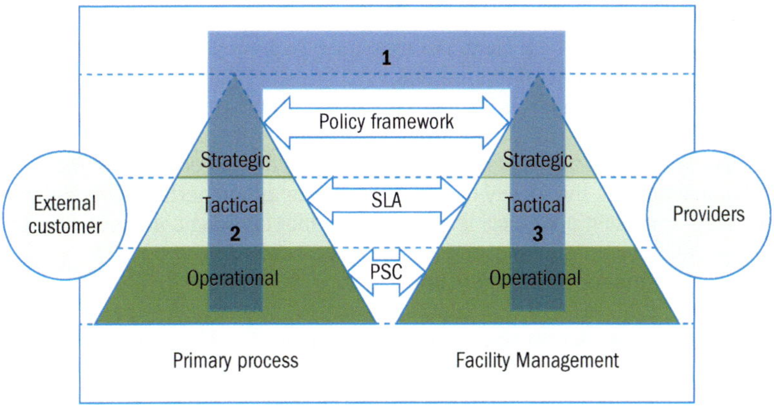

This information comes from various sources, such as customer satisfaction surveys, service evaluations, the facility management information system (FMIS), facility account managers or benchmarks. The facility manager coordinates this with his facility staff (pillar 3). Then, budget holders (paying customers) can select from the offer. They are responsible for the costs of what they consume (externally and internally). These budget holders inform their employees (benefitting customer) - in pillar 2 - about what they have chosen within the established policy frameworks. In this ideal situation, end-users with complaints about the services level will not contact the facilities service desk, but their line manager. This prevents facility professionals

Policeman

from having to take on the role of policeman who continuously sells 'no' to requests that fall outside the frameworks or agreements.
As already mentioned, Figure 4.2 describes the ideal situation. In practice, dissatisfied users often do not complain to their manager, but do so to the facilities service desk.

Account management

Account management, or customer management, is essential for the smooth running of this alignment process. Account management occurs at all three levels of the organisation, but not always as a separate function. Especially in larger mature organisations, facility account management is more formalised; in smaller organisations, account management is one of the tasks of the facility manager.
Strategic account management is a task that the facility manager will never outsource because it is inextricably linked to the company's culture.
Account management at the tactical level of organisations is mainly focused on process design, defining the service concept and performance management. In the case of integrated outsourcing, tactical account management is also outsourced (demand management organisation; see section 1.2.3). In such cases, the account manager of the provider is in direct contact with the budget holder (paying customer), without the intervention of the in-house facilities team. If only a few services are outsourced, the provider's account manager will be in contact with the in-house facilities organisation and the contract manager. All communication with budget holders goes through the in-house account manager.
All executive employees are customer managers. At the operational level, they are the eyes and ears of the facilities organisation. They should know like no other how to optimise the satisfaction of the benefiting customer.

Building and maintaining customer relationships through account management consists of three steps:

- Step 1: getting to know the internal customer: his work processes, the nature and needs of the (external) customers of the customer, what is needed for him to be successful (determines the added value of facility management), the facility needs, bottlenecks and important agreements;
- Step 2: making and fulfilling agreements about products and services to be delivered. Communicating about the agreements and possible adjustments to these is very important;
- Step 3: evaluating the service based on customer satisfaction surveys and discussions with customer representatives.

It is especially important that the account manager has a clear understanding of the customer's needs and that he or she is able to translate the customer's core business into the optimal range of facilities products and services. However, this is far from being the standard: in practice, supply-driven rather than demand-driven operations are still common.

4.2.2 Roles and activities of the client, facility management organisation and provider

To understand the complexity of matching supply and demand, the facility manager must have an understanding of the activities and roles of the client, facility organisation and providers at all hierarchical levels. The nine-grid (Figure 4.3) shows the actions of facility management at different hierarchical levels and from different roles. On the vertical axis, there are three hierarchical levels of the organisation, from top to bottom: the strategic, tactical and operational (see also Figure 1.1). Horizontally, the three parties involved are listed: the organisation (client), the facility management division (commissioning) and the provider (delivery). The client determines the demand for the hard services, soft services and resources (e.g. office supplies). Facility management is responsible for making these available and maintaining them. As services are increasingly being outsourced, a distinction is made between commissioning and delivery. In this model, the provider delivers services and resources. However, it is essential to note that the provider can also be an in-house supplier, as is the case with in-house sourcing.

FIGURE 4.3 The 9-grid for matching demand and supply

Facility Management

	Organisation = demand	Facility Management	Provider
Strategic level:	Mission,vision, goals and strategy of the organisation	FM policy that matches the organisation, demand and requirements set by society	Aligining with modern developement, topicality of services
Tractical level:	Organisational structure, design of primary process and support processes	Structure of the facility management organisation	Aligning with and cooperating with external and internal suppliers
Operational level:	Operational Management	Executing tasks and having them excecuted, steering and examining; suitable services and provisions	Quality checks - measuring results

Next, we will explain the matrix, starting from the top left to the bottom right.

The strategical level of an organisation

Top managers - members of the board of directors, director(s) - determine the mission, goals and strategy of an organisation. These policies are, in fact, the DNA, the soul, of the organisation.

Mission statement

The mission indicates why and for whom the organisation exists. The catchphrase term 'right to exist' is often used in this context. A good mission statement gives direction to all decisions that need to be taken and goes beyond objectives; it describes the desired identity and ultimate goal.

Vision

The vision describes the future dream of the organisation in a stimulating way, as it were. An essential part of this is the idea of what and how the organisation should contribute to society. The vision concretises the mission.

By defining clear goals, it is possible to measure whether the intended results of the mission and vision can be achieved. Think of growth, profit, sales area and number of branches, customer satisfaction, quality, innovation or market share. The goals are expressed in measurable units, such as profit/loss, turnover, costs, numbers, per cent, length of stay (hospitals), study succes (universities) or the net promoter score (NPS, see section 5.3.1).

Strategy

How the organisation wishes to achieve its goals is called the strategy. The approach involves concrete actions set out in a schedule with the expected milestones and measurable results.

Facility Management at strategic level
What is vital to the top of the organisation (described above) is the most considerable influence of the functioning and sourcing of the facility management division. To be of added value, FM's mission, vision, positioning and objectives (the facility policy) are to be derived from the company's mission, vision, goals and strategy. We call this 'alignment'. The positioning of facilities organisations is based on the degree of strategic alignment with the core business (Figure 4.4):

- A passive facilities department is supply-driven. It provides services based on its assessment of what is needed. The facilities strategy (if any) be aimed at the (perceived) client, but no dialogue with the client takes place.
- A reactive facilities department provides services that are tailored to the customer's needs, but there is no strategic alignment yet.
- A proactive facilities department enters into a dialogue with the client. It bases its facilities strategy on the client's strategic objectives.
- In the case of a business partnership, the client strategy and facilities strategy partly coincide. Strategic objectives are jointly determined.

FIGURE 4.4 The positioning of in-house FM-organisation

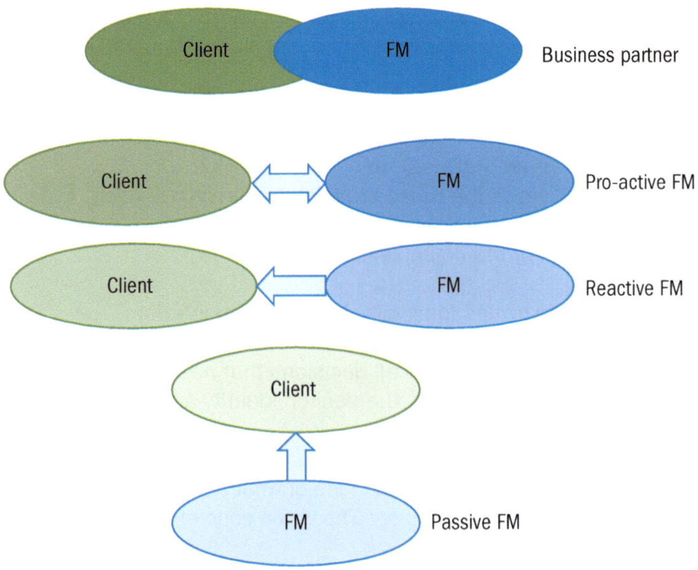

Passive and reactive facilities are of limited added value to the client. As explained in Chapter 1, facility management delivers added value to the client. This effect can be significantly improved if coordination takes place at the strategic level. In that respect, passive and reactive FM-departments are of limited value to organisations.

The added value of facility management for the organisation and the preferred form of sourcing are determined at a strategic level. There are three principles when it comes to sourcing: make, buy and cooperate. All are valid; it is a matter of selecting the best option for the organisation (client). The choice of in-house or outsourcing is often made with the aid of a decision tree. The principles are as follows:

Sourcing

Make, buy and cooperate

- *Make*. The facilities organisation carries out all the services itself unless there are convincing arguments and benefits to be gained by outsourcing a particular service to an external party.
- *Buy*. The facilities organisation outsources services unless there are convincing arguments and benefits can be gained by performing a particular service itself.
- *Cooperate*. The facilities organisation outsources the execution and management of facilities services integrally to a single provider. The client and provider work together in the form of a partnership.

FM PROVIDER PERFORMANCE RESEARCH, 2018, P. 45 (HOSPITALITY GROUP, 2019).

Open-ended contracts allow for more innovation and continuous improvement

The satisfaction of clients with an open-ended contract for innovation and continuous improvement is demonstrably higher than with short-term leases. For example, clients are more satisfied with the extent to which the provider responds to innovations and improvements without having a business interest of its own. Besides, the implementations of proposed changes will take place earlier, and the provider will share more knowledge about innovations and best practices in the market with the client. This proves that a partnership pays off. The longer the engagement, the better providers understand their clients and the more they know what is essential for business success. As providers gain trust in the long-term relationship, they dare to experiment more and implement innovations sooner, leading to more satisfied clients.

On all of these grounds, frameworks are determined for the structure of the service provision: sourcing, the service level, the budgetary frameworks and the indicators for measuring the accomplishment of strategic objectives. With integral outsourcing of facilities services (such as IFM, TFM, IFS), the provider is managed based on performance indicators as defined in the policy frameworks (see Figure 4.2).

The provider at the strategic level
The market for commercial facilities services is extremely competitive. To keep its pole position, the provider must keep abreast of and adapt to the latest developments in its field of expertise - in terms of working methods, materials and tools used. To distinguish themselves from competitors, providers of facilities services highlight the core values that characterise how they provide services and the unique selling points for their products and services. In a tender process, they have to empathise with the mission, vision and core values of the potential client to make an offer that is as appealing as possible. A commercial account manager aims to achieve turnover by closing or extending (additional) assignments and contracts.

● www.ziggo.nl/zakelijk/inspiratie/in-zaken/mas/aanbestedingen/

How did the MAS Cleaning Company impress the Rijksmuseum Amsterdam?

Delivering tailor-made solutions, being involved; these are beautiful terms, but they don't immediately make you stand out if you try to win a tender. Moreover, it is not always possible to make a distinction based on price, because more and more often, the budget is more or less fixed in advance. In that case, you can only quote a few per cent higher or lower.
Director Oumaima el Mouden explains how they managed to make a difference at MAS: 'We find out what our potential customer is doing and match our offer. What does their operational management look like, how do they want to differentiate themselves from others? When a company is open 24 hours a day, we adjust our schedules. And if they work with self-directed teams, we will also experiment accordingly. This way, our employees become part of the customer's team, and that is well appreciated!
The cleaning market is competitive. MAS won the Rijksmuseum's contract through a tough European tender. How did they make the difference? According to the facility manager of the Rijksmuseum, MAS was able to positively surprise them. 'MAS came up with a bonus: new company clothing, in the same style as the clothing of our catering staff.' Instead of being disturbed by the often-colourful clothing cleaners wear, the Rijksmuseum perceives MAS almost as its employees. According to the facility manager, this is not the only reason why MAS won the tender. The proactive approach convinced them too. 'If we have an important event, I'm sure that the division manager will visit us in person in advance to check that everything is perfectly in order.'

The organisation at a tactical level
Tactical level
Setting up the organisation and designing and preparing primary processes (the execution and the production) takes place at tactical level. Here, departmental heads or managers of business units lead staff employees. Also, the planning is made, monitored and kept up to date at this level.

Facility Management at a tactical level
On a tactical level, the relationship between the facility management departments and the department heads or business unit managers (paying customer) is at stake. The agreements made with these managers of business units and facilities departments must fit within the policy frameworks set (see Table 4.1) but may be tailored to the wishes and specific requirements of budget holders (aying customer). The challenge is to make clear agreements about the intended final objectives (what), but to leave the execution (how) to facility professionals.
At this organisational level, the strategic choice for (out)sourcing is translated into programmes of requirements and service level agreements (SLA). Proper management of services is guaranteed at this level, amongst others by measuring results. Based on the latter, if necessary, the services are adjusted - with the ultimate goal of satisfying the customer.

Service level agreements (SLA)

4

The provider at the tactical level
The responsibility for managing the execution of services with the fundamental concepts of cost, result and response time belongs at the tactical level. In consultation with the commissioning party (facility management organisation), the services are regularly evaluated and adjusted. Evaluations are based on (periodic or real-time) performance measurement and performance reports in which at least costs, quality and customer satisfaction are displayed. The provider is also responsible for managing any subcontractors.

The operational level
On an operational level, the aim is to provide customer-oriented service. The benefitting customer must be served at his or her convenience, but within the set frameworks (service level agreement). Whether that service is provided by the facility management unit or by an external provider makes no difference to the benefiting customer. All staff employees are ambassadors of the facilities organisation, with a leading role for services. Above all, a clear framework, excellent communication (account management) and expectation management are essential for smooth operations. The benefitting customer wants convenience and pleasant experience. His/her satisfaction is influenced by visibility, availability, clear ordering instructions and compliance with agreements. Critical tools for managing the expectations of customers are a product and service catalogue (PSC), information on the self-service portal, or a user-friendly app.

Product and service catalogue (PSC)

This section has outlined the interplay between supply and demand and the parties involved. In the following sections, we will look at activities that are preconditions for setting-up the organisation of facilities services, starting with positioning, identity and image (section 4.3) and the establishment of facilities policy (section 4.4).

4.3 Positioning, identity and image

Thinking about positioning, identity and brand image started in the fifties of the last century. Organisations became aware that the perception of the organisation by external stakeholders influences the success and failure of organisations; subsequently attention was paid to the image of an organisation. Initially, the main focus was on stakeholder-communication.

Later on, the understanding arose that the organisational culture and the employees are important carriers of the corporate image. Thus, increasingly attention was paid to behavioural aspects of identity and the extent to which identity influences the functioning of the organisation itself. In the 1990's the concept of corporate branding was introduced. Branding is an integrated approach that links the concepts of brand positioning, identity and image. Positioning relates to the question: who do we want to be? Identity refers to the question: who are we? And image is about the question: how do others perceive us?

The organisation's brand positioning is expressed in its mission, vision and core values (what the organisation is good at) and its brand values (what the organisation wants to display).
A few examples:

- The mission of integrated FM provider ISS is: 'service performance that facilitates our customers' purpose through people empowerment' (ISS, 2019).
- The mission of caterer Eurest is: 'Inspiring people and creating moments of happiness, that's our mission!' (Eurest Compass Group, 2019).
- The mission of the facilities consultancy Hospitality Group is: 'creating better places' (Hospitality Group, 2019).
- The mission of the Royal Albert Hall is: 'to give you unforgettable memories of fantastic experiences' (Royal Albert Hall, 2019).
- The mission of the facilities company of University Medical Center (UMC) Utrecht is: 'we make a professional and appropriate contribution to the well-being of our patients, visitors and employees by facilitating patient care, education and research based on our expertise'.

Vision

The vision relates to how the organisation sees the future developments in the market and to its proposition. A vision identifies and motivates the organisation's course in broad terms.

Core values

The core values of an organisation are the special qualities in which the organisation wants to excel. Core values mainly have a task in internal communication: they describe what is expected of the employees. A few examples of core values:

- Cleaning company EW facility services: 'hospitable, enterprising, self-directing, together, people-oriented' (EW facility services, 2019).
- Schiphol Passenger Services: 'inspiring, sustainable, efficient, reliable and hospitable' (Schiphol Passenger Services, 2019).
- Group4Security: 'integrity and respect; security and the best service; innovation and teamwork' (G4S, 2019).
- CBRE: 'RISE: Respect - Integrity - Service - Excellence' (CBRE, 2019).

The brand values are a translation of the core values into what the organisation wants to communicate to its customers. In the service industry, the core values and brand values are close to each other. These values become visible and tangible in the identity of the organisation.

Identity

Identity is the DNA of an organisation, which is expressed in branding, communication, behaviour and performance.
Branding uses symbols to represent the body of thought and identity of an organisation. The logo is the most well-known symbol. Besides, the location,

the building (architecture), workplace concepts, the corporate design, the uniforms and the use of colour can be seen as symbols.
Communication encompasses all target groups, both internal and external. Internal communication is particularly important and is often underestimated. After all, all employees must contribute to the desired identity in their actions and behaviour.
Performance is not only about how services are delivered but also about the actions of the organisation concerning its environment, and the organisation's ambitions regarding corporate social responsibility. Behaviour relates to the attitude of employees: tone and style of communication, speed of responses, customer orientation, creative search for solutions, etcetera. Because in the service provision behaviour is inextricably linked to service provision, this touches the core of the performance of the organisation. It is not only about executive facility staff: all layers of the organisation are involved, from management to service engineers.
Providers need to align their performance and behaviour with the identity of the client. In the selection of providers, clients must take into account the identity of providers and the extent to which they match.

4

Image

Image is the appraisal that customers and other stakeholders have of the organisation, in other words: its reputation.
If the corporate image is more favourable than the corporate identity, this poses a risk, because there is a chance that the customer will lose confidence in the organisation.
If the company's image is less favourable than its identity, the organisation does not make effective use of its opportunities. Thus, it will try to influence perception in a positive direction. Figure 4.5 shows the image of in-house facilities organisations in the Netherlands. The percentage indicates how many per cent of the respondents (fully) agree with the statements on the horizontal axis.

FIGURE 4.5 The image of Facility Management

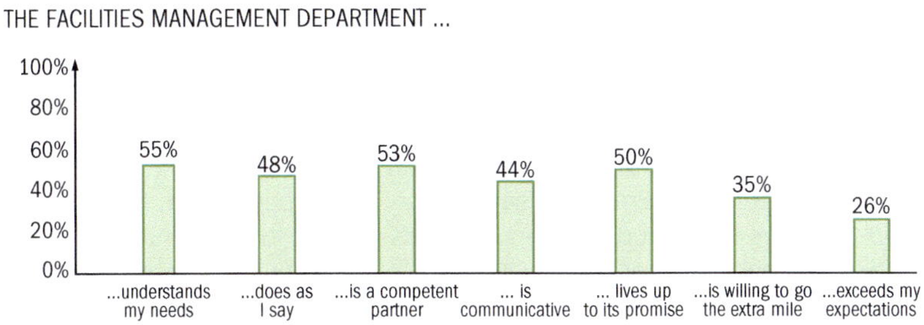

Many facilities organisations face the challenge of improving their image. One of the reasons for this is the customer's perception of various services, as explained in the Kano-model (see Figure 4.6).

In new organisations (or merging organisations), no image has yet been formed. The organisation at this time must use its branding, means of communication and behaviour to create an image. If the image and identity match, the perception of the customer, and the desired image coincide. However, even in such a situation, a favourable image must be actively preserved.

4.4 Facility policy

Facilities policy

As outlined in section 4.2, the facilities policy must be aligned to the course of the organisation to contribute to the achievement of the organisation's objectives. If the organisation adjusts its path because it identifies opportunities or threats in the market, the facilities organisation will also have to adjust its course (reactively). New developments in the facilities field can also lead to a change of plan, to seize opportunities that will help the organisation in an innovative way (proactively). It is a dynamic process in which many parties are involved. Beckers & Roelofs (2016) describe seven steps to arrive at a facilities policy plan. The plan itself (the eighth step) comprises all the building blocks that resulted from steps 1 to 7. Next, an example is given for steps 1 to 7 from the facilities policy plan of the facilities service organisation (FSO) of Saxion University of Applied Sciences.

Dynamic process

Step 1: Determine the context

Client

In this first step, the facility manager describes the client for whom he provides the facility services (in terms of vision and goals). This is the starting point for the facility strategy. Furthermore, this will give clarity on the building or all branch buildings within which the services will take place (the spatial framework).

EXAMPLE OF STEP 1

Saxion is a university of applied sciences with the ambition to become the best university of applied sciences in the Netherlands. Saxion's motto is: 'Get ready for a smart world!' Saxion has over 26,000 students and 2,800 employees (over 2,100 FTEs). Demographic developments (fewer young people, therefore fewer students, thus less budget) are the most significant challenges for the future. The Saxion 'True North' strategy focuses on increasing the adaptability of the organisation; cost control (lean); strengthening the technology profile to distinguish itself from other universities of applied sciences. In summary, the Saxion strategy comprises six pillars: (1) a standardised educational model (Saxion Education Model); (2) a decisive organisation; (3) focus on living technology; (4) internationalisation; (5) research & entrepreneurship; (6) external orientation & profiling. In addition to strategic indicators derived from these pillars, student satisfaction is an essential strategic indicator. Student satisfaction is related to the study success of students and Saxion's position in the rankings and therefore, to the influx of students. Employee satisfaction is also paramount. Satisfied employees are more productive and are less likely to look for another employer. Employee retention is very important in a market in which talent is scarce.
Saxion has subsidiaries in Enschede, Deventer and Apeldoorn and owns a total of almost 107,000 m2 GLA and rents 9,700 m2 at external locations.

Step 2: Describe yourself as a service provider

Raison d'être

In step 2, the facility manager mainly looks at the raison d'être of the facilities organisation (mission) and how it can best support the primary process of the client. The facility manager indicates the role of the facility organisation as a service provider (vision) and the priorities in the service provision (value proposition).

EXAMPLE OF STEP 2

Saxion's Facility Service Organisation's mission is: 'We give every customer a good feeling'. FSO offers inspiring learning, research, work and accommodation facilities for students, staff and visitors in a hospitable environment.

Clients see FSO as a professional, responsible and progressive department that creates a 'hospitable Saxion'. FSO follows a customer excellence value strategy. They strive for the right balance between looking at the customer's needs (customer intimacy) and the smart, fast and effective design and implementation of the service (operational excellence). In this way, FSO aims to be one of the top 3 Dutch universities of applied sciences in the facilities industry in terms of quality and costs.

Saxion's strategic principles are embedded in FSO's strategic vision (see image). The FSO strategy consists of four goals: close to the customer; efficient and effective operations; agile and vigorous service organisation; support of high-quality decision-making.

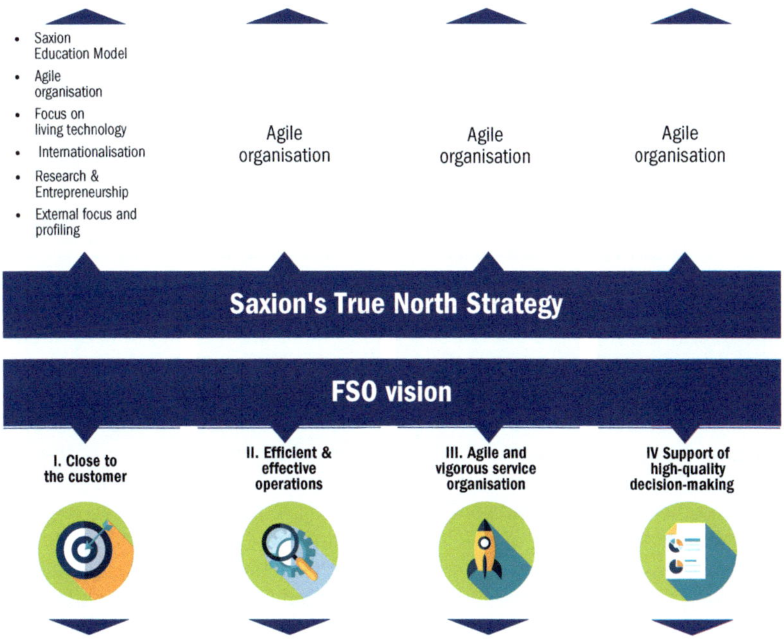

Step 3: Make an inventory of the customer requirements

Customer requirements

The facilities client is diverse (see Table 4.1). Make an inventory of what the defining customer, paying customer and benefiting customer consider necessary.

EXAMPLE OF STEP 3

Concerning the defining client (Executive Board and directors): in line with Saxion's mission, FSO will become a leader in three areas: data-driven facility management, sustainability and procurement management. In this respect, FSO actively cooperates with the education sector to involve students in the developments. FSO is actively engaged in education, research and relevant

networks. This makes it possible to experiment with or apply new insights, technologies and developments in the service offered.
As far as the paying customer (academy management) is concerned: FSO is an expert business partner for Saxion academies and services. FSO has knowledge and understanding of the primary process and shows responsibility for its activities. The (inter)active relationship with the customers is aimed at understanding the need and making it specific in terms of appropriate service provision. FSO anticipates making things better, more innovative and more efficient for its customers.
As far as the benefitting customer (students, employees and visitors) is concerned: FSO ensures that students, employees and visitors feel comfortable in Saxion's buildings. Facilities are safe, clean and comprehensive, the range of services on offer suits the target groups and the atmosphere is pleasant and hospitable. The facilities contribute to well-being, which promotes study and work results. Students and staff are aware of the services provided by FSO and perceive services as reliable, easily accessible, flexible and professional.

Step 4: Set up facilities services
Based on the customer's wishes, the facility manager determines the offer of facility products and services at the desired level of service, and how these can best be organised (sourcing).

EXAMPLE OF STEP 4
The organisation opts for a variation on the directive model, whereby operational services are outsourced unless there are reasons to keep them under one's management. The figure below shows how FSO organises the matching between supply and demand. In the service chain, FSO distinguishes between activities of the front office (direct interaction with the benefiting customer) and the back office (which take place out of the customer's view). Managers are integrally responsible for the entire chain to ensure efficiency and effectiveness. Suppliers are contracted based on partnership and are part of the chain. An example: FSO cooperates with ENGIE (installer) according to lean principles (see Chapter 5.4.4). ENGIE and the internal real estate and technology team have a joint week start meeting. In this, they review the previous week and together determine the priorities for the coming week and beyond.

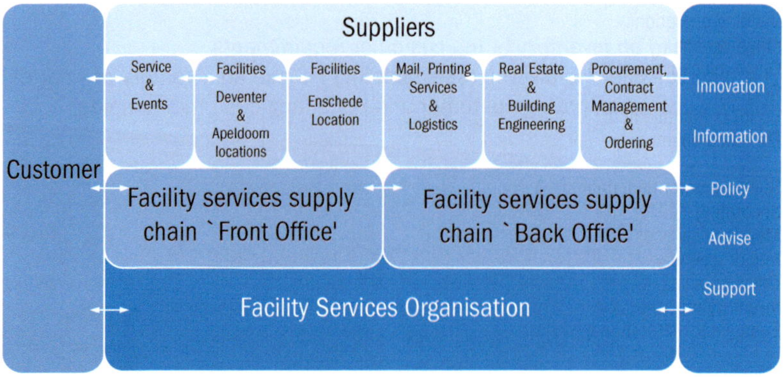

Step 5: Establish management and control

Next, you determine the management and the control of the facility processes. This involves how the facility manager directs the services.

Management and control

EXAMPLE OF STEP 5

The FSO strategic objectives have been elaborated into measurable results that give direction to the design of the facilities services. Essential indicators are quality, cost, safety, ethics and process efficiency. Objectives for improvement are linked to these indicators, such as a 10% improvement in quality compared to previous measurement results. FSO uses continuous improvement (lean) as a standard way of working. This ensures a 100% focus on customer demand and reduces waste in the processes. It is regularly tested whether customers experience this as such. The overall performance is benchmarked against other universities of applied sciences. The figure below shows that implementation and management are separated to achieve FSO's intended agility.

Strategic goal	Resulting in	Results for 2019/2021
Agile and vigorous service organisation	• Teams with process responsibility and ownership.	• Teams can set up, implement and improve services according to frameworks. This promotes speed and prioritises execution.
	• Innovative through intense collaboration with academies, partners and suppliers.	• Applied innovation of governance (e.g. planning, consultation structure, external partnerships). • Annually three innovation/improvement projects in progress in cooperation with education & external suppliers.
	• Decisive and supportive leadership.	• FSO Management is visible to employees (Gemba-management). • Decisions are made and communicated quickly. • Employees are open about their work. Even in the event of mistakes. • Every employee has a development agreement
	• Engaged, committed and flexible employees.	• Open and proactive communication about FSO and Saxion • Employees contribute to improvement initiatives. • Increased employee satisfaction.
	• Separated execution and management.	• Organisational design with a focus on execution (make it happen). • Supply chain manager takes care of the conversion of customer demand to service adjustments.

The performance of suppliers is actively managed (continuous improvement), in terms of agreed performance indicators and their contribution to innovations.

It is also essential to organise the processes management itself: how, how often and with whom do you consult? FSO has opted for a weekly start meeting with all managers. This enables FSO to quickly respond to developments, to increase cohesiveness and to eliminate unnecessary consultations.

Step 6: Determine how you communicate with customers

When it is clear which facilities services the facilities organisation has to offer its customers, the next step is to determine how the offer is communicated to the customer. This involves setting up a storefront in which the facilities organisation presents its range of services and how the mutual

Communication

communication between facilities organisation and the customer takes place.

EXAMPLE OF STEP 6

As far as the Executive Board and the directors are concerned: for these client contacts, the frameworks for the FSO strategy are determined each year. Every quarter, the FSO director reports on agreed performance indicators in management reports.
As far as managers are concerned, supply chain managers have structural contact with academy management and managers of services to determine the (future) need for services. This is how they fulfil their role as a business partner. The net promoter score (NPS) is used to determine to what extent managers see FSO as a business partner.
As far as students, employees and visitors are concerned, the facilities service point functions as a one-stop-shop for complaints, wishes, requests for information and malfunctions. They can be reached digitally, by telephone and physically. The products and services catalogue provides a complete overview of all services and conditions. Employees can place (online) orders via the order desk. Employee satisfaction is measured in the annual employee satisfaction survey (ESS). Student satisfaction is measured in a national student satisfaction survey (SSS). In this survey, FSO also experiments with specific measurements for a particular feature, for example, for catering facilities. An app measures the experience of guests over a more extended period, so that based on trend-data, insight is gained into patterns in use and possibilities for improvement.

Step 7: Review and revise

Of course, there is no stopping the world around the organisation. In this last step, the facility manager is alert to whether the facility organisation is still doing the right thing (effectiveness), and how well it is doing (efficiency).

Up to date

If the facility plan is no longer up to date, the process starts again at step 1, but now within the changed context.

EXAMPLE OF STEP 7

The Saxion strategic plan has a duration of four years, just like FSO's strategic plan, and is designed as an improvement A3. FSO's strategic plan is translated into specific annual objectives with related measures. The progress of the goals is continuously monitored. As such, the progress and topicality of the strategic plan are annually reviewed.

4.5 Facility marketing and expectation management

The marketing of in-house FM organisations differs from that of consumer marketing. While consumer marketing is mainly aimed at stimulating the

Facilities marketing

demand for a product or service, internal facilities marketing is aimed at increasing efficiency and cost awareness among consumers. After all, the purpose of the facilities organisation is not to increase its turnover - read: costs for the client. Facilities marketing aims to strengthen the visibility and image of facility management by clearly communicating the added value for the organisation. Commercial FM-providers aim to increase turnover and market share with existing and potential customers. Moreover, they give substance to the brand and to increase the attractiveness as an employer using employer branding.

Employer branding

As mentioned before, the facilities organisation has various types of customers (seeTable 4.1 with main groups of customers), but also within these groups, there are differences. Administrative staff have different needs in terms of facilities than salespersons, and directors will expect a different service level policymakers. A facilities organisation dedicated to the customer wants to tailor its services to the specific needs of these different internal customers.

4

4.5.1 The FM marketing mix

There are several models to shape facility marketing. We discuss the 6P model and the SIVA model.

Marketing mix

With the 6P model, a specific marketing mix is created for each customer group. The mix consists of *Product, Price, Place, Promotion* and *People*. In the FM-industry, the sixth P for *Process* is essential for delivering services at the right quality-price level.

6 P's

There are multiple ways to specify the 6 P's.

Products and services

Although many facility products and services are offered to all users, a distinction can is made between service levels. In the company restaurant, for example, everyone will be able to enjoy a regular lunch. Still, the management will often also make use of luxury boardroom lunches. Differences in service level can be large and small. Examples are the types of lease cars, the functionality of copiers, and the options of the coffee machine.

Price

The *price* may be used to increase cost awareness. By charging a higher rate for a higher level of service, the internal customer can make a cost-benefit conscious assessment. If internal customers can also purchase facilities services elsewhere, pricing can be used to proof market conformity.

Place

The key to *place* is the accessibility of facilities services and the ease with which they can be obtained. Do you have to call four different numbers to speak to a specific facilities employee, or can you contact a facilities service desk with all your questions?

Promotions

Promotions can also be understood as communication. The way in which the facilities organisation communicates must correspond to the purpose and target group. Examples of communication tools used by FM are the product and service catalogue (see also section 4.6.2), newsletters/mailings, information meetings and announcements on the intranet or notice boards (e.g. in the lift) and in brochures.

People

People is perhaps the most decisive factor for customer satisfaction. A service-minded attitude and professionalism of employees are necessary for a facilities organisation committed to delivering quality (see also section 3.4). But quality should not depend on the individual employee; it is unacceptable for one employee to provide a completely different level of service than another.

Process

Without process management, it is not possible to control and guarantee a consistent quality of service (see section 3.3).

SIVA-model

Another model that can help determine the facilities marketing mix is the SIVA-model. The abbreviation stands for: *Solutions, Information, Value* and *Access*. While the 6P model is about a marketing strategy based on the product or service, the SIVA-model takes the customer's needs as its starting point. This is why the SIVA-model is very suitable for service-oriented, customer-oriented organisations.

Solution

Solution is about solutions that meet the needs of the customer. It is about the result that facility management delivers, not about the characteristics of the service. For example: 'every customer feels welcome by our hospitable reception', instead of: 'our reception is open from 07:30 to 18:00'. By presenting the service as a solution, the customer recognises that facility management meets his needs.

Information

Information. The internet provides fast and easy access to information, which makes the (facilities) market more transparent. Customers can use reviews to get a picture of the experiences of other customers easily. It is therefore important for in-house facilities teams as well as for commercial providers, to provide clear information to the customer. And if the claims are substantiated with factual data, customers are more willing to accept.

Value

Value. Value is about the relationship between price, quality and added value of the solution offered for the customer. It is up to the in-house facilities organisation or external provider to demonstrate the added value it delivers.

Access

Access. For the customer, the support and services must be quickly and easily accessible through channels appropriate to the service: a mix of 'click-call-face'. Click stands for digital accessibility; call for accessibility by telephone; face for personal contact (for example, a service desk).

4.5.2 Customer satisfaction

Facility marketing is an opportunity to highlight the added value of facility management and to strengthen the image (both for internal and external providers). Facility marketing is also crucial for managing and positively influencing customer expectations. After all, it is the 'fate' of the facility manager that the customer takes most of the services for granted. In general, customers will not be surprised by the availability of a workplace, tasty coffee, and sanitary services. Because the customer assumes that all these facilities are available and that they have a certain level of quality, deviations will be particularly noticeable in the negative sense. Without attention to facilities marketing, the emphasis is quickly on customer dissatisfaction based on complaints received.

The relationship between a customer's satisfaction and his expectations can be described as follows. The customer is satisfied if the experience at least matches the expectations he/she had in advance.
Suppose you go to the cinema, but you are slightly late. The previews have already started the moment you still have to buy your ticket. You expect that you will soon have to hurry to find your place in the dark, where there is a good chance that you will disturb other people. But now the cashier says: 'Unfortunately, the doors are closed, but the main film hasn't started yet. I'll give you tickets to the aisle of a favourable row and call a colleague to show you your place. The employee will call a colleague who kindly guides you to the right place and wishes you a pleasurable movie night. In this example, your expectations are exceeded by the service offered. The result is that you are satisfied. But now for the tricky part. The next month, the same

situation occurs. Again, you are late for the movie, but now you expect that it will not be a problem. Relaxed you are at the box office, where an employee informs you that the doors are already closed and that it is no longer possible to buy tickets for this show. Your expectations were high because of the previous experience, and thus, you are highly dissatisfied. This example underlines the importance of influencing the customer's expectations. If the employee had emphasised that for once he can make an exception to the rule (for whatever reason), but that this is not the policy, expectations would have been realistic the second time. Even then, the client would be disappointed but would understand that this had to do with a too-tight schedule. What's more, if the cashier hadn't reacted harshly, but had kindly recommended another film, then the result might still be a satisfied customer.

Kano-model

As indicated in the introduction to this section, this principle described here does not always apply, as the Kano-model illustrates (Figure 4.6). The model was developed in the eighties of the last century by the Japanese professor Noriaki Kano. The model helps to identify customer expectations, priorities and specific needs. Also, it helps us understand how quality and satisfaction are interrelated.
The model consists of two axes: the functionality (ranging from not implemented to fully implemented) and the degree of satisfaction that this implies for the customer. When mapping needs, you can distinguish between three different factors (or categories of services): the must-be factors (basic factors); the performance factors; and the attractive factors (so-called wow-factors). Providing for some customer needs can lead to higher or neutral satisfaction, while not meeting other customer needs can lead to lower or neutral satisfaction. For priority-setting, facility professionals need to understand which services are decisive for customer experience, and what are so-called basic factors that require a minimum standard to prevent dissatisfaction.

FIGURE 4.6 Kano-model

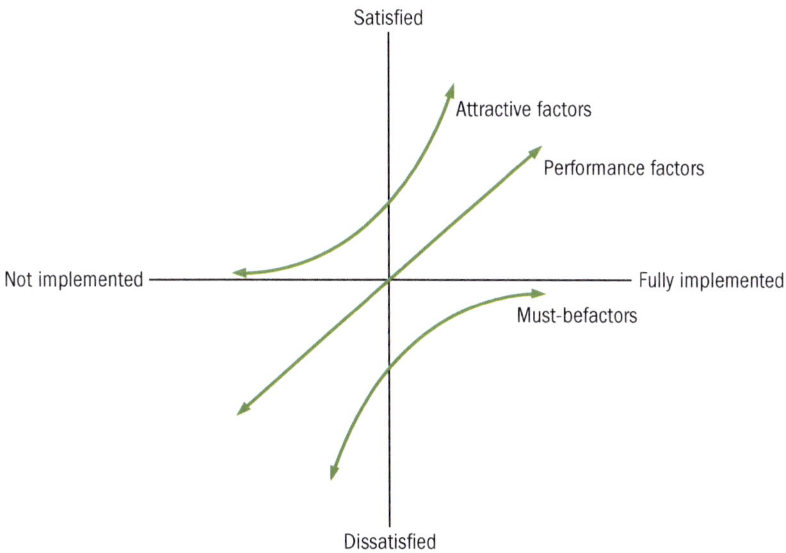

'Heterogeneous groups of users, by definition, have different wishes and needs. It is naive to aim for an average rate of confirmation of expectations.'

— Herman Kok, principal lecturer in Facility, Health and Innovation Management at Wageningen University, www.facto.nl, 6 February 2019

Basic factors

Basic factors (must-be) are factors that must be provided for as a minimum standard; these are also referred to as hygiene conditions. These are requirements and needs that customers take for granted but at the same time, which they are fully justified to expect. If the minimum standard is not provided for, then by definition, the customer is dissatisfied. This is why basic factors are called dissatisfiers. Take, for example, a hotel room. The availability of a bed as such, will not satisfy the guest. And, it is not necessarily true that more beds lead to happier hotel guests. On the other hand, hotel rooms without beds will inevitably lead to dissatisfied guests. Cleaning and security are also examples of basic factors. An organisation that promotes itself as an employer with clean bathroom facilities and a safe workplace is likely to get frowned upon, as this is what workers may expect.

Dissatisfiers

Performance factors

For performance factors, the rule says that the more they are present, the more satisfied the customer is. Conversely, a customer may also become less satisfied if these factors are present to a lesser extent. Services of this type are called satisfiers. The size of the workplace can be a performance factor. A tightly dimensioned workplace results in lower employee satisfaction; and employee satisfaction increases with more spacious layouts. Catering is also an example of a performance factor. The fuller the product range and longer the opening hours, the higher the level of user satisfaction will be. Conversely, a limited range and limited opening hours lead to less satisfaction.

Satisfiers

Wow factors

Delighters

Wow factors (delighters) are things that the customer does not expect and with which he/she is pleasantly surprised. Because customers do not expect these items to be present, they will not be dissatisfied when these are unavailable. Some examples are fresh fruit at the reception desk, a bottle of drinks for visitors on departure or a nice bouquet on the counter.

Customer expectations are continually changing as previous experiences feed them. This is why what distinguishes today is tomorrow's standard. Performance factors can even turn into basic factors, like, for example, free wifi.

4.5.3 Managing expectations

Managing customer expectations is the key to customer satisfaction. A facilities organisation has several tools at its disposal to manage customer expectations.
The first tool is to make clear agreements about the services to be delivered as well as their quality. To this end, an important role is reserved for facilities account management, as described in section 4.2.1. The agreements between the facility manager and the client about the facility products and services to be delivered are recorded in an SLA (see Table 4.1). But with an SLA, the facility manager does not influence the

expectations of the customers. After all, customers are usually not involved in determining the SLA, so it is very doubtful whether they know what that SLA entails. To inform end-users of precisely what they can expect in terms of facilities and services, the SLA is converted into a product and service catalogue (PSC). The PSC contains all the information that is important for users of the work environment; it is a kind of showcase for all facility products and services. For each of these, practical information is presented, such as delivery conditions, throughput and delivery times, ordering units, conditions, procedures and costs. By indicating the prices of these products and services, even if they are not charged, the customer becomes aware of the costs.

Product and service catalogue (PSC)

A PSC, for example, states how lunch is booked and what the costs are, how malfunctions must be reported, what the service hours of the company restaurant are, how the service desk can be reached and what the opening hours of the building are.

The facility manager can use a PSC to ensure that the beneficiary customer (the customer at an operational level) has realistic expectations of the services. A customer with realistic expectations is more easy to satisfy. An example: a customer that is informed that lunch catering must be requested at least 24 hours in advance, will not be dissatisfied if told that last-minute bookings are limited to standard sandwiches.

A PSC can be made available in the form of a guide, but nowadays many organisations place it on the intranet (see screenshot of a digital PSC) or in an app. The advantage of this is the direct link with the integrated workplace management system (IWMS). A customer searches in the PSC to find out how he can submit a request, place an order or report a malfunction. Next, the appropriate digital form can be opened, which will be processed in the IWMS after submission. A PSC thus becomes the central point of access for all the services offered within the organisation. The task of the service desk then shifts from registering notifications and informing customers, to managing the PSC and monitoring the processes.

Integrated workplace management system (IWMS)

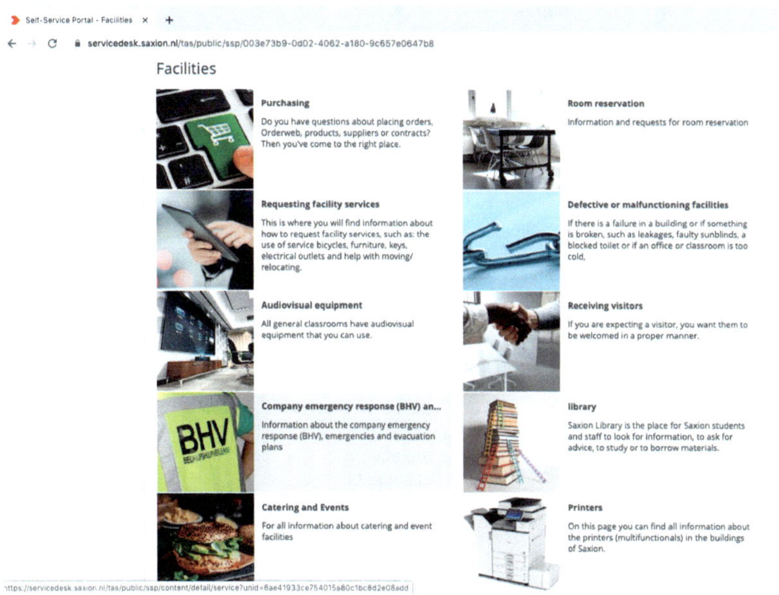

4.5.4 The SERVQUAL-model

SERVQUAL-model

When dealing with dissatisfied customers, it is essential to know what the root cause is. With the help of the SERVQUAL-model (an abbreviation of SERvice QUALity; see Figure 4.7) from Zeithalm, Berry and Parasuraman, it is possible to look at causes of dissatisfaction among customers in a more detailed way. In line with what we discussed before, this model is based on the presumption that dissatisfaction results from a discrepancy between expectation and experience. Helpfully, this is a diagnostic model that also indicates how these gaps can arise. And once the causes are known, the solutions are clear.

The first 5 gaps

Discrepancies between customer expectations and reality can emerge on several levels:

1 *Difference between what management thinks the customer expects from the service and what the customer expects from the service.*
2 *Difference between what the customer expects from the service and the way the management translates those expectations into policy.*
3 *Difference between the policy of the organisation and the way the staff provides the service.*
4 *Difference between the promises the organisation makes to the customer and the quality it can deliver.*

4

FIGURE 4.7 The SERVQUAL-model gap 1 up and to 5

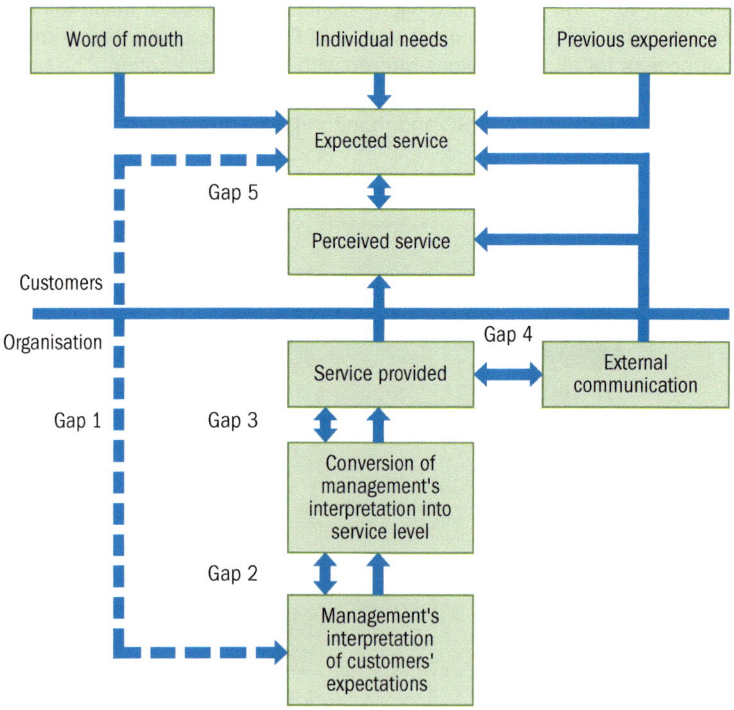

These four gaps together expose the quality of service as experienced and assessed by the customer. The SERVQUAL-model shows that the expectations of the customer is influenced by:

Expectations

- experiences of other users (word of mouth);
- personal need (what the customer wants);
- own experiences in the past (both with the organisation and the industry);
- communication (does the external communication match the wishes of the customer and does the communication match the service provided?).

The SERVQUAL-model is used in response to a complaint received: the customer is not satisfied with the service offered or the service level. This is the fifth gap:

5 *Difference between what the customer expected from the service and what he experienced.*

The investigation of the complaint in gap 5 starts with an examination of gap 1, gap 2, gap 3 and gap 4 to find the cause. In 2002, in addition to the SERVQUAL-model, Zeithaml and Bitner described nine services attributes on which the customer assesses the service. Gap 5 occurs when the customer judges one of these aspects to be insufficient:

1. Reliability
   - Are agreements adhered to the service provision?
   - Is there adequate assurance that the services will be provided?
2. Responsiveness
   - How are any comments about the services handled?
   - If situations change, can the service be adapted accordingly?
3. The expertise of the service provider
   - Is the service provider a professional partner?
   - Does the expertise of the service provider match the wishes of the customer?
4. Customer focus
   - Is the service provider proactive or reactive?
   - Are the interests of the customer also the interests of the service provider?
5. Communication
   - Is communication clear and unambiguous?
   - Was the communication conducted at the right level?
6. Credibility
   - Does the service provider have such a high level of trust or confidence built up that it is believable?
   - Is the assigned role (partner, customer-service-provider) performed accurately?
7. Safety/confidence
   - Can the services be carried out under the right conditions?
   - Can the customer have the service carried out without supervision?
8. Empathy, understanding the customer
   - Is the service provider customer-oriented?
   - Are customer developments taken into account in the execution of the service?
9. Facilities
   - Are the resources/facilities used in line with the customer's wishes?
   - Have the resources/facilities been used appropriately?

The additional 8 gaps of Facility Management

Almost every facility organisation outsources one or more services. With various parties involved in the organisation of the services coordination processes become more complex, and miscommunication is more likely to occur. This miscommunication can result in eight additional gaps, which can be visible and noticeable for the customer. Hillenius (2013) has supplemented the original SERVQUAL-model to include the thirteen gaps in facility management. The first five gaps are derived from the SERVQUAL-model. They are described from the perspective of a facility management organisation. The other eight gaps are additional and are partly related to the services provided by a third party through outsourcing (Figure 4.8). These additional gaps are:

Thirteen gaps of facility management

4

6 Unworkable work instructions.

Work instructions, protocols and policy are often made at a certain distance from the shop floor. This distance can mean that the proposed working method is not feasible in practice. The reason may be that the required expertise or the required resources are lacking, that the estimation of the necessary time is not correct, or that internal circumstances hinder proper execution. In those cases, the protocol or job description is not well attuned to the working method used in practice.

FIGURE 4.8 Gaps 6 up and to 13 of Facility Management

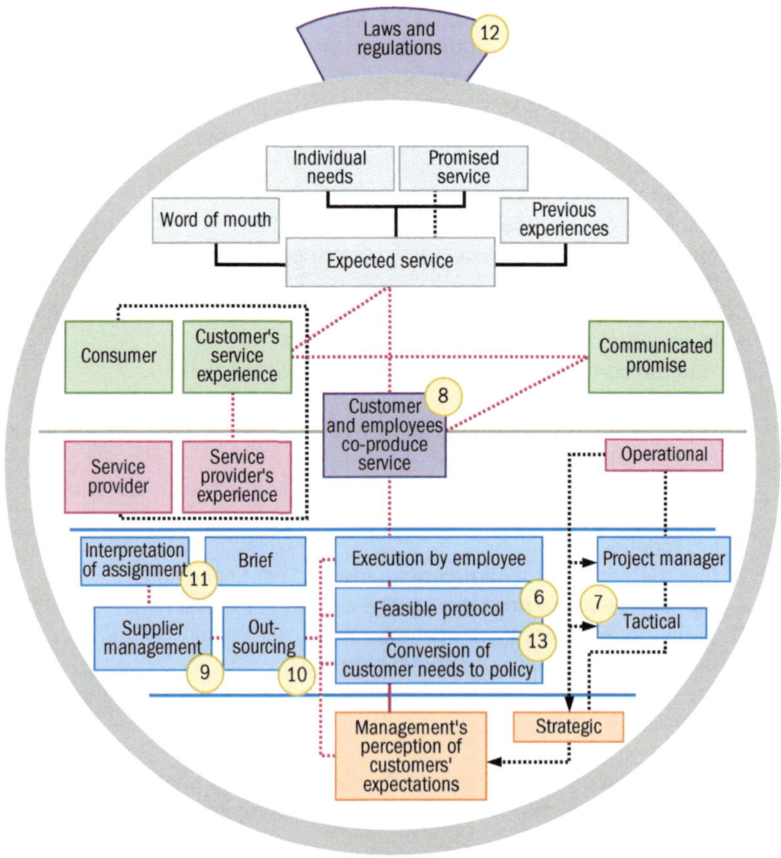

7 Conflict of interests.
In the SERVQUAL-model, it is assumed that the customer is in charge of the service. Still, the facility manager has to deal with stratification of customers (defining customer, paying customer, benefitting customer) and their interests (see Table 4.1). The quality of the service may be compromised as a result. Professional facilities account management can contribute to preventing differences in customer experience and to making well-informed choices (see section 4.2.1).

8 Unpleasant interaction between customer and employee.
Even if the employee strives to do his/her best, this does not mean that the service provided is appropriate for the given situation. In services, employee and customer are jointly responsible for the performance, and precisely because of this interaction, things can go wrong. Excellent listening skills and professional communication (also with angry customers) are essential to prevent misunderstandings and dissatisfaction.

9 Incompatible cultures.
Each organisation has its own identity, culture and core values. It is important that the selected contractors (providers) align the way in which they perform the services with the identity, culture and core values of their client. Especially when the client comes into contact with the hired service providers, the client runs the risk of a negative experience among users. Therefore, the programme of requirements must specify what added value the supplier must offer. Still, often the client limits himself to a general description of this. If suppliers are selected based on a quotation, the client can include mandatory references.

10 Unclear or incomplete specifications (brief).
The client must be aware that outsourcing service provision requires a careful approach. First of all, the client must know what he wants. Next, the procurement department must have a clear view of the functional and technical requirements imposed on a product or service. Information from facility information systems (FMIS) can be used to this end. However, the information in the FMIS is not always complete. It often mainly reflects the incidents, especially in organisations that work in an incident-driven manner. In other words, is registers complaints and errors instead of the overall performance.
If a supplier uses its own systems in the execution, the question arises as to who owns the data. If the client chooses another supplier in the future, does it still have access to its data? And can new suppliers see this data for quotations?
But even when the specifications are precise, a cooperative attitude on the part of the client is essential for proper implementation. Both parties must be willing to invest in the collaborative relationship and take the time to get to know each other to ensure that the contract is properly tailored to the requirements. In reality, such is not always the case.

11 Misinterpretation of the specifications (brief).
The organisation itself does not always compile the specifications or brief; often, an external consultancy firm is called in to do so. The more parties involved, the higher the risk of misinformation or incomplete sharing of information. The commissioning party can prepare a brief based on

incorrect (outdated) information. As a result of which, essential matters may be overlooked, or the service level is not adequately described. Sometimes the consultancy firm misinterprets the service level desired by the client. In addition, the supplier may misinterpret the specifications. This may lead to incorrect starting points in the quotation and incorrect work instructions and protocols, as a result of which the work is not carried out correctly. By including the desired service level in internal SLAs as standard, the client can prevent aspects from being overlooked or misrepresented. Before setting-up services, suppliers should carefully check the specifications. Modifications at this stage are still relatively easy to implement; they are much more drastic (and costly) if the supplier has already started to provide services.

12 The customer's wishes do not comply with laws and regulations.
The government adopts laws and regulations that also influence how suppliers provide their services. The legislation is not static but continually changing. Legislative changes and new rules, for example, about working conditions, can conflict with the wishes of the customer and therefore harm his perception of the service. Excellent communication and the provision of information by the organisation is vital to foster understanding among customers.

13 Project groups fail to take the customer-perspective into account.
Every organisation uses projects for large and small improvements on an annual basis. In business process redesign projects, it is essential that the organisation not only looks at the business result but also at the effect on the customer experience. Many project groups only work from an organisational perspective. They only involve the customer at a late stage in the decision-making process. To prevent a tunnel vision, policymakers can mention the intended effect on customer experience as a required result of projects.

4.6 Procurement

Facility procurement

Indirect procurement

The simplest definition of procurement in general is selecting and contracting suppliers of products and services that are invoiced. Facility procurement is a form of indirect procurement. This is the sourcing of all services and resources that are not purchased for the primary production processes directly, but that are needed to maintain and develop the core business operations. Thus, the goods and services classified under the umbrella of indirect procurement are commonly bought for consumption by internal stakeholders (business units or functions) rather than the external customer or client. FM-procurement includes amongst others accommodation, maintenance, lease cars, office supplies, energy, and workplace design. Moreover, it encompasses all services outsourced by the facility management organisation, such as maintenance, cleaning, catering, security and landscaping. It is essential to strive for the best value for money when purchasing facility services. Though it is part of the facilities focus area, FM-procurement is often organised as a separate team or department of the facilities division.

4.6.1 Outsourcing

In practice, we see that the responsibility of the Procurement Department increases with higher outsourcing ratios. Many organisations choose to outsource this task to specialised procurement agencies, especially in the case of (complex) tendering processes (see section 4.6.4), as they lack the necessary expertise to do so themselves. Larger organisations (>500 employees) usually do have this expertise at their disposal.

Outsourcing ratio

Procurement, including facility procurement, cannot be viewed in isolation from the organisational strategy. Whether the facilities organisation chooses to make, buy, or cooperate, depends on the facilities strategy (see also section 4.2). And the facilities strategy is a derivative of the organisational strategy. Whether outsourcing is the right choice depends not only on the corporate strategy but also on the culture and challenges facing the organisation. The degree of interdependence with the primary process also plays a role because this partly determines the extent of the impact and risks of outsourcing (see Figure 4.9).

Impact and risks

FIGURE 4.9 FM's impact on the primary process

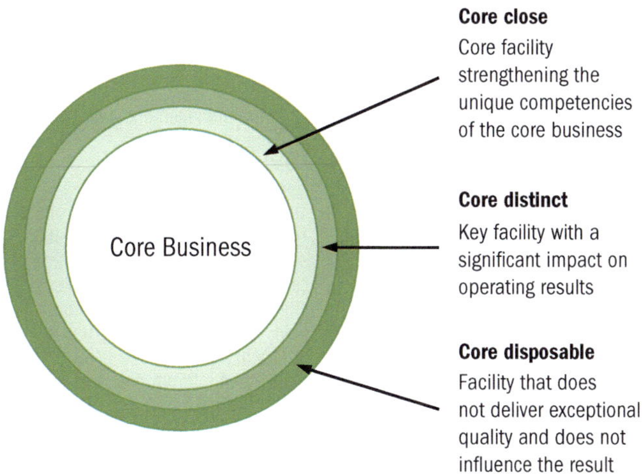

There may be several reasons to carry out services yourself (make) or to purchase from a supplier, or to choose a combination of these. Combinations of in-house and outsourcing in the care and cure sector can be:

- Suppliers supply semi-finished food products that are used by nutritionists or room service to create meals that comply with dietary requirements and the customer's wishes.
- Employees wash the personal linen of clients and patients; the cleaning of the bedding (bedding, towels, tea towels) is outsourced.
- Apartments of residents or patient rooms are cleaned by hospital staff and circulation areas and public areas by an external cleaning company.

Table 4.2 shows the benefits and consequences of in-house operations and outsourcing.

TABLE 4.2 In-house or outsourcing: considerations and consequences (based on Beckers & Roelofs, 2016)

	In-house sourcing	Outsourcing
Advantages	In-house facilities expertise	Access to expertise if the (facilities) organisation does not have the expertise, or if it is expensive to keep specialist knowledge up-to-date.
	The facilities organisation is independent of suppliers, retains its autonomy and sense of control.	The facilities organisation brings in a specialist and therefore, higher quality and possibly lower production costs.
	Facilities staff are colleagues who know the culture and other members of the organisation well so that the family atmosphere is maintained.	Enables an organisation to focus on core business.
		The FM-organisation can focus on strategic alignment and is not distracted by the issues of the day that operational activities bring.
		Outsourcing offers more flexibility to respond to changes in demand.
		Transparent costs due to monthly invoice
		Lower fixed costs (for staff, overheads and payroll administration)
		The organisation can utilise the supplier's economies of scale, which can lead to financial benefits.
Effects	Company recruits and retains facility staff.	Contract management is required to manage contracts.
	The company is a payroll administrator.	Outsourcing can lead to vendor dependency.
	The company itself is responsible for the education and training of its staff.	The lower fixed costs are also offset by hidden costs: transaction costs and costs of financial administration (invoicing) and contract reviews.
	The company itself handles replacement in the event of illness.	The risk of the costs of additional work (in the case of effort-based contracts)
		Insufficient internal benchmarking, which leads to: • lack of reliable (financial) control data; • lack of understanding of indirect revenues and savings.

Pros and cons

Table 4.2 does not mention 'cheaper'. Contrary to what is often said, outsourcing does not necessarily lead to lower costs. Certainly not if the costs of the outsourcing process itself (the so-called transaction costs) and the costs of invoicing and contract management are considered. There are pros and cons of outsourcing, as you can see in Table 4.2. Outsourcing is, therefore, a process that must be completed with care, and which requires a great deal of time and hence, costs. A decision tree can help in going through the decision-making process (see example in Figure 4.10).

FIGURE 4.10 Example of an (out)sourcing decision tree

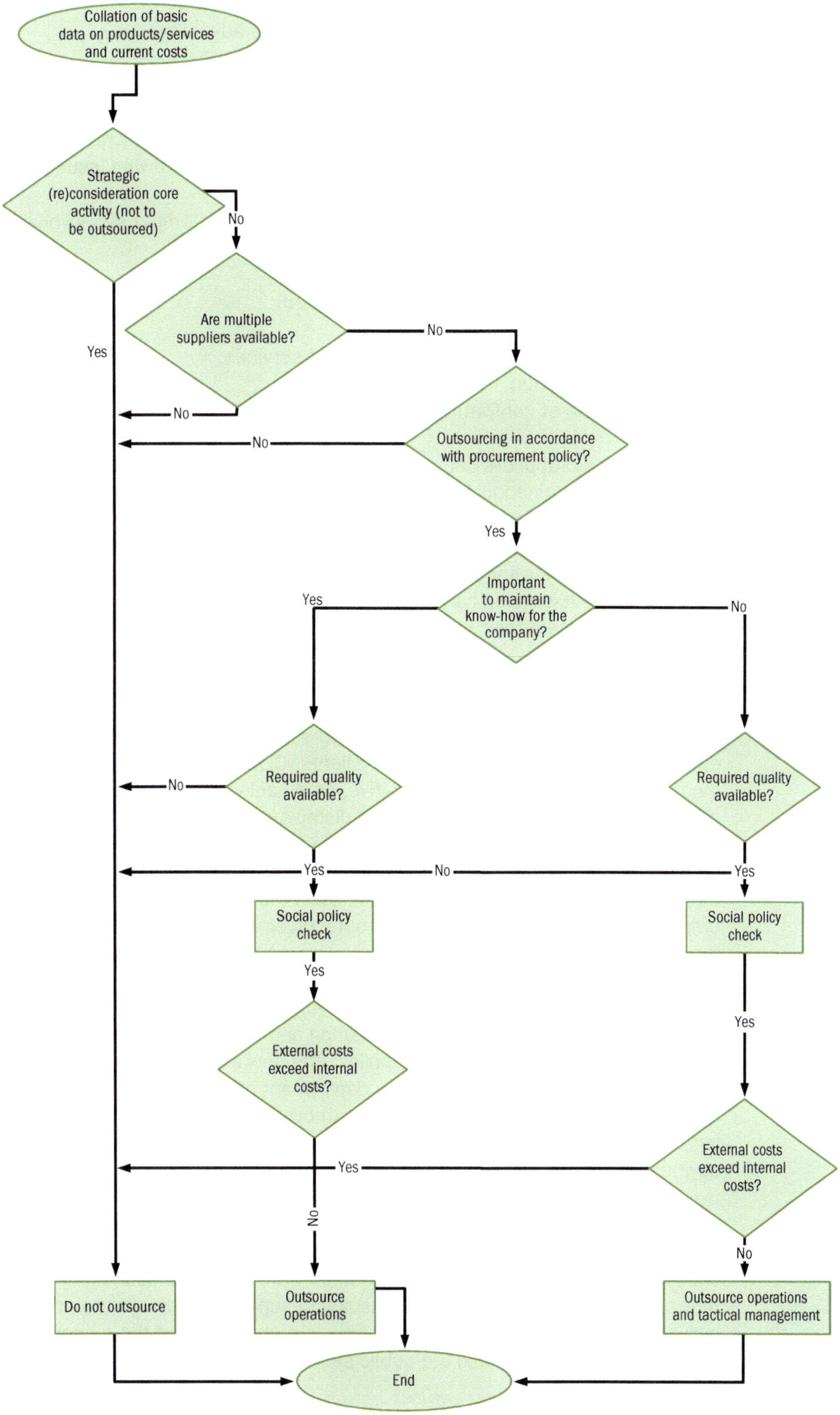

4

Assessment criteria

The assessment criteria for outsourcing are:

- The impact on the primary process (added value and risk).
- The qualitative added value: can the supplier do better?
- Economic added value: is it economically more advantageous?
- The social considerations.

Because the impact on the primary process is a vital assessment criterion, various parties have a say in the outsourcing of facility services processes. When outsourcing, careful consideration is given to the following issues, among others:

- the impact on the various stakeholders (including the employees involved);
- the effect on the public image;
- the process of matching supply and demand;
- the development of partnership (if intended);
- the fulfilment of customer needs;
- maintaining control over the execution of services;
- ensuring innovation;
- ensuring market conformity.

In some cases, suppliers can offer better quality than in-house service provision. As explained in Table 4.2, such is the case when state-of-the-art expert knowledge is needed. Furthermore, suppliers sometimes offer better value for money. They can achieve economies of scale, and have a pool of dedicated and well-trained staff to compensate sick leave quickly. Furthermore, outsourcing can increase the flexibility of organisations to downsize or upscale their business processes.

Transfer of undertakings

VAT

As far as economic added value is concerned, for companies that are bound by a collective labour agreement (CLA), outsourcing work can be cheaper. Such is the case if the CLA of a supplier prescribes lower wages and employment benefits compared to the clients' CLA. However, organisations must be careful not to be overly positive. An organisation that wants to outsource security, for example, cannot dismiss employees with a permanent contract of employment just like that. The legislation on protection of employment act can also apply to outsourcing. According to this Act, employees follow work and, in principle, retain the right to their terms of employment. Therefore, when a client outsources activities for the first time, the schedule of requirements often also requires the service provider to take over the employees involved in the outsourced activities. In subsequent outsourcing of the same activities with labour as the most decisive production factor (such as cleaning, catering and security), the protection of employment act usually no longer applies. The new service provider then has the freedom to give preference to the deployment of its employees. Another important financial consideration is the turnover tax (VAT). Suppliers are legally obliged to charge VAT. This is no problem for the private sector. In the payment to the tax authorities, they are allowed to deduct the VAT they have paid from the VAT that they have charged to their customers. That is why they pay close attention to whether the offer is inclusive or exclusive of VAT. Public authorities, such as healthcare, education and cultural organisations, do not charge VAT to their customers. Therefore, they are unable to deduct any VAT they pay. The price these organisations pay to suppliers is always the price inclusive of VAT, which is much higher. This is probably one of the reasons for the lower outsourcing rate in the public sector.

Social considerations arise from corporate social responsibility. An organisation can, for example, outsource the maintenance of the grounds to the green service of the social work agency. As a result, outsourcing now creates employment for people with a distance to from the labour market. Social considerations may lead to the decision to bring services under management again.

As described in section 1.5.2, more and more organisations are opting for small strategic facilities departments with highly educated facilities specialists (the so-called management organisation) who are capable of setting out the organisation's 'facilities strategy' and who can outsource the necessary coordination to the market.
Regardless of whether the facility manager outsources a service or not, s/he remains responsible for the quality of the service provided and the working conditions under which they are carried out. And so, s/he will have to keep his or her finger firmly on the pulse. Is the service provided following the agreements in the contract? Is the customer satisfied? In other words: a facility manager can outsource the execution of a service and the contract management, but never the supervision of the quality to the supplier of the services covered by the contract.

4.6.2 The procurement process

Procurement process

In essence, the procurement process is a cyclical process consisting of several phases (see Figure 4.11).

Scope

Phase 1 is about making a sourcing analysis and the procurement strategy. Here, the scope of the project is determined and the set-up of the procurement process. A stakeholder and risk management are used to map the wishes and needs of all parties involved, and to gain insight into the market. This involves checking whether the programme of requirements takes account of future developments of other business units. For example, the HRM department is increasingly involved in facility purchasing issues relating to the working environment because it is closely interwoven with HRM policy. In this phase, a distinction is made between requirements (need to have) and requirements (nice to have). The type of product or service (see section 4.6.3) also influences the purchasing strategy. The result of this first phase is a strategy document that indicates the framework of the project and for all subsequent phases. The goal is to arrive at a choice for a solution that is genuinely beneficial for the organisation.

Need to have

Nice to have

Strategy document

Phase 2 concerns the actual purchasing process. A programme of requirements (POR) clearly describes the expectations and responsibilities of the intended supplier. As a rule, three to five suppliers are invited to submit a quotation. The selection of the provider is made based on the criteria laid down in the POR. It is about achieving the most attractive combination of quality, sustainability and cost. It is also possible that none of the offers will qualify as positive. In that case, it must the requirements need to be adjusted, or other suppliers will be approached. After selection, the terms and conditions are negotiated. After agreement, the agreements are laid down in a contract. Contracts represent an important part of the procurement process; they are an essential part of a procurement transaction. The contract specifies which goods or services the supplier provides at what price and on what terms. As a rule, reference is made to the POR, to avoid differences in interpretations during the execution.

Quotation

Contract

Quality requirements

Contracts are subject to the following quality requirements:
- the contract meets the needs of the organisation;
- the provisions in the contract are enforceable;
- the organisation benefits from the contract;
- the contract complies with the legislation and, if the parties have subscribed to the Code of Responsible Market Conduct, also with the provisions of the code (see section 2.7.2);
- the contract is complete and unambiguous.

Contract register

All contractual details must be clear to both the supplier and the customer. Besides, all provisions (where relevant) must be clear to the organisation. In any case, this requires a contract register, i.e. a system in which necessary contract data is stored. Also, think of online-ordering and the administrative procedures necessary for processing invoices under the agreements in the contract.
Once the contract has been ratified, the tactical purchasing phase can be completed and operational purchasing can begin.

Phase 3 is crucial because it concerns the implementation of the contract stipulations. Especially in the case of a change of supplier for services, it is essential to make agreements about the transition period. How and what should the previous supplier transfer and when? And what does the new supplier need to be able to start up while the current supplier continues to provide services? You can imagine that this can be a delicate process. The client asks the supplier whose contract is terminated to help its competitor take over. Phase 3, therefore, requires the formulation of a clear consultation structure, explicit authorisations and tight control. If things are going well, then the implementation will be virtually flawless. If too little attention is paid to the implementation, chaos is lurking, resulting in dissatisfied customers.

Contract management

Finally, *phase 4* is the management phase, which consists of operational procurement (ordering and invoicing) and contract management. Associated activities are performance management, procurement portfolio management and supplier management. Contract management is managing, implementing and monitoring compliance of both parties with contracts for purchased services and resources. Contract management refers to the physical archiving of contracts and the management of relevant supplier data. In the implementation, the arrangements and the way of working agreed upon are communicated. Consistent contract management enables organisations to optimally benefit from the services provided during the term of a contract.

A contract manager's day-to-day work consists of carrying out periodic contract evaluations and measuring performance. The supplier performance assessment is an essential tool for this (see also Chapter 5). If the organisation opts for a partnership with a supplier, it is necessary to jointly evaluate the results achieved, allowing suppliers to commit themselves more to the strategic goals of the organisation. Given the dynamics of organisations (see Chapter 2), it may be desirable to make interim adjustments to agreements.

All lessons learnt from phase 4 can deliver input to phase 1 of a subsequent procurement process, with which the organisation prevents it from making the same mistakes over and over again.

At the operational level, purchasing only consists of ordering the desired items under the terms and conditions laid down in the contract. For example, a team can order office supplies directly without the intervention of a buyer. The delivery of orders and payment of the invoices are monitored for contract-compliance. After-care refers to the evaluation of the internal customer's experience with the supplier. What goes well, what can be improved? Are new requirements identified? These questions are part of the periodic meetings of the contract manager and the supplier.

FIGURE 4.11 The procurement cycle

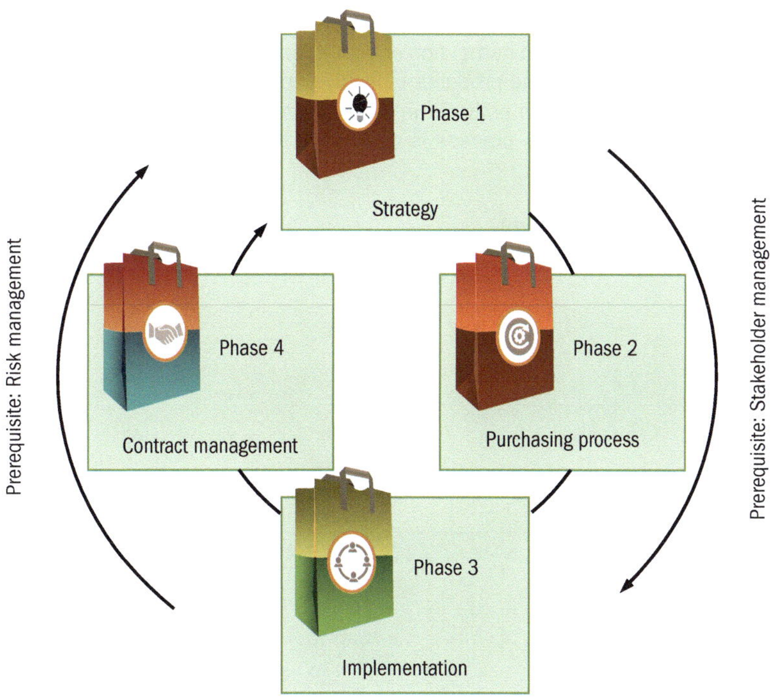

Correct identification of needs, drafting the specification, searching for the best suppliers and applying optimal purchasing conditions have the most significant influence on the cost level of facility purchasing. Within these steps, the first step is the most important: determining what the company needs. If demand is unknown, products and services purchased are not optimally effective and will not be fully appreciated. In addition to a competitive price and responsible market behaviour, the buyer, therefore, also pays attention to quality and added value for the customer. In addition to the investment costs, the buyer must also take account of the total cost of control. Contrary to what you might think, buying at the best possible price is not the main objective of the buyer. It is also crucial that buyers allow suppliers a healthy margin to ensure their business continuity.

Total cost of control

4.6.3 Supplier management

A purchasing organisation can have many suppliers, but not all of them are equally important. Everyone will understand that stationary involves far less risk than buying a new printing line for the repro department. There is a big difference in costs and in how difficult it is to obtain the product (the supply risk). Therefore, a different supplier policy applies. To help buyers determine the best strategy, Kraljic (1983) developed a model for purchase classification (see Figure 4.12). On the one axis, the supply risk is depicted. Situations with high supply risk occur from, for example, scarce materials; government instability or natural disasters; complex logistics; or few suppliers. The other axis represents the profit impact. The profit impact is high when the item adds significant value to the organisation's output. For example, because it makes up a high proportion of the production (like fresh fruit for a juice maker) or has a high impact on quality (such as the cloth used by a high-end clothing manufacturer).

Relative importance

In the matrix thus emerging, Kraljic distinguishes four groups of products: leverage items, strategic items, non-critical items and bottleneck items. The relative importance of the products is highly dependent on the industry in which the organisation is active. The relative size of the facilities budget compared to the primary process is decisive in this respect.

FIGURE 4.12 The Kraljic matrix

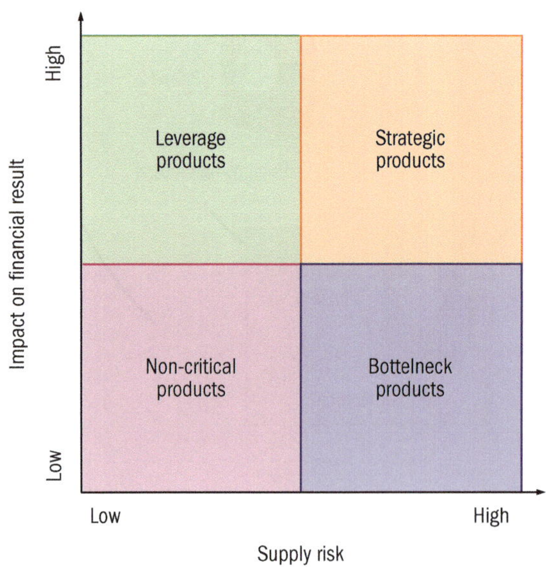

Next, we will discuss these products:

Leverage products

1 Leverage products make the difference for a company because of low supply risk and a high impact on the financial result. These are products for which, however, several suppliers are available and for which there is the possibility of using alternative products. The buyer thus has a strong position in negotiations with suppliers. An example of a leverage product for a university of applied sciences is energy. The strategy that the buyer

can apply here is *competition*, to make maximum use of the competitiveness in the market. The purchasing activities focus on tendering contracts and regularly reviewing contract agreements.

2 Strategic products have a major impact on the financial result, but also involve a great deal of risk. It is difficult to change suppliers for these products because there are often few suitable suppliers. For example, ICT is a strategic product for a university of applied sciences. A good match with the supplier is important, so an obvious strategy is *cooperation*. Procurement activities are focused on careful supplier selection and building a partnership with added value for both parties.

Strategic products

3 Non-critical products. An example of a non-critical product for a university of applied sciences is office supplies. The purchasing strategy focuses on *simplifying the purchasing process* and reducing the number of suppliers. Purchasing activities focus on standardisation, master agreements, fixed order and invoice processing processes.

Non-critical products

4 Bottleneck products. These products can only be purchased from a limited number of suppliers and the organisation relies heavily on this product. In negotiations on products of this type, the supplier has a great deal of power. A suppliers-market can result in waiting lists, high prices and poor service. An example of a bottleneck product for a university of applied sciences is the supplier of textbooks. The purchasing strategy aims to secure delivery by creating a *preferred position* with the supplier. At the same time, through *developing alternatives*, supplier-dependence can be reduced.

Bottleneck products

4.6.4 Public procurement

Government procurement involves a high risk of corruption because of the enormous financial turnover. The complexity of many procurement processes in which businesses interact very closely with politicians and civil servants contributes to this. The personal interests of public officials are often not the same as national interests. Such a conflict of interest increases the risk of corruption even more. Most countries, therefore, regulate government procurement at least to some extent.
Public procurement rules govern the way public authorities and their agencies purchase goods, works and services. For fair competition in an open and transparent market, laws and regulations require the procuring authority to issue public tenders if the value of a bid exceeds a threshold value. Government procurement is also the subject of the Agreement on Government Procurement (GPA), a plurilateral international treaty under the auspices of the World Trade Organisation (WTO).

European tendering

European tender

A European tender is a tender procedure under European legislation and guidelines. The starting point of the European procurement directive is that every bidder throughout the European Union must have an equal chance of winning a contract from public authorities. Public procurement rules are set out to create a level playing field for businesses across Europe. These rules govern the way public authorities and public utility operators purchase goods, works and services. They are transposed into national legislation and apply to tenders whose monetary value exceeds a certain amount. For tenders of lower value, national rules apply. Nevertheless, these national rules also have to respect the general principles of EU law. The principles of the European Union Directives on Public Procurement are:

- Objectivity. There are no preferences before or during the tender procedure and there is openness about the imminent contracts.
- Transparency. The procedures and rules must be clear in advance.
- Non-discrimination. Objective selection criteria must be applied.
- Proportionality. There should be no disproportionate requirements.

Aim

The aim of these procurement guidelines is an open European internal market for all companies within the member states of the European Union. In the Netherlands, the directive has been translated into the Public Procurement Act. This Act obliges central government, local government (municipalities, provinces and water boards), bodies governed by public law (universities, independent administrative agencies) and particular sectors such as energy supply, to award contracts above a certain amount (threshold value) to market parties by means of an EU-wide tendering procedure. The procurement guidelines do not apply to commercial companies, but they do affect them. After all, governments buy from commercial companies. And, if they want to have a chance of winning a contract, these companies are therefore obliged to follow the tendering procedures.

Procurement procedures

In a European tender, two types of procurement procedures can be followed: an *open procedure* and a restricted procedure. The contracting party is free to choose between the open procedure and the restricted procedure with prior selection. In specific circumstances, other options are available, including a negotiated procedure without prior notice, a competitive dialogue, a design contest, an electronic auction and an innovation partnership.
Those procedures are subject to conditions. For example, the project to be tendered must be so complicated that the contracting authority cannot make a selection without dialogue with economic operators.

Open procedure

In an *open procedure*, any interested economic operator may submit a tender. The open procedure is a one-round tender procedure. This procedure starts with a contract notice to be published in the Official Journal of the European Union (OJEU) and on the web by Tenders Electronic Daily ('TED'). In this procedure, the selection of candidates and the award of the contract are carried out in a single operation; there is no question of pre-selection. This means that all tenderers who meet the eligibility requirements must have their bids assessed in terms of content. It is therefore not necessary that one bidder has better references and higher turnover than another party; it is sufficient to meet the requirements. The contract is awarded on the best price/quality ratio, the lowest price, or based on cost-effectiveness. A choice for the best price or lowest costs must be justified in the tender. Negotiation is not permitted.
In practice, it appears that an open procedure is mainly used for the procurement of leverage and non-critical products. Leverage products can be purchased from multiple suppliers, but the costs are high. These include hospital beds, building installations and legal services. Leverage products generally compare well in a transparent market with multiple suppliers. This is why the strategy focusses on purchasing as economically as possible by, for example, achieving volume discounts through joint purchasing. The open procedure with the lowest price as the award criterion is the most appropriate procedure for these products. For non-critical products, such as

office supplies and transport, there are sufficient suppliers, their value is low, and alternatives are available. These products must be purchased as cheaply and efficiently as possible, to optimise the purchasing process, for example, through e-Procurement. These products, too, are best procured through an open procedure with the lowest price as the award criterion. In the case of joint purchasing, it may also be possible to achieve volume discounts.

Advantages of a public procedure are:
- much competitive pressure on bidders;
- the ability to bid throughout the market;
- objective, transparent and non-discriminatory.

The disadvantages of an open procedure are:
- a strict description of the procedure, a lack of freedom in the organisation as to how the procedure is organised;
- a relatively lengthy procedure with prescribed minimum time limits, particularly compared to national procedures;
- the risk of (over)submission of bids;
- high procedural costs (among other because of the high number of bids), both for contracting authorities and for tenderers;
- no possibility to negotiate;
- no benefit from prior experience with suppliers.

Restricted procedure

In a *restricted procedure*, a separation is made between the selection and the award, as two different parts of the process. This procedure also starts with a contract notice (via Tenders Electronic Daily) in the EU Official Journal (OJEU). The candidates (tenderers participating in the procedure) and the tenders are evaluated in two rounds. In the first round, the most suitable candidates are selected from all applications (usually five) on the basis of pre-determined and communicated selection criteria. The selected candidates are invited to submit a bid. In the second round, these tenders are evaluated. The contract can be awarded on the basis of the best price/quality ratio, the lowest price or the lowest costs on the basis of cost-effectiveness. Selection on the basis of the best price-quality ratio is the standard. If the choice is made to award the standard on the basis of the lowest price or the lowest cost, this choice must be justified in the tender documents. Here too, negotiation is not permitted. In practice, it appears that a restricted procedure is mainly used for strategic products and bottleneck products. Strategic products are usually complex products tailored to the specific needs of an organisation, for which there are few suppliers in a non-transparent market. These products represent a high procurement risk for an organisation. Entering into long-term partnerships with suppliers in which knowledge and experience are exchanged is generally wise as changing suppliers is often difficult. The restricted procedure or, where possible, the competitive dialogue are the best procedures. Bottleneck products represent a relatively low value in monetary terms but are nevertheless susceptible to supply risks due to a supplier"s monopoly position. The risks of supply in the short and longer term will have to be properly assessed, as suppliers have a lot of bargaining power. The best procedure for purchasing this type of products is also to use the restricted procedure (see also Figure 4.12).

Advantages of a restricted procedure are:
- competitive pressure on candidates;
- possibility of notification for the market-wide tendering;
- attractive procedure, especially for suppliers;
- objective, transparent and non-discriminatory;
- limited number of bidders, resulting in a relatively low administrative burden.

Disadvantages of a restricted procedure are:
- the procedure is very narrowly defined, and there is little freedom to set up the procedure;
- lengthy procedure with prescribed minimum time limits, exceeding the length of the open procedure;
- relatively high procedural costs, including for tenderers, but much lower than the open procedure;
- no possibility to negotiate;
- no benefit from own previous experience with suppliers.

The purchasing portfolio of central and local governments is comprehensive. Contracts in which the tendering procedure for facilities products and services is applied are divided into three categories:
1 *Supplies*, such as tickets for business trips, lease cars, security equipment and printed matter;
2 *Services*, e.g. cleaning, catering, reception and courier services;
3 *Works*, e.g. the construction of new buildings and the execution of renovations.

Thresholds have been set for each of these categories, each for a period of two years. If the total purchase in a particular class does not exceed the threshold in a given year, the obligation to use European public procurement does not apply. Different thresholds apply for different types of contracts. And lower thresholds apply to the central government than to other bodies. Table 4.3 shows the threshold amounts for the period 2020 up to an including 2021.

TABLE 4.3 Public procurement thresholds according to the category 2020-2021

	Supplies	Services	Works
Special sectors (railways, airports, energy, military, etc.)	€428,000	€428,000	€5,350,000
Central government authorities	€139,000	€139,000	€5,350,000
Sub-central government authorities	€214,000	€214,000	€5,350,000

Source: www.ec.europa.eu

The purchasing profession has become more sophisticated and, partly as a result of increasing outsourcing, it is also becoming strategically more important. For example, procurement is essential in areas such as sustainability and circularity (sustainable and circular procurement), innovation, new ways of working and new media.
For more information on public procurement and European tendering directives, see https://ec.europa.eu/info/index, for current tenders see: Tender Electric Daily (TED) via https://ted.europa.eu/.

Federal procurement in the United States of America
In the United States of America, federal procurement is regulated by many public procurement laws and regulations. All federal laws related to public procurement have been implemented in the Federal Acquisition Regulation (FAR), codified in Title 48 of the Code of Federal Regulations. The FAR sets forth the uniform policies and procedures for acquisitions by all federal departments and agencies and implements or addresses nearly every procurement-related statute or executive policy. In doing so, the FAR reaches every stage of the acquisition process.

The FAR establishes several basic methods of contracting. The three most common are:

1 Sealed bidding
A sealed bid is a document enclosed in a sealed envelope submitted in response to an invitation to bid. The five steps in the sealed bidding process are (1) preparation of invitations for bids (IFB); (2) publishing the invitation for bids; (3) submission of bids in sealed covers; (4) evaluation of bids; (5) awarding the contract. Typically, an IFB includes a description of the product or service to be acquired, instructions for preparing a bid, the conditions for the purchase, delivery, payment and other requirements associated with the bid, including a deadline for bid submissions. Each sealed bid is opened in a public setting by a government contracting officer, at the time designated in the invitation. All bids are read aloud and recorded. The agency then awards a contract to the lowest bidder who is determined to be fully responsive to the needs of the government.
Sealed bidding is a preferred method to award contract goods and services with clear and specific government requirements.

2 Contracting by negotiation
Contracting by negotiation is used in many federal procurement actions, even though it is more time consuming for all parties involved. This method is applied for the procurement of highly technical products and services, of which the value exceeds $150,000. The first step is a Request for Proposal (RFP). In a typical RFP, the government will request a product or service it needs and solicit proposals from prospective contractors on how they intend to carry out that request, at what price. Proposals in response to an RFP can be subject to negotiation after they have been submitted. All other procedures do not permit negotiations with bidders.

3 Simplified procedures
To reduce the administrative burden of awarding procurements simplified acquisition procedures (SAP) may be applied below a threshold value USD 150.000 as of 2014. SAP allows informal quoting and competition procedures for simple, small-dollar-value purchases. For example, quotes may be submitted in oral rather than written form, and contracting officers may directly compare quoted prices. Items commonly purchased through this program include office supplies, computer software, and groundskeeping services. Also, the winner may be chosen directly by a contracting officer rather than a source selection team.

4.7 Quality management

The facility manager is ultimately responsible for the quality delivered to the customer, regardless of whether the sourcing of services. But what is clean, tasty, safe? It seems that every user appraises quality differently. That is precisely why the added value of facility management is challenging to measure.

Quality

By definition, quality is to what degree the consumer's (product) requirements and wishes are met. Measuring the quality of service is essential because:

- when the facility manager can objectively determine the performance, s/he can demonstrate to the paying customer that the facility performance meets the requirements;
- by measuring the performance, the facility services can be checked and optimised.

This means that even if the facility manager does not outsource, s/he has to do (some kind of) contract management - even if it is within the organisation. A professional organisation is always in control.

In control

4.7.1 Types of quality

The quality experienced by the customer occurs during and after the service. Grönroos (2015) distinguishes between three factors:

- technical quality (what);
- functional quality (how);
- relational quality (who).

Technical quality

The technical quality refers to the service or product; to what is delivered. In the case of cleaning, for example, technical quality is the extent to which the result of the cleaning meets expectations. In the case of a pre-ordered lunch, the critical factor is whether the food is served to order, at the right temperature, at the right location.

Functional quality

In functional quality the how-question is central: how did the service process happen? What was the response time? Is the environment where the service is enjoyed pleasant? Think about temperature, the comfort of the furniture, light and design. If the customer takes part in a conference call, while someone is drilling into the wall next to him or her, the perceived functional quality is low.

Relational quality

Finally, the relational quality refers to the person providing the product or service, which is also an essential factor for the experienced quality. If there is to be intensive contact between the customer and the service provider, the employee must be friendly, sympathetic to the customer, competent and capable. If the customer feels well treated, the relational quality is likely to be good. Of course, the opposite can also happen. Unkind, uninterested interaction can result in a negative perception of quality, even if the output provided is still excellent. Relational quality is also expressed in timely and appropriate management reports.

4.7.2 Standardisation

Quality management systems have in common that they are building on the so-called Deming-cycle: plan-do-check-act (PDCA). Quality management is the cyclic process of setting goals (plan), providing the service (do), measuring whether the performance meets the objectives (check) and - if not - adjusting the process (act). Figure 4.13 illustrates this continual process.

Quality management

FIGURE 4.13 The Deming-cycle, or PDCA-cycle

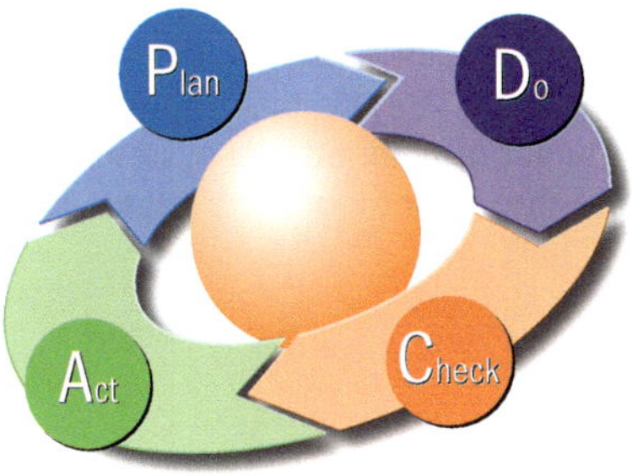

4

PDCA FOR ONBOARDING

We give an example of the onboarding process for new employees. The SLA dictates that on the first day at work, a new employee receives a warm welcome by the service desk, along with a key card, safety instructions, a mobile phone, laptop, and a workstation (plan). The service desk coordinates the onboarding process (do). To know if the service level is following the agreement, amongst others, we must check if everything is available at the start of the onboarding process. A simple checklist will do (check). If it turns out that some things are missing, a manager will look for causes and will make adjustments accordingly (act). If all goes well, the next measurement shows improved performance.

In 1987, the European Committee for Standardisation (CEN) introduced international standards for quality established by the International Organisation for Standardization (ISO). ISO 9001 is a general standard for quality management systems. It provides a framework for setting up a quality management system. If an organisation wants to be ISO certified, it must set up, describe and implement a quality management system. It comes down to this: (1) say what you do, (2) do what you say, (3) show that you do what you say. ISO standards aim for a systematic application of the PDCA-cycle (see Figure 4.13). These standards mainly focus on the working method of the organisation. They aim to increase predictability and controllability of processes and to minimise risks. The result of the process is less important. In other words: ISO-certified chocolate does not mean it is tasty.

ISO 9001

ISO 9001 and its certification are sometimes associated with unnecessary bureaucracy. However, this is not due to the standard, but to the assumption of some organisations that drawing up detailed procedures and extensive regulations leads to more quality. Fortunately, there is a growing awareness that compact plans and dashboards are more effective than a cabinet full of quality manuals that nobody reads. The advantage of certification is that it enables an organisation to prove to stakeholders that its quality management is in order.

There are various quality management systems, such as EFQM, TQM and Six Sigma. Which method is used by an organisation depends on the preference and the nature of the company. Some industries use a sector-specific model. Organisations that work with a quality management system must also ensure the quality of the system itself. They do this through quality assurance. Quality assurance is the set of planned and systematic actions that are necessary to demonstrate that the product or system complies with the requirements set.

Quality assurance

4.8 Environmental management

More and more organisations are aware of the effects of their business operations on people and the environment. As a result of news reports on climate change, the Paris Climate Agreement and government ambitions concerning the transition to a circular economy came to be formulated (see section 2.4.5 and 3.2.6). If the built environment is to meet the objectives of the Paris Climate Agreement by 2040, the energy consumption of the built environment worldwide must fall sharply compared to the current average. This is a firm but necessary ambition, in which facility management has a crucial role. As mentioned earlier, the facility manager is a catalyst in making business operations more sustainable. A substantial part of the CO2 emissions is related to user behaviour and the energy efficiency of the building and its installations. Also, the facility manager has a substantial say in whether or not to separate waste flows, because s/he negotiates contracts with waste processing parties. As the example below illustrates, there is a need for FM-professionals to take a leading role.

Paris Climate Agreement

● climateactiontracker.org

Climate Action Tracker

The Climate Action Tracker (CAT) is a consortium of a group of three climate-research organisations, that monitors the progress of 32 countries in meeting the Paris treaty goals. These 32 countries together account for 80 per cent of the world's greenhouse gas emissions. As the 2100 warming projections show, current policies will result in a 3.1–3.5 degrees Celcius global warming – not nearly enough to meet the 1.5°C temperature goal of the Paris Agreement. However, not all countries are equally inadequate, as the next graph shows. The only two countries that are currently set for meeting the Paris Agreement targets are Morocco and Gambia.

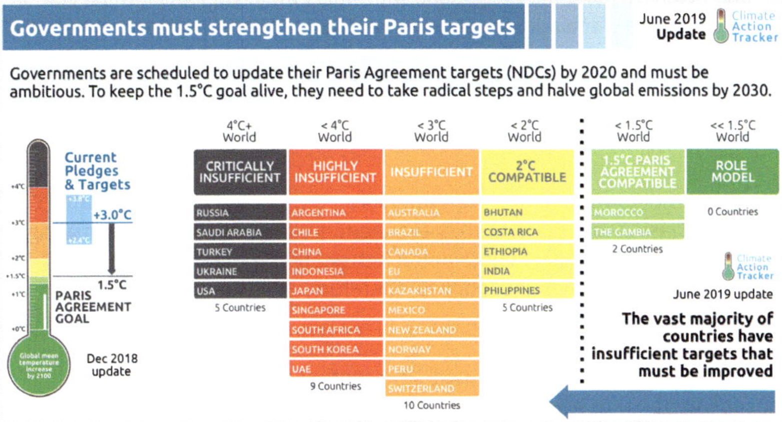

But why wait for governments to take action? Facility management professionals around the world by their own initiative can, and must, make a difference.

4

4.8.1 Environmental management systems

Environmental management system

An environmental management system is the basis for an organised approach to ecological care that complies with the applicable standards. Which legal standards apply depends on the organisation's activities. For example, different standards apply to the chemical industry than to an office organisation. ISO 14001 is one of the standards of the ISO 14000 series and is used worldwide to set up and certify environmental management systems. In the same way, as for the certification of quality management systems, certification helps an organisation to demonstrate that its environmental management system is in order. This can be important if sustainability is a selection criterion in purchasing or outsourcing processes. Sustainability measures must be seen in connection with each other and the overall business operations. An environmental management system supports this integrated approach.

ISO 14001 describes the critical elements for setting up, implementing, maintaining and improving an environmental management system. The environmental management system is based on the plan-do-check-act cycle (PDCA cycle; see figure 4.13), and consists of formulating ecological policy, determining the environmental aspects and effects, drawing up environmental objectives and programmes, implementing an environmental management system and evaluating the progress aimed at the continuous improvement of environmental performance.

To set up an environmental management system, an organisation goes through seven steps:

1 To carry out a so-called baseline measurement;
2 Identifying the ecological risks for the organisation;
3 An analysis of the environmental risks as a basis for a management method;

4 To draw up an environmental or sustainability policy, containing ambitions and objectives and the obligations with which the organisation must comply;
5 To compile a manual of procedures, indicating how the organisation will implement the environmental management system;
6 To apply the procedures described in step 5;
7 Measuring whether the objectives have been achieved and adjusting the environmental management system accordingly, if necessary.

Many internal environmental care measures are related to facility matters (e.g. purchasing and building maintenance). In addition, the facility manager can promote and support sustainable behaviour through interaction with end-users. The facilities company is, therefore, a logical place for the environmental affairs coordinator.

When it comes to making the built environment more sustainable, it is necessary to know how sustainable a building is. This provides insight into the improvement opportunities with the most significant positive contribution to sustainability performance. To this end, specific assessment methods have been developed, of which BREEAM and LEED are the best known. These aim to reduce the adverse effects of construction and development on the environment.
BREEAM and LEED have many similarities. They help to raise awareness amongst owners, occupiers and designers, and to successfully adopt sustainable solutions cost-effectively. Moreover, they provide market recognition by certifying buildings, giving them a quality mark established by an independent assessor. Such a label is particularly interesting for organisations that want to demonstrate their sustainable approach. For example, this will concern real estate investors and parties who commercially rent out real estate. The sustainability of a building can be an essential consideration for a potential tenant, for example, because it has lower energy costs or because it fits in with the sustainable image of the tenant. Certification can also be attractive for building owners who use a building themselves. A proven sustainable building visualises an organisation's sustainability policy to stakeholders and can thus contribute to the desired image.

Certifying

BREEAM

BREEAM is an assessment method to determine the sustainability performance of buildings. The abbreviation stands for Building Research Establishment Environmental Assessment Method. This method was developed and introduced by the English research institute BRE but has been adapted to various real estate markets, such as the Dutch BREEAM-NL. There are different labels: for the area, for development and renovation, for buildings in use and demolition and disassembly. BREEAM In-Use is most relevant for facility managers. It consists of three parts:
1 Building Asset: this section deals with the sustainable performance of the building;
2 Building Management: this is about how sustainably the building is managed;
3 Occupier Management: this section looks at the sustainable use of the building.

A score can be achieved for each component. The rating for BREEAM In-Use consists of scores for the various sustainability categories: management, welfare and health, energy, transport, water, waste, land use and ecology, and pollution. BREEAM awards a maximum of five stars, based on the score on components. BREEAM can also be used for self-assessment to identify opportunities for improving sustainability performance.

LEED

LEED is the abbreviation for Leadership in Energy and Environmental Design. LEED was established in 1998 by the US Green Building Council (USGBC). It is the most widely used sustainability tool in the United States since LEED is often chosen for government projects. LEED is a scoring system that uses a checklist. LEED also has various quality labels: for building design and construction, for interior design, management and maintenance and area development.

LEED assesses the following elements: development of sustainable locations, innovation and design, water savings, energy efficiency, use of materials, indoor climate quality, accessibility and awareness (education). Based on the overall score a Committee awards a basic, silver, gold or platinum certificate.

4.8.2 Corporate social responsibility

Corporate Social Responsibility

Organisations are devoting increasing attention to sustainable enterprise (see also section 2.7). Due to the rise of sustainability thinking within organisations, environmental objectives are increasingly becoming part of the general policy about corporate social responsibility (CSR). Corporate social responsibility is a management concept that supports businesses to integrate social and environmental concerns in their business operations and interactions with their stakeholders. CSR is generally understood as how a company achieves a balance of economic, environmental and social imperatives ('Triple-Bottom-Line Approach'); at the same time addressing the expectations of shareholders and stakeholders (United Nations Industrial Development Organization, 2019). It is about finding a balance between people, planet and profit. This balance appears to lead to better results for both the company and society.

Balance

Balanced approach

An excellent example of the triple-bottom-line approach is the renovation of the a.s.r.-head office in Utrecht (see case). This resulted in a more sustainable, cost-efficient and much more pleasant building for end-users.

- www.gebouwvanhetjaar.nl

Sustainable renovation

With its 86,000 m2, the head office of a.s.r. in Utrecht is one of the largest office buildings in the Netherlands. The a.s.r.-building dates from the seventies. In 2010 it was decided to carry out the sustainable renovation. The aim was creating a transparent, energy-efficient, durable and cost-efficient office, in which it is pleasant for employees to work and in which external parties also feel welcomed.
The building has been completely stripped and fitted with a double facade that has been laid out more generously around the building in a distinctive

design. The double facade creates space for green winter gardens, more openness, maximum daylight, efficient ventilation and reduced energy consumption.
The spacious entrance hall with its high atrium is connected to the brasserie, coffee bar, restaurant and underground meeting centre, creating an attractive meeting area with various formal and informal work and consultation spots. Users highly value these.
A high-quality image has been achieved with relatively limited financial resources. Materials with a long lifespan have been chosen, and old furniture has been reupholstered to reduce waste. The building has a BREEAM Excellent design certificate. The transition from energy label G to energy label A+ equals 50% more energy efficiency.

There is no such thing as a little corporate social responsibility. Sustainable business operations require an integrated approach and therefore, also a sustainable facilities policy. An excellent example of an issue that brings together the factors people, planet and profit, is the indoor climate. Research has shown that a pleasant climate increases labour productivity by 10 to 15% and even reduces absenteeism by a quarter. A research report by the Dutch Labour Inspection shows that in one in five companies, the employees do have complaints about the indoor climate. In about one in four companies with climate-related issues, the indoor climate is a continuous source of dissatisfaction among employees. So, there is much to be gained from improving the indoor climate. For example, sick leave alone can be reduced by 0.6% by replacing the filters in air conditioning systems twice a year.

Many companies go further than required by legislation and regulations. Some organisations even incorporate sustainable practices into their corporate philosophy, such as Ben & Jerry's and The Body Shop. These companies actively contribute to the realisation of the UNESCO Sustainable Development Goals (SDGs) (see Figure 4.14).

FIGURE 4.14 UNESCO Sustainable Development Goals (SDGs)

The sustainable development goals (SDGs) provide the framework for sustainable development worldwide up to 2030. These goals offer all countries a challenge in terms of environmental (planet), social (people) and economic (prosperity) dimensions. Coherence between the themes forms the starting point for the SDGs: policy aimed at one aspect can also have an (unexpected) impact on other goals. A core principle of the SDGs is that even the most disadvantaged should be included - leave no one behind (LNOB). On 25 September 2015, all UN member states committed themselves to these goals. The conditions for achieving the goals in 2030 are peace and cooperation.

Global Corporate Responsibility RepTrak 100 is a worldwide survey that ranks global companies based on their corporate social responsibility (CSR) reputation. In late 2018 it collected the opinion of over 230,000 people from the general public across the 15 largest economies. According to their ranking, Lego and Danone are leading in CSR.

ISO 26000

ISO 26000

To arrive at an integrated CSR approach, organisations often use the ISO 26000 guideline. The ISO 26000 provides companies with a normative framework for determining their social responsibilities and advises on the structural implementation of CSR. This ISO standard cannot be used for certification. Still, its purpose is to help companies make their business operations more sustainable.

Generic CSR principles

ISO 26000 has four main components: (1) an organisation examines (2) together with its stakeholders (3) which social themes (CSR core subjects) (4) it will address (implementation).
Figure 4.15 shows their interrelatedness.

ISO 26000 defines seven principles of social responsibility. These principles form the basis for every decision and every activity of an organisation:

1 accountability (taking responsibility and being accountable);
2 transparency;
3 ethical behaviour;
4 respect for the interests of the stakeholder;
5 respect for laws and regulations;
6 respect for international standards of conduct;
7 respect for human rights.

ISO 26000 specifies the importance of the environment for the organisation and the importance of the organisation for the environment. Therefore, the guideline describes how an organisation can identify which stakeholders represent the environment and how it can find out what they expect from the organisation in the area of CSR.

CSR Core subjects

In addition to the CSR principles, ISO 26000 identifies seven CSR core subjects that need to be addressed in every organisation. These core subjects are divided into 37 CSR issues. Organisations must decide for themselves whether and to what extent a particular CSR issue is relevant and which issue must be given priority.

FIGURE 4.15 Schematic overview of the ISO 26000

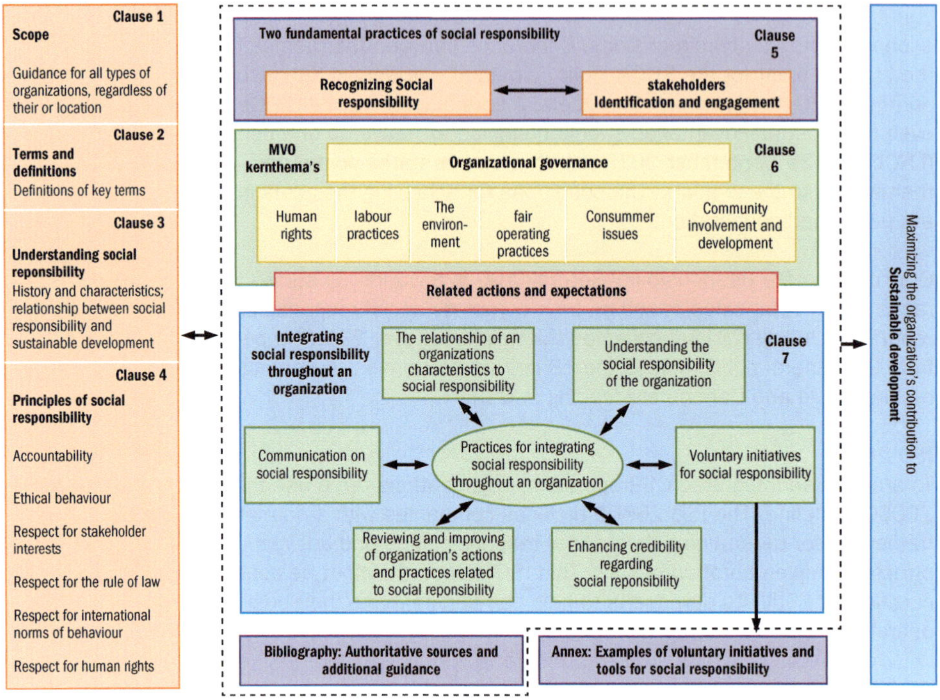

ISO 26000 provides guidelines for each CSR issue to be able to implement this, namely:

1 the management board of the organisation;
2 human rights;
3 working conditions;
4 environment;
5 fair business practices;
6 consumer issues;
7 involvement in community development.

4.8.3 Sustainable behaviour

The facility manager can come up with such beautiful, sustainable plans; corporate social responsibility depends on sustainable human behaviour. Think of reducing the use of paper by not printing everything as a matter of course, reducing energy consumption by switching off the computer and lights when leaving a room and of separating waste. Reducing CO2 emissions by using public transport for business trips is also a good example. Ultimately, CSR is a combination of technology and services on the one hand, and sustainable behaviour on the other hand.

FMN DGBC Dutch Green Building Council

In September 2011, FMN and the Dutch Green Building Council (DGBC) launched a code of conduct for sustainable end-users as part of a joint effort to stimulate and accelerate the sustainability of the work environment and buildings. Because the end-user can exert a great deal of influence thereon, the **code of conduct** encourages end users to take responsibility.

Those who endorse the code of conduct are committed to the following:
1 Our organisation is responsible for making our workplace more sustainable.
2 We make our employees aware of the contribution they can make by changing their behaviour.
3 We impose sustainability requirements on the premises we use.
4 We only develop new premises if there is no existing alternative.
5 We take responsibility for any premises we leave behind.
6 We look at our accommodation costs as a whole so that there is room for investment and sustainability measures.
7 We pursue an active environmental policy.
8 We measure and improve the physical work environment of our employees.
9 We share and promote our knowledge and experience on the sustainability of our workplace.
10 We actively pursue or develop in the short term, a policy to make our workplace more sustainable.

Changing behaviour is a laborious process. New routine only lasts if its effects are visible. In other words: if the practise pays off in one way or another. To ensure the success of the CSR policy a facility manager will, therefore, have to be particularly attentive to change management. The Triad-model (Poiesz, 1999) is instrumental in this respect. This model assumes that people will only display specific behaviour (e.g. business trips by public transport) if three conditions are met at the same time: **Triad-model**
1 The person is motivated to show the intended behaviour (motivation).
2 The person has the necessary knowledge and tools to be able to exhibit the behaviour (capacity).
3 The circumstances are favourable to be able to exhibit the behaviour (opportunity).

This defines the three main conditions for sustainable behaviour: *motivation, capacity and opportunity*. For example: if the employee wants to reduce CO2 emissions by making business trips by public transport, he may be given a travel card (opportunity). However, if he does not know how to get from stations to the place of destination in a comfortable and sustainable manner, then this will not lead to the desired behaviour. And if an employee has the necessary knowledge (he can do it) and the opportunity (access to public transport) without the motivation, he will not show the desired behaviour either. Just as if he is motivated and has the necessary knowledge, but the opportunity is not offered (for example, he is expected to travel by leased car). For a behavioural change, all three conditions must, therefore, be met at the same time.

A score of 1 indicates that the condition has been fully met; 0 means that nothing has been achieved yet. The probability of displaying specific behaviour is calculated by multiplying the scores of the three conditions. The result of this is shown visually, as in the left example in Figure 4.16. The larger the volume of the pyramid, the more likely the desired behaviour is. To obtain a high total score (i.e. a large volume) all three factors must be scored at a maximum. If two factors score high and one of the factors scores low (see the pyramid on the right in Figure 4.16), there is a weak chance of achieving and sustaining the desired behaviour.

FIGURE 4.16 The Triad-model

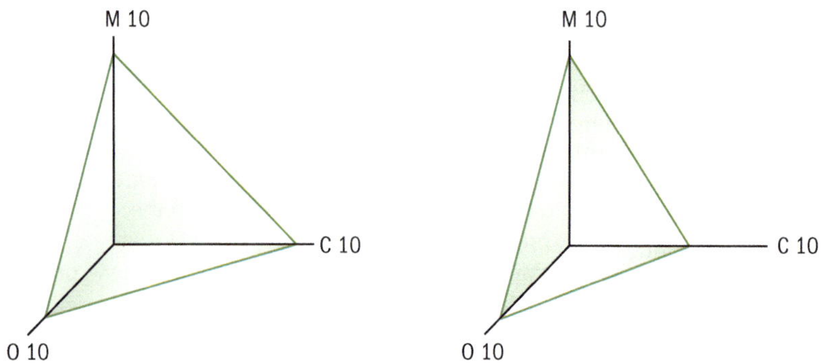

Of the three main conditions, opportunity (O) should be addressed first. Motivating and providing knowledge or tools is of no use at all if employees subsequently do not have the opportunity to show different behaviour. The motivation (M) of employees can be increased by displaying the results. For this, the facility manager needs several quick wins (quick results), especially in the beginning. The immediate results increase the commitment of the employees involved to continue. Also, quick wins can be used to convince the less motivated employees of the usefulness or the necessity to participate. In general, remuneration is an excellent way to achieve sustainable behaviour, and that does not always have to cost money. After all, employees consider the appreciation of managers to be at least as important. It is motivating if ideas and suggestions from employees are taken on board and if they are widely shared.
Capacity (C) in the sense of knowledge is achieved through education, instructions, demonstrations and the like. Tools must be practical in nature. They should, as it were, give a helping hand to refresh the memory or to make sustainable behaviour easy. An example of this is the sending of a push message: 'Don't forget to turn off your computer'.

4.9 Risk management

A risk is a chance that damage or loss will occur as a result of a given uncertain event. Risk management comprises the coordinated activities to manage and control an organisation about risks (Figure 4.17). The risk management process is the systematic approach of identifying, analysing, evaluating, treating and monitoring risks in a specific context. Also, the advice and communication during such a process are part of it. Risk management
It is crucial how risk management is embedded in an organisation. According to one of the principles in ISO 31000, risk management is only useful if it is an integral part of the organisation's operations.

FIGURE 4.17 The risk management process model

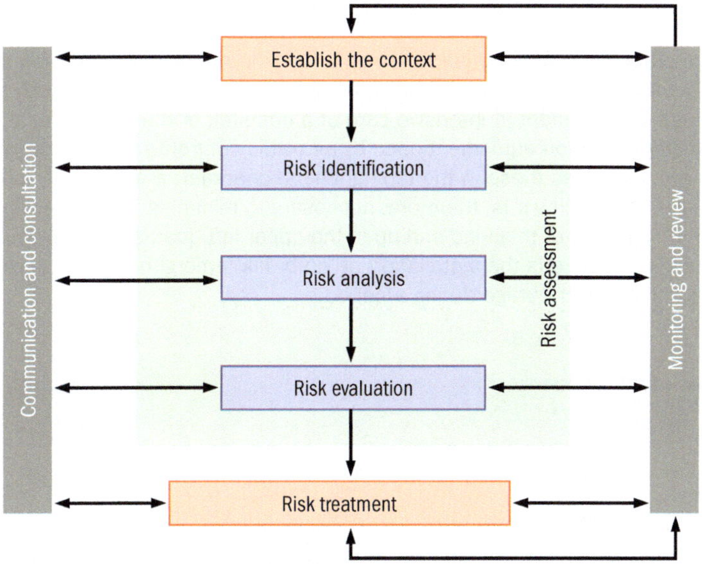

4

4.9.1 Types and definition of risks

Risks occur at many levels: of the individual, of people, of the workplace, of the location and the environment (Reitsema, 2014). For each level, examples are given, starting with the environment and ending with the individual:

- Environmental risks are, for example, natural disasters, but also an attack such as in the tram in Utrecht, in Zaventem Airport and on London Bridge. The activities of surrounding companies also create environmental risks, for example, heavy industry, the transport of hazardous substances in the vicinity, but also the presence of an event location. This results in specific environmental risks for a facility manager to take into account.
- Examples of risks at the level of a location are fire, power failure, a strike or a violent attack.
- An example of a risk associated with the work itself and the workplace is the incidence of physical and psychological complaints.

- Risks at the level of other people (colleagues or public) include bullying at work, sexual harassment and violence against employees.
- Finally, there are risks associated with individual characteristics. Think of physical, psychological or mental limitations, but also ignorance (e.g. concerning safety regulations) or carelessness.

Risk management is intended to protect the organisation from disasters, calamities, vandalism, terrorism, crime and the like. At least as far as the work environment is concerned (see also section 3.3.5). Fraud prevention is the responsibility of the finance department or security/auditing department.

4.9.2 Probability and effect

Risk matrix

Risks can be mapped out in a risk matrix (Figure 4.18). The horizontal axis shows the likelihood of occurrence (once in the ...). The vertical axis shows the severity of the consequences (impact). Both the probability and consequences are characterised in qualitative terms.

In the practical example in section 4.9.3, the occurrence of a power failure is discussed. The consequences can be quite severe and harmful, such as the loss of equipment in intensive care of a hospital, or the loss of radar and communication equipment used by air traffic control to navigate flights. Figure 4.18 places these in the top right-hand quadrant: a significant impact at a high probability. It is, therefore, necessary to minimise the probability, so that incidents of this kind end up in the upper left quadrant. An example of such a measure is the installation of 'no break' emergency generators to cope with a power outage during a period.

FIGURE 4.18 Risk matrix

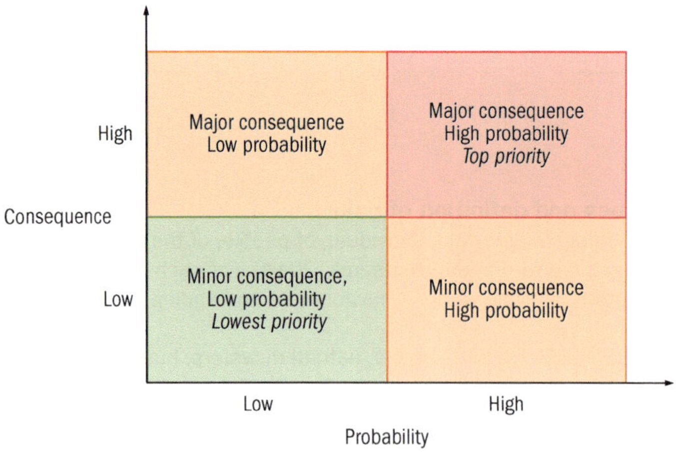

The ISO 31000 defines risk as the effect of uncertainty on the achievement of objectives. This definition defines risk as a neutral concept (opportunities and threats). This is in line with risk management as the foundation of doing business. 'Doing business is taking a risk', i.e. dealing with uncertainties that can have positive and negative effects on the business results and thus the success of the organisation.

A risk analysis is carried out periodically to identify the risks, for example, every three years. To keep the data up to date, a limited annual update can be done. It is not necessary to cover all risks. As stated above, taking risks is inherent in doing business. Depending on the probability and the consequence, risk can be tackled in four ways, whether or not in combination (Figure 4.19). The four T's of risk management are:

Risk analysis

- Terminate: the removal of one or both of the factors chance and effect or both;
- Treat: the reduction of one or both of these factors in likelihood and impact;
- Transfer: mitigating risks to insurers;
- Tolerate: when the risk is less than the established acceptable loss for the organisation.

FIGURE 4.19 The four T's of risk management

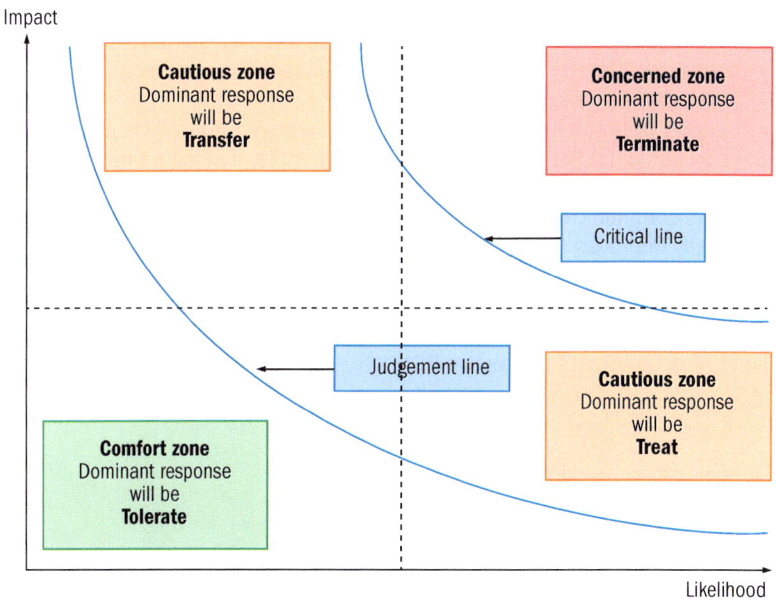

Because how the organisation deals with the risks it runs, affects the organisation's continued existence, it is up to the management to determine the risk profile for the organisation. The facility manager then has the responsibility to manage the risks with regard to the building and (work) environment in such a way as to avoid doing too much (which leads to unnecessary costs) as well as too little (which can lead to unacceptable consequential damage).

Risk profile

Risks that play a role in a work environment are caused by, for example, fire, burglary and power failure. The consequences of a fire can be considerable. If no fire safety measures are taken, there is a good chance that a fire will occur. The risk is therefore high (high probability × high impact). Risk management aims to reduce the probability of a fire occurring and takes measures to ensure that if a fire does occur, the effect is limited. This can be done, for example, through the installation of fireproof furniture, fire doors and sprinklers. An escape plan and fire drills (part of the in-house emergency response plan) help to ensure that people can leave the building safely.

Besides, the financial consequences of risks can be outsourced to an insurer or limited by a mutual guarantee fund. After all these measures, the risk of fire is acceptable. This example, therefore, involves prevention, reduction and outsourcing (Figure 4.19). The remaining risk is accepted.

4.9.3 Business continuity management

Business continuity management (BCM)

Limiting consequential damage

Closely related to risk management is business continuity management (BCM). BCM refers to what needs to be arranged to ensure that the business-critical processes can be continued at a pre-agreed level, in the event of a calamity or crisis. It, therefore, focuses on preventing or limiting consequential damage (see Figure 4.19). In the event of the loss or failure of a location, for example, the employees must be able to continue their work elsewhere; be it an alternative (fall-back) location, or home office. Other examples are the outbreak of an epidemic (or pandemic like covid-19) or a strike. In this case, too, the critical processes must continue, but incidents of this kind require a different approach. ISO 22301 Business Continuity Management offers tools for setting up a business continuity management system. Organisations that have set up BCM have access to:

- a strategy for getting out of a crisis, complete with priorities and instructions on where the resources are;
- support from top management, so that it is clear that BCM is just as good a goal for the organisation as, for example, making a profit;
- crisis management for the organisation and IT;
- crisis communication, because good communication during a crisis greatly improves the position of the organisation;
- exercises, complete with evaluation and points for improvement.

The practical example shows the impact of power outages on industries and illustrates the need for effective BCM.

● www.fosterfuelsmissioncritical.com

Loss of power

In today's world, even a short-term loss of power can lead to significant losses in a variety of industries. During recent natural disasters like Hurricane Irma, we saw the effects of power outages on businesses. Nearly two-thirds of Florida's electricity customers lost power during Hurricane Irma. A significant number of companies were still without power for days after the hurricane hit. These unplanned power outages from natural disasters and other events have a considerable impact on businesses operating in the areas impacted. Power outages can occur at any time, and if companies in the following industries are unprepared, they may experience significant downtime and loss of revenue:

- **Data Centres:** Data centres are facilities that contain an organisation's IT operations and equipment. These centres store, organise, manage and process a company's data. When a data centre goes down due to a power outage, a loss of mission-critical data can occur. Companies rely heavily on data stored in their data centres; losing this crucial information could affect everyday operations long-term.

- **Medical Facilities:** Patients' lives depend on the proper functioning of medical equipment, even in the event of a power outage. Typically, hospitals have backup generators to make sure everything continues to run. These generators require continuous maintenance and testing to ensure they will work in the event of a power outage. Patient data and lives are all on the line when the power goes out. And without working telephone lines, people will not be able to call emergency services for help.
- **Financial Corporations:** Companies involved in the stock market can lose out on millions of dollars in a fraction of a second if a power outage occurs. Because transactions happen every second, even a short-term disruption can have a drastic impact on businesses.
- **Military Operations:** Power outages can cause valuable weaponry and equipment to become useless in the event of an attack. With non-functioning equipment, military personnel could be left defenceless.
- **Retail and Wholesale:** Companies in the retail industry suffer lost sales revenue with unplanned power outages. For larger retail companies, a few minutes of downtime can lead to tens of thousands of dollars in lost revenue. To understand the possible impact, we can look at Amazon. In 2016, Amazon's website was unavailable for 13 to 15 minutes. They estimated that this short lapse lost the retailer $2,646,501 in revenue. Although Amazon's site did not go down due to a power outage, this serves as a perfect example of how downtime can affect a business.

4.10 Information management

This chapter discusses the many areas to which a facility professional should pay attention in managing the facility organisation. Information is needed for all these processes. Information management is hence the subject of this last section. There are various types of information. In this book we differentiate between operational information and management information. Operational information is information that is needed to keep the processes going and is also called process information. Management information is information that is needed to evaluate and adjust or improve the service.

4.10.1 From data to information

You can record data and still have no information. Information consists of composed and structured data. To be able to use data for management purposes, it must be organised in such a way that it actually tells you something (see Figure 4.20). A facility example: a stack of prints of all the reservations for meeting rooms is of no value to the manager of the meeting centre. These are just data. It only becomes information when he knows what the average occupancy rate is per part of a day and whether there is a rising or falling trend.

FIGURE 4.20 From data to information

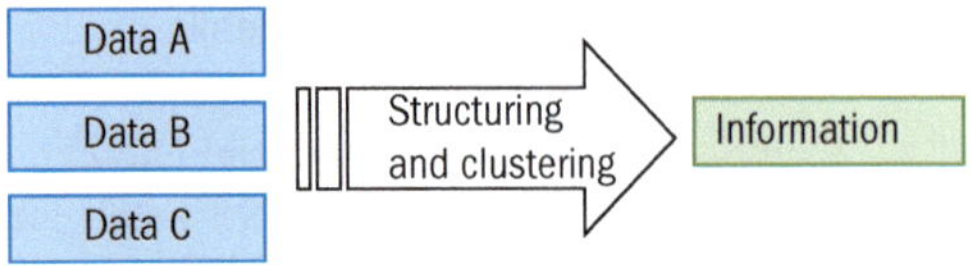

So, a manager needs information to be able to plan, steer and control. The manager's information needs are linked to the hierarchical level at which he or she operates. At the strategic level, relatively little (high-level) information is used to cover a wide range of matters. At the operational level, there is a need for extensive (detailed) information on relatively few subjects. For the management of the company, for example, one table may be sufficient to show last month's sales figures. Although it contains little information, the table does show the total production result. At the operational level, spreadsheets are often used: for each machine, for example, the operating hours, maintenance intervals and the speed of production are recorded. Other examples of information at operational level are the average occupancy rate of workplaces, average customer satisfaction with cleaning, number of complaints about indoor climate or downtime of installations. Table 4.4 highlights the difference between operational information and management information.

TABLE 4.4 Operational information and management information compared

Information	Operational level	Management level
Use	To ensure that day-to-day operations run smoothly, for example, information about events (location, persons, attributes).	To support decision-making regarding the future (e.g. identifying potential bottlenecks)
Period	Short term	Medium to long term
For what?	To control the process	To be able to assess and adjust the process (e.g. KPIs)
For whom?	Intended for operational staff	Designed for managers at the operational, tactical or strategic level in the organisation

Figure 4.21 shows the connections between the various information flows. As you can see, information from top to bottom is increasingly specific, and the information that goes from bottom to top is becoming less detailed. For this process to properly function, it is necessary to appoint process owners who are responsible for the process management. It starts with the goals of the organisation on a strategic level. The objectives of the organisation are translated into strategy and policy. The policy gives direction as to how processes are configured. At the tactical level, consideration is given to related performance requirements for the operations. Control measures help to ensure that the right things are done in the right way. At the operational level, information is collected about the performance of the process (e.g. about speed, sustainability, quality). This information is provided in the form of process reports. At the tactical (management) level, data from many different processes are clustered in management reports, which indicate to

what extent the organisation is on track in accomplishing its policy. Finally, this tactical management information is even further condensed to a strategic dashboard with only a few indicators to monitor the organisation's performance. In Chapter 5, this is discussed in more detail.

FIGURE 4.21 Process diagram

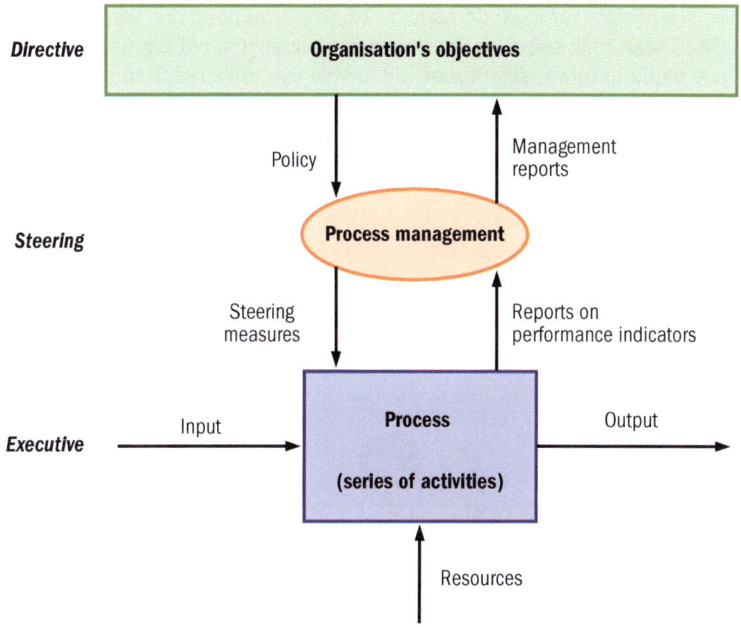

4.10.2 Computer Aided Facility Management software (CAFM)

To be able to offer an excellent service it is essential to be 'in control'. The question is: who does what, how is it done, what does it cost, and what is the result? For a sound management system, an organisation needs an integrated set of tools that support its working methods. Computer Aided Facility Management Software (CAFM) is used to support work processes and management processes. Examples of CAFM are Facility Management Information Systems (FMIS) and Integrated Workplace Management Systems (IWMS). In this way, CAFM supports being in control of the facilities organisation. CAFM is also used for the registration and handling of questions, reports and complaints, both by operational facility staff and for self-service by the end-users. CAFM has a modular structure, in which it is possible to expand the standard FMIS or IWMS with options that meet the wishes of the organisation.

Computer Aided Facility Management Software (CAFM)

Examples are modules for meeting room reservations, visitor registration and inventory management. CAFM is a stand-alone software system, which is ideally suited for integrated deployment. CAFM then plays a central role in the exchange of information with other systems or devices, for example, with smart sensors and the building information management system (BIMS). Smart sensors are able not only to detect an event but also to communicate it (through the Internet of Things). An example: a sensor

Building information management system (BIMS)

monitors the temperature in a specific room. As soon as the temperature exceeds a minimum or maximum setpoint, the sensor communicates this with the system. This automatically creates a new work order for checking the installations. Thus, complaints can be solved before they are reported. The number of (automatic) reports may increase, but the number of reports of complaints will decrease. We give another example. If printers generate automated orders for new toners faster than average, the system will notice this. Based on this signal, you can consider whether the capacity of the printer is sufficient for local demand, or whether you need one with a higher capacity. These may seem straightforward applications, but big data analysis leads to new insights that will enable you to increase the quality of service and reduce costs. With smart systems, it is possible to make ever-better predictions about customer behaviour, even in environments that are subject to change.

Well-known suppliers of FMIS/IWMS systems are Archibus, FAMAS, NPQ Prequest, Topdesk, Planon and Ultimo. Figure 4.22 shows an example of modules such a system can include.

FIGURE 4.22 Example modules of an FMIS/IWMS

Let's give an example of the functionality of CAFM. Suppose a customer wants to schedule a meeting. He needs a conference room for twelve people with a U-shaped set-up, a beamer, and a selection of refreshments. The meeting will be attended by four external guests who come by car, two of whom need a charging station. The customer contacts the service desk or logs on to the self-service portal and makes the request outlined above. The service desk assistant will look into the conference room management module to see which meeting rooms are available. In consultation with the customer, a room is booked. The service employee then checks whether a standard beamer is available. If not, the system reserves a beamer. It sends a message to the technical support that a beamer must be ready at

the discussed time and location. Also, a request is sent to the conference service staff for refreshments and required set-up of the tables.
In the self-service portal, the customer enters the date, time and requirements concerning capacity and facilities. The reservation system then displays the available conference rooms that meet the criteria given. The customer reserves his preferred option, indicates the desired layout of the space and selects refreshments from a list of food and beverages. External guests are registered in the visitor registration module (either by the customer himself or by the service employee), and parking spaces are reserved for guests arriving by car, with or without a charging station. All reservations are confirmed by email. Service desk employees (and in case of self-service also customers) can check the status of the requests in the FMIS/IWMS.

For the customer, working with CAFM has the advantage of one-stop-shopping, and in the case of self-service, that he can arrange things at any time without delay. For the facilities organisation, this way of working means that the operational services are perceived as more customer-friendly and transparent. On a tactical level, management information can be retrieved from the system, such as the average occupancy rate of the meeting rooms and the average time between reporting and resolving a malfunction. Due to the central registration of all requests, notifications and reservations, work processes can be effectively coordinated.

The following points of attention are essential in the implementation of CAFM:

- The work processes must be well structured and organised (see section 3.3.2).
- The organisation must accept working with CAFM. A system that does not fit in with the work processes of the organisation is ineffective, even if it is an excellent system. Therefore, always involve future users in establishing the programme of requirements and in the selection of a system.

● www.facto.nl

The Boijmans Van Beuningen museum in Rotterdam now works with an FMIS. Given a major renovation and the opening of the new Boijmans Van Beuningen depot, the facilities department, the IT department, the technical department and the security department felt the need for tighter registration of incoming requests and complaints. Without a system, everything becomes elusive, and you are not in control. It was necessary to get a better grip on the work and to be able to set priorities.
The selection process started with an inventory of requirements and wishes and the preparation of a programme of requirements. After the pre-selection, two suppliers were invited to demonstrate
their system. The choice was made based on objective criteria and the expected quality of the cooperation.
The final decision was followed by a kick-off meeting with representatives of the museum and the supplier. Of course, there was already a significant degree of paperwork. But in such a meeting, you discuss what you expect from each other and how you want to collaborate and communicate. Building the relationship is crucial; however, an often-overlooked success factor.

Young Professionals in Facility Management

Mathijs Gallé, Senior Manager Facilities, Corporate Solutions/IFM, JLL Services, Brussels Area, Belgium - graduated Bachelor International Real Estate and Facilities Management at BUAS in 2015

Making the difference daily

While still getting my bachelor's degree in International Real Estate and Facilities Management at Breda University of Applied Sciences, I started working for the Corporate Solutions division at JLL Services. JLL is one of the global leading Real Estate and Facilities Management firms. With them, I got the opportunity to work for many international corporate clients, like Honeywell, the Royal Bank of Canada and Procter & Gamble, in cities like Frankfurt, Amsterdam and Brussels.

For my clients, I manage all day-to-day workplace services and engage in their strategic portfolio planning. My role is to align all operational services with ongoing capital investment projects over multiple years. Within my area of responsibilities, my team and I cover a wide variety of services like Building management and occupancy, technical building installation maintenance and repairs, soft- and security services, car fleet management and long-term site master planning.

What I love about working in corporate Facilities Management is an international context and the ability to impact all of my client's employees. My objective is to provide them with the best possible workplace to transform their experiences into something that contributes to their productivity.

My advice to everyone that is starting their professional career is: keep an eye out for any chance that comes along. Accepting challenges can genuinely bring you far and help you develop to become the best version of yourself.

Summary

- The tasks of the facility manager are broad and differ per organisation and market sector.

- The facility manager has clientele at all levels of the organisation, each with their interests and needs. We call this the stratification of the client. At the strategic level, these people are referred to as clients (defining customer), at the tactical level as customers (paying customer), at the operational level as consumers or end-users (beneficiary customer).

- The matching of demand and supply takes place at all hierarchical levels, with the client organisation on the one hand and the providers on the other. The 9-grid shows the activities of facility management for the different roles in matching demand and supply.

- The positioning, identity and image of an organisation are interrelated. The *positioning* is expressed in the mission, vision and core values. The positioning of facility management can be divided into passive, reactive, proactive and business partnership. *Identity* is the 'DNA' of an organisation, which is expressed in branding, communication, behaviour and performance. *Image* is the value judgement of customers and other stakeholders about the organisation.

- Facilities policy is derived from the organisational strategy; in the facilities policy, the organisation's spearheads and objectives are translated into a facilities services concept and matching facilities priorities and goals.

- Account management is customer management. It is an activity that aims to translate the customer's wishes and future developments into the facility services offered at the desired level of service. Building and maintaining relationships with customers consists of three steps: getting to know the customer, making and keeping agreements about products and services to be delivered and delivering the agreed services.

- A service level agreement (SLA) is a tactical agreement concerning the products and services to be delivered to the business unit or department, the service level and the costs to be charged. The SLA must fit within the frameworks laid down in the master agreement at the strategic level. At the operational level, the agreements are communicated to end users by means of a product and service catalogue (PSC).

4

- The 6 P's of facility marketing are product, price, place, promotion, people and process. The 6P model is used to determine the marketing strategy for a product or service. Another model is the SIVA model. SIVA is the abbreviation for solutions, information, value and access. The SIVA model takes the customer's needs as its starting point for the marketing strategy and is very suitable for customer-oriented service organisations.

- Facility purchasing refers to all services and resources that are not purchased directly for the primary processes. The procurement process consists of two phases (tactical and operational procurement) which together comprise eight steps. Facility purchasing by government organisations is subject to European procurement rules. Not all products purchased are of equal importance to an organisation (industry-specific). We distinguish between leverage products, strategic products, routine products and bottleneck products.

4

- Quality is the degree to which the characteristics of a product correspond to the requirements and wishes of consumers. In service provision, a distinction is made between technical quality, functional quality and relational quality. Quality management is a continuous process aimed at quality control, quality assurance and quality improvement.

- Usually, a facilities organisation engages in the outsourcing of services. Because more parties are involved in the service organisation (at least the internal customer, the facilities organisation and the supplier), coordination processes become more complex. The chance of communication problems and dissatisfied customers thus increases. The SERVQUAL model is a tool for discovering sources of customer dissatisfaction; 13 gaps can be distinguished when adapted to the facilities organisation.

- Facility professionals set an example in the field of sustainability and socially responsible behaviour. Sustainable business operations require an integrated approach and a sustainable facilities policy, but sustainable operations depend on human behaviour. To evoke and sustain customers' green behaviour, three conditions must be met: motivation, capacity and opportunity.

- The purpose of risk management is to ensure the continuity of business operations (business continuity management, BCM) in the event of a calamity or crisis, at a pre-agreed level. FAFS is part of risk management. The policy is determined based on a risk analysis.

- CAFM is a software application with a database for supporting work processes and generating management information for the facilities organisation, e.g. a Facility Management Information System (FMIS) or Integrated Workspace Management System (IWMS). By linking CAFM to the building management system (BMS) or Buildings Information Management System (BIM), and by using smart sensors, new application possibilities have emerged.

Questions and Assignments

4.1 Who are the paying, determining and benefitting customers within your university or the company you work for?

4.2 New coffee machines have been in use at a large distribution centre for a month now. With these machines, it is no longer possible to choose for soup. The paying customer has decided so because the canteen always serves a soup of the day during lunch and because by omitting this option in the vending machines, costs can be saved. The distribution centre works in three shifts. The night shift has complained several times to their team leader about the lack of soup, which is sorely missed during the nightly breaks. The team leader refers the matter to the facility manager and says that he does not understand why the manager has chosen this machine. Can you explain what goes wrong here?

4.3 Explain why marketing for internal facility organisations has become increasingly important over the years.

4.4 The facility manager, together with the purchasing agent, compiles a brief to replace all the office furniture in their organisation. The facility manager is well prepared and has a long list of requirements for the new furniture to meet. The buyer discusses the list point by point and makes a distinction between nice to have and need to have. Why does he or she do so?

4.5 Sustainability has become a principal core value for organisations. The facility manager has translated this into a sustainable facility policy. For example, sustainability criteria have been drawn up for the purchase of catering and cleaning products, and a project has been started for the separate collection of waste. Agreements have been made with the waste processor, various types of collection bins have been placed, and clear instructions have been given to the users. However, few people have yet switched to actually separating waste streams. Paper still ends up as residual waste, and advertising material wrapped in plastic film in the paper bin, and in the company restaurant customers do not even bother to separate the organic waste. Can you explain what's happening here? What are the tasks and responsibilities of the facility manager to make the project a success?

4.6 Give an example of a leverage product, a strategic product, a routine product and a bottleneck product for an energy company and an airline.

4.7 Mention three factors that determine the image of an in-house facilities department.

4.8 Identify two aspects of the technical, functional and relational quality of reception services for a leading law firm in a business district.

4.9 List three facility services that are outsourced within the school where you study or the organisation you work for. For each service, give a reason why you believe this service is outsourced.

4.10 Security gates at a school's entrance are more common at a vocational training school than at a university. How would you define risk = probability x effect for a vocational training school compared to a university?

n Excellence
HUBLOT
CHANEL
GIVENCHY

5
Quantifying Facility Management

In this chapter, we will address the following questions:
- Why is performance measurement important to the facility manager?
- Where are information sources for the facility manager?
- What are the most critical indicators for a facility manager?
- How may a facility manager measure indirect results?
- What is performance management, and what is its importance for a facility organisation?
- How is the facility budget established?
- What is the planning and control cycle?
- How can a facilities organisation assess its market conformity?

◄ Through the eyes of the facility manager:
'A good example of combined functionality: a hotel lobby with a shopping mall. But a challenge to smoothly manage this varied business activity.'

Expert Column

Soft and hard data go together like Yin and Yang

These days, the availability of data is overwhelming. The abundance of data affects how facility professionals can measure, analyse and steer to achieve its goals. The options are numerous, and it is, therefore, essential to make choices about the subjects to be included in performance management. After all, each indicator involves the use of systems, measuring, data analyses and reporting. And thus, time and money.

There are roughly four reasons for data within Facility Management to have insight:

- to better understand the client/customer/end-user, for example, to adapt services or customer communication accordingly;
- to provide insight into performance, for example, based on KPI's for steering and internal reporting;
- to achieve better decision making, for example about the volume of warehouse stock based on consumption;
- to better anticipate future scenarios, e.g. for new accommodation based on occupancy rates.

It is interesting to note that facilities services have a 'hard' objectively measurable side, but also a 'soft', more subjective edge. Hard data involves, for example, budgets and realisation, service level agreements and quality measurements. The soft, more subjective data, concerns aspects such as expectations and perception of service provision, connection to needs, and sensory impressions such as smell, colour and sound. It is up to the facility manager to monitor both types of data. After all, as can be read in this chapter, this 'softer' side of service provision has a high degree of effect on perception.

The 'soft and hard' components of performance management are, just like 'Yin and Yang', complementary values and inextricably linked with each other.

Jacqueline Bakker
Regional Director Central Europe at Euroflorist

5.1 Introduction

As you learned in Chapter 4, organisations are managed based on a strategy with (medium to) long-term goals. If the organisation were a ship, those goals would be the destination or the point on the horizon. The strategic objectives are elaborated into actions in the short term, for example, in an annual plan. Such a plan breaks down the route to the final destination into stages. For each step, it is possible to determine what needs to be done to get to the next point. While sailing, a skipper regularly determines his current position to the outlined route, to decide whether or not he is still on the right track. Similarly, in organisations, it is important to measure the extent to which objectives are achieved. And just as with a ship, it may sometimes be necessary to change a course (temporarily), to slow down or accelerate it, because circumstances do require this.

Steering based on measurements sounds logical, and in theory, objectives must be phrased in a 'SMARTI' way: SMARTI

- specific: or unambiguous;
- measurable: in numerical values or to be expressed in yes/no;
- acceptable: understandable and obvious;
- realistic: relevant, meaningful and feasible; , supported in the organisation and accepted;
- time-bound: to be expressed as much as possible in terms of throughput, delivery or end time;
- inspiring: a stimulating goal that incites action.

In practice, objectives can be vague, open to multiple interpretations and without a clear timeframe and therefore not measurable. You can imagine that in such a situation, people may have a different idea of the final destination and that they see different routes towards it.
However, good governance is more than achieving goals. Every organisation runs business risks, such as financial loss, the risk of staff turnover and liability risks. It is therefore essential, in addition to the objectives, to accurately monitor and cover the risks. To remain in the metaphor of a ship at sea: the ship must be seaworthy to be able to offer a safe crossing, and the crew must be well-trained. Furthermore, the vessel must have sufficient supplies of food and fuel. And malfunctions in radar, communication and navigation equipment must be corrected before accidents occur.

Measuring provides insight into the strengths and weaknesses; into what is going well and what needs to be improved. If an organisation has systematically set up this process, then the organisation is 'in control'.
As you now know, the facilities policy is derived from the strategic plan of the organisation. The facilities mission describes the added value it provides for the organisation. The facilities organisation also develops a strategy that helps to achieve the organisation's strategic goals (in the medium term). The facility manager makes agreements with the board of the organisation about the performance that he must deliver in the year to come (annual plan). In general, these agreements include standards for customer satisfaction, the service level and the costs for the service. Depending on the added value that the facilities organisation wants to achieve, additional performance agreements are made. Organisations differ in the extent to which facility performance indicators are established for specific performance areas. The

most important thing about measuring is that the facility manager or supplier can demonstrate his added value to the organisation.

In control

How facility manager can be in control, and what this involves, is discussed in this chapter. The first part of this chapter deals with the systems at the disposal of a facility manager: process management, performance management and financial management (see Figure 5.1).

FIGURE 5.1 The focus of this chapter

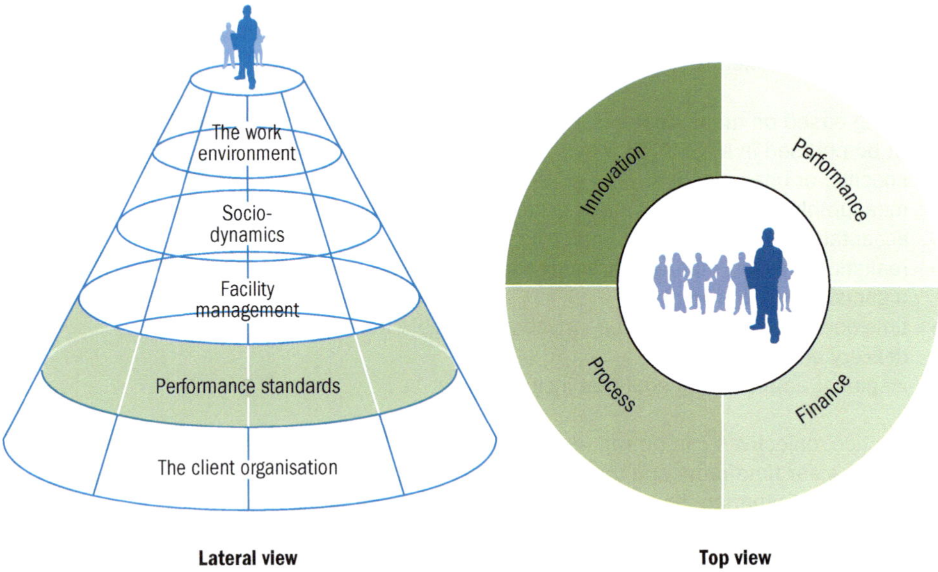

Information is needed to control processes and organisations. Section 5.2 explains the meaning of information and where facility managers retrieve it. Section 5.3 describes how the facility manager uses dashboards to steer. Section 5.4 deals with performance management. We discuss various models that can help to monitor and improve the quality systematically (performance). To finish this chapter, in section 5.5, we look at financial management and the establishment of the facilities budget.

5.2 Monitoring requires information

Before exploring what and how it is measured, we consider the essence of information (see also section 4.10). The occupancy rate of a building, for example, can be expressed as a ratio (with a numerator and a denominator) and is, therefore 'information'. The numerator shows the number of employees present. This is just a fact; it does not mean anything in itself. The denominator represents the number of workplaces, also a given. Only when you combine these data do you get meaningful information about the occupation: the number of employees per workstation. This information can be compared to a standard or to a trend in time.

In Chapter 1, we discussed the three hierarchical layers in organisations. From top to bottom, these are the strategic, tactical and operational level. Schematically, these layers are depicted as a pyramid. A pyramid is narrow at the top and wide at the bottom. Besides the number of people involved, this symbolises the amount of information used in the organisation. At the strategic level, little but comprehensive information is used because it covers the entire organisation. At the operational level, much specialist information is available about a specific part of the organisation, for example, of a particular process.

All the information used at the three levels of the organisation is derived from the information required by top management. In other words, the demand for information always comes top-down. This means that every manager or controller receives a request for information from a superior. However, information is built up from bottom-up. All specific, specialist information at the operational level is reported to a higher organisational level. This report is a derivative of all operational data and information, but combined and condensed. Otherwise, the report would be far too extensive, too detailed and difficult to understand.

Top-down

Bottom-up

We will give an example of information in a cleaning company. Much information is needed to clarify what is mutually expected when outsourcing cleaning. Specifications, therefore, include, among others, the following operational information:

- the number of employees and the number of FTEs for whom the environment must be kept clean;
- the workplace policy, e.g. clean desk;
- the average number of users and the distribution of use, such as peak days;
- the cleaning times, such as day, morning or evening cleaning per type of room;
- the opening hours of the building;
- the method of reporting, which states who is responsible for the planning and assessment;
- the settlement method - fixed price per square metre or based on the actual costs incurred;
- the cleaning frequency (input specification) or the degree of cleanliness to be maintained (output specification);
- the specification of cleaning activities per object and space or the degree of cleanliness;
- specialist cleaning on demand;
- how to deal with building mutations, vacant spaces;
- waste management, with indicative volumes.

From this detailed information, a (bottom-up) report can be compiled for the tactical management level. Such a report contains other information and combines several activities - think of the pyramid. Examples of this periodic information are:

Report

- trends, or changes in cleaning activities;
- work on post-calculation compared to the budget;
- deviations from the average number of complaints and reports.

At the next higher organisational level, this information is combined with that of other services. It thus provides insight into satisfaction with, for example,

Aggregation

the extent to which the working environment enables users to 'really enjoy their work'. The structure and flow of management information are, therefore, bottom-up. This bottom-up process is characterised by a decrease in the amount of information and by an increase in scope; the so-called aggregation of data. Figure 5.2 illustrates this principle as the aggregation pyramid.

FIGURE 5.2 Pyramid of aggregation

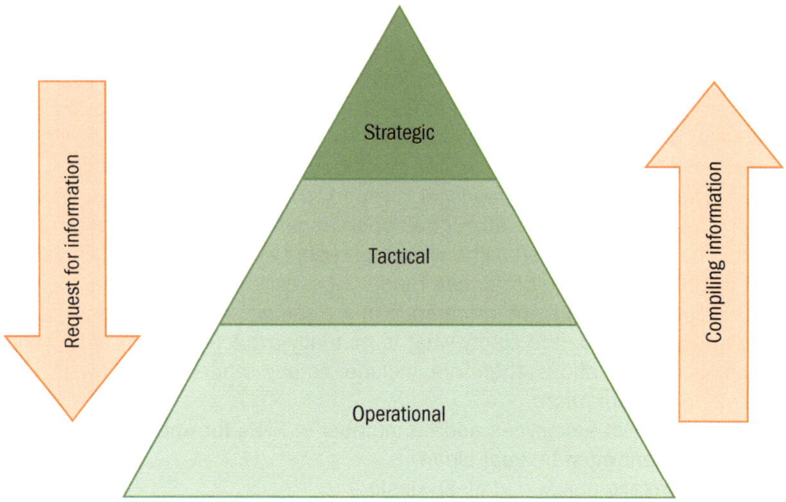

5

5.3 A facility manager's dashboard

Classify the information

The aggregation of information is by no means simple. To be able to do this successfully, it is necessary to first classify the information. A subdivision generally used in business administration is quality, quantity, time, and cost. A simplified and structured overview that allows for the recognition of deviation of standards and long-term trends is called a 'dashboard'.

In practice, FM's leading dashboard indicators are:

- customer satisfaction;
- service level (quality);
- costs.

Dashboards

All services at the operational level use dashboards that indicate costs versus budget, customer satisfaction and achieved service level. At an operational level, extensive information is collected on smaller units, for example, on the execution of work processes. The information from these different dashboards is aggregated to higher-level dashboards for the operational manager, to coordinate activities and make adjustments if deviations occur. At a tactical level in the organisation, the operational management information is clustered. The facility manager will look at the developments in all departments for which s/he is responsible. Here it is assessed whether the organisation performs up to standard. At a strategic level, this management information is compared with external developments to determine whether the course is still appropriate or adjustments to the policy are necessary (see Figure 5.3).

FIGURE 5.3 The relation between dashboards

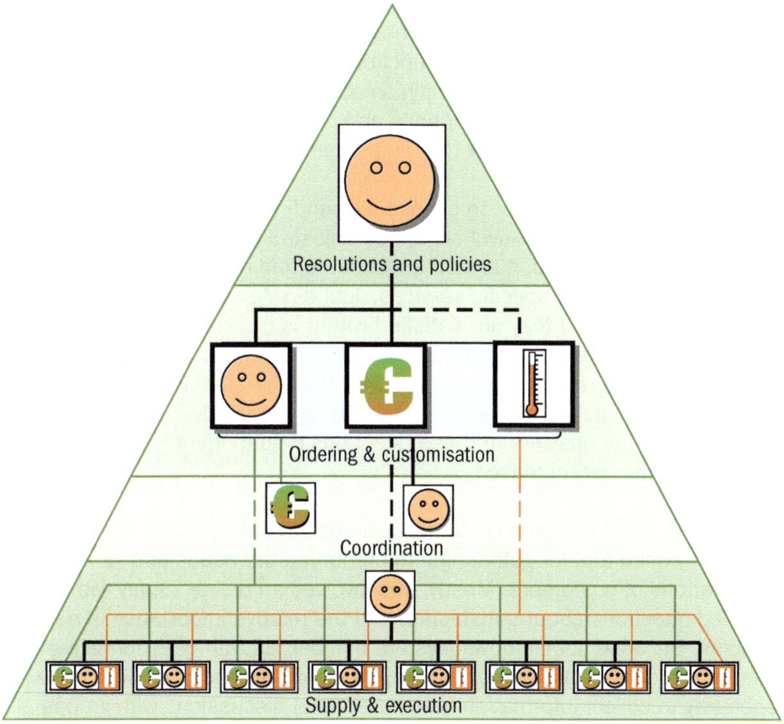

These leading dashboard indicators of customer satisfaction, service level and costs will be explored in succession. However, in (almost) every organisation, additional performance areas for facility management apply. These are often related to the added values of facility management. In section 5.3.4, we will discuss performance management related to productivity and vitality; image; sustainability; and flexibility (adaptability).

5.3.1 Customer satisfaction

Customer satisfaction

The most important indicator is customer satisfaction. After all, if the customer is not satisfied with the facilities services, then the facilities company loses its right to exist. In such a situation, the image of the facilities company is damaged and inadequate (see section 4.3).

Customer needs and satisfaction

Defining customer satisfaction is not easy. This has to do with the stratification of the customer, as mentioned earlier, into the determining customer (client), paying customer (customer) and the benefitting customer (end-user). Each of these customers has its drivers for satisfaction:

1 *Client.* As a rule of thumb, you could say that the decision-maker is only satisfied when the service in general is in line with the strategic policy of the organisation; provided that the customer and end-user are also satisfied.

2 *Customer.* Business unit managers or department managers are mainly interested in the extent to which their effectiveness is supported, and the extent to which employees are satisfied. Examples are the fulfilment of agreements, the average response time, and how the facilities organisation fulfils its advisory function.
3 *End-users.* The users have an opinion about the quality of operations based on their personal experience and preferences. Costs and effectiveness play a smaller role in the satisfaction of end-users.

Customer needs are subject to change. Depending on changing needs of the customer, the service level should be adjusted regularly (see Figure 5.6). For facility professionals, it is also essential to find out why individual customers do not use specific services. Just asking the visitors to the company restaurant if they are satisfied with it is not enough. It is equally interesting to find out why others never use it. Is it because of poor quality? Because the assortment is too expensive? Because opening hours do not match work schedules? Or because they prefer the sandwich shop a few blocks away? The answers to these questions help to identify how satisfaction may be increased.

In the service industry, the customer satisfaction survey (CSS) is a commonly used tool for gathering information on the needs and expectations of customers. Mostly, this involves an online survey with detailed questions about satisfaction and the relative importance of service aspects. Because people nowadays are bombarded with questionnaires and feedback requests, the willingness to participate is generally low. A more open way to collect information is to have panel discussions with groups of customers. A critical comment we would like to make, however, is about the value of customer satisfaction surveys. When a company conducts an annual customer satisfaction survey, it is a random snapshot of what a person thinks based on past experiences. Because someone's perception fluctuates over time, this snapshot is not very informative. Moreover, satisfaction is often measured on a multitude of sub-aspects. This does not provide any insight into the impact of the workplace on people's performance. Big data opens up new possibilities in this regard.
In the customer journey of a working day or visit to an institution, users experience service at various touchpoints (see also section 3.4.3). Each of these touchpoints is an experience of the quality of service. Real-time feedback methods (such as the happy-meter in section 3.3.9) can be used to measure a person's experience at the moment. Collecting this form of feedback is called experience sampling. If used over a more extended period, a vast amount of data is collected. With an analysis of these big data, relations between the workplace experience, the environmental conditions and the user performance can be revealed and used to demonstrate the effectiveness of the facilities organisation and its suppliers.

The facility manager has one more important source of information: the complaints made by customers. Dealing well with customer complaints is a critical contribution to a successful organisation. This applies to both profit and non-profit organisations. Adequate handling of complaints has a significant impact on customer satisfaction and the organisation's image. Facility professionals should regard complaints as free-of-charge advice from which the organisation can learn. It is, therefore, a pity that the majority of the complaints do not reach the facility manager. Figure 5.4 shows that no less than 60% of dissatisfied customers do not make a complaint.

FIGURE 5.4 The economics of complaints

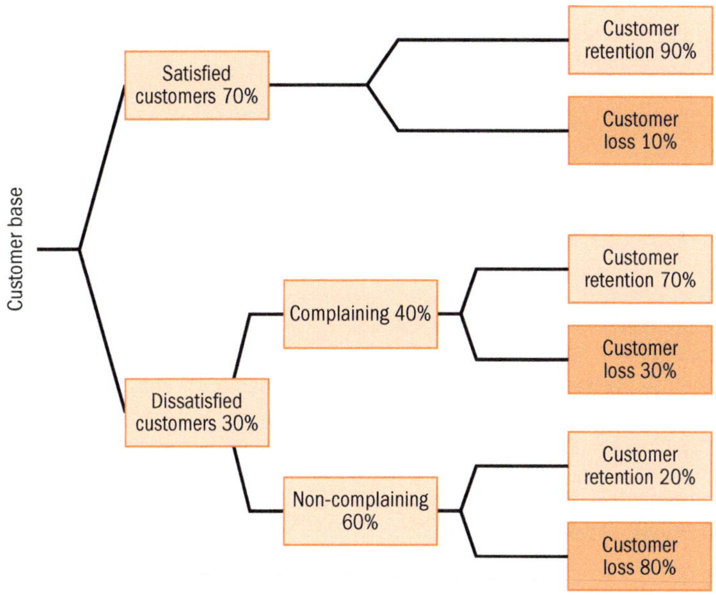

Of the 60% of dissatisfied customers who have not issued an official complaint, 80% ultimately give up and no longer use the services. Of the customers who do complain, 70% are retained as customers. It is therefore vital to lower complaint-thresholds and to take every complaint seriously. Recruiting new customers is many times more expensive than retaining existing ones. You may assume this only applies to providers, but the in-house FM-organisations can also lose customers. Just think of all students that go to a supermarket during their breaks, instead of using the canteen.

Effective complaints management enables organisations to:

- understand its customers;
- learn from its customers;
- strengthen customer relations;
- create goodwill (and prevent anti-advertising);
- retain customers;
- retain employees.

The same applies to employees as to customers: dissatisfied employees are more likely to look for a new challenge than satisfied employees. When experienced employees leave the company, valuable knowledge about the company and its customers is lost. Moreover, this entails extra costs for recruiting and training new employees: on average one and a half times the annual salary. A high appreciation by employees is prerequisite for both customer satisfaction and a positive financial result. Information about (the cause of) absenteeism, staff turnover (how long people on average stay in service) and the experience of the work environment can be used as an indication for staff satisfaction. And concerning the work environment: in every building, you will find crowded places and places where employees prefer not to sit. You could say that people vote by behaviour. Likewise, if employees massively choose to work from home, dissatisfaction with the work environment may be one of the causes.

Integron conducts annual research into the facility experience in the Netherlands among approximately 4,500 end-users. The study covers: building, cleaning, catering, workplace, climate, security, reception, service desk, printing services, and waste management. The 2019 survey showed that the overall appraisal score for facility services increased from 7.1 to 7.5. The rating for cleaning also improved from 6.8 to 7.2. End-users were most satisfied with the reception desk (7.8), the service desk (7.6) and the print services (8.0). In addition to appraisal, the study also provides insight into the relative importance of these domains. This study shows that employees consider the service desk and the workplace to be the most important. Combining the importance and appreciation, this results in a ranking that is led by sanitary cleaning, followed by the climate of the workplace, the quality of the solutions offered by the service desk, and service desk' responsiveness.

Customer satisfaction cannot be considered in isolation from customer expectations since what the customer expects is a significant factor in its satisfaction. Fortunately, by adequate communication, facility professionals can influence this expectation. If, for example, hot meals will no longer be served in the company restaurant, it is very important to explain why this decision is taken. Furthermore, all information on the notice boards, the intranet and the products and services catalogue must be timely updated (see also section 4.5.3).

Net promoter score

Net promoter score (NPS)

One way to measure the reputation of facilities organisations (both commercially and in-house) is the net promoter score (NPS). The NPS uses only one question: 'How likely is it that you would recommend organisation X to a friend or colleague?'. The customer can answer this question by giving a rating between 0 and 10.
Based on the number, the customer classified into one of the following three categories:

1 Detractors (opponents or critics). These respondents give a score between 0 and 6 and are most likely dissatisfied with the organisation.
2 Passives (passively satisfied). These respondents give a score of 7 or 8. These people are happy with the organisation.
3 Promoters. These respondents give a score of 9 or 10. They are very enthusiastic about the organisation.

The NPS gives insight into the percentage of customers that highly recommend the service (promoters), the percentage on the 'threshold' between being enthusiastic and not being enthusiastic (passives), and the percentage of detractors. The NPS is calculated by subtracting the percentage of detractors from the percentage of promoters; the higher the NPS, the larger the group of promoters in the organisation (see Figure 5.5).

FIGURE 5.5 Calculating the net promotor score

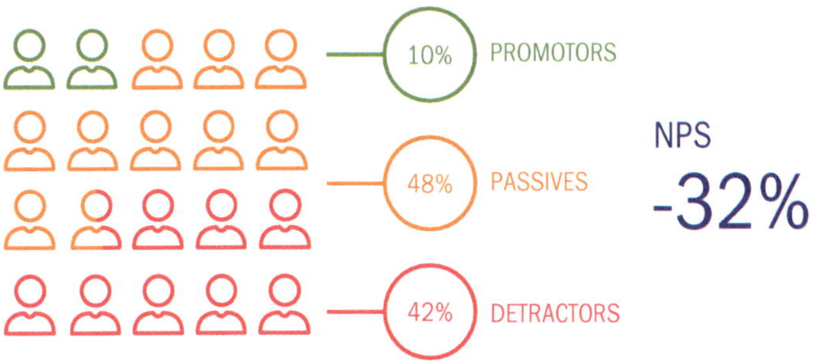

5

It is evident that organisations want to reduce the number of detractors. However, the NPS also indicates that satisfied customers do not do much for the reputation of the organisation. Only when people belong to the category of promoters will they actively contribute to a positive image for the organisation. Especially in the hospitality and leisure industry, this is a frequently used fast indicator for customer satisfaction. Alternatively, this question can be added to a more extensive questionnaire used for other purposes.

Positive image

For facility professionals in the Netherlands, there is still a considerable challenge when it comes to enthusing end-users about the services offered. Research by Integron (2018) showed that the average percentage of detractors (42%) is still considerably higher than the percentage of promoters (10%). From our experience, we dare state that the situation is not much different in other countries.

5.3.2 Service level

Service level

Instead of service level, the terms 'quality' and 'contract compliance' are also used in this context. As this is somewhat confusing, we will use 'service level' in this chapter. The service level gives an objective indication of the quality of the amenities, expressed in terms of range, the intensity of use, responsiveness and use of materials. These are standards that articulate the customer's expectations. Examples of a high level of service include a prompt response from a service desk, an extensive range of fresh products in the restaurant and extended opening hours for a building. Determining service levels is part of planning the organisation (tactical level).

Figure 5.6 links the relative importance of a service or service attribute to the service level. If the relative importance and the service level are low, then the service in question is well coordinated. The same applies to a high level of importance and a high level of service. That is why the quadrants in the figure depicting low-low or high-high combinations are coloured green.

FIGURE 5.6 Service level related to importance and need

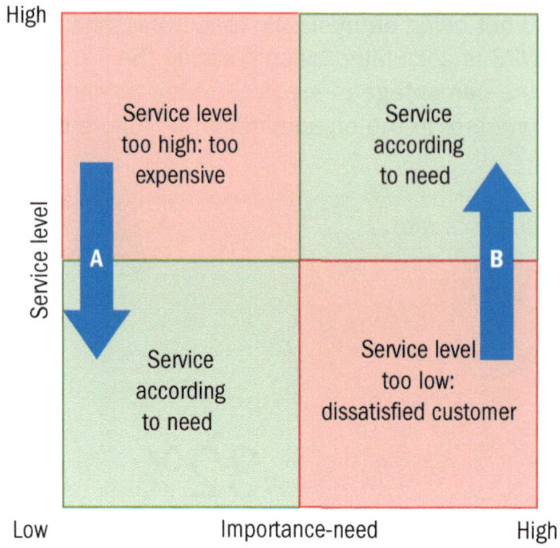

5

Mismatch

The red squares indicate a mismatch between service level and importance. Services that are placed in the top left quadrant have an excessively high service level. This leads to unnecessarily high costs and does not result in extra customer appreciation. Here, services levels can be lowered without compromising customer satisfaction; hence cost-savings are possible. The lower right quadrant contains the services with an inferior service level, while the customer attaches great importance to them. It seems that these services are economic, but an insufficient level of service is the reason why the customer is dissatisfied.
A mechanism that increases satisfaction without increasing the bottom-line costs is as follows. By lowering the service level above left (and thus the costs; arrow A) the service level of these services that matter to the customer can be upgraded (arrow B). This way, within budget, the service levels are adequately matched to the demand.

Research has been conducted into the influence of hard and soft service aspects on customer satisfaction. Hard aspects are, for example, techniques, products or instruments. Soft elements include behaviour, attitude, sensory perception and delivery time. Customer-oriented organisations are increasingly discovering that it is not so much the product or brand that makes the difference in the market. How customers are served decisively determines customer satisfaction and loyalty.
Behaviour and communication are so important in meeting human needs that their impact sometimes outweighs the hard aspects. However, for none of the aspects, the quality must fall below a minimum standard. Thus, within a certain margin, the hard elements can be adjusted downwards in terms of service level if the soft aspects are reinforced at the same time. The result is at least the same level of customer satisfaction at possibly lower costs, as in general, the hard aspects outweigh the costs of the soft factors.

Service operations are frequently outsourced so that the facility manager cannot control the quality of delivery directly. Supervision of quality control is transferred to the supplier, for example, the property manager or the

catering manager. This way, the performance of external suppliers is vital for customer satisfaction. How then does the facility manager keep track of and control over the quality of service? At the operational level, this is monitored by checking whether the service level delivered complies to the programme of requirements.
Checklists are used for this purpose, in which an auditor indicates whether agreements are met. Tablet computers are often used, as part of a software application that can register, weigh and analyse the scores to arrive at an overall score. To inspect the quality of service delivery mystery guests can also be used, especially for monitoring the behaviour and attitude of service providers.

Supplier performance assessment

Supplier performance assessment can be carried out to assess whether the suppliers are fulfilling the contractual arrangements made. This assessment includes adequate delivery and compliance with agreements. However, a supplier assessment should focus on building the relationship with the supplier, not on 'settlement of scores'.

Checked

In general, it is not wise to have the quality checked by the person who also performs the service in question, because this person is biased. Which is understandable, because only a few people will spontaneously report that they have made any errors. Control should, therefore, be carried out by an independent party that is commissioned jointly by the client and the service provider. An example of this is the performance of HACCP checks in the catering industry (see section 4.7.2). However, there is an exception to this rule. When the following circumstances apply, it is acceptable to be self-checked:

Commodity

- The result of the inspection can be determined objectively. In such a case, we are talking about a commodity, a straightforward product, service or article.
- There are strict, verifiable guidelines that the party must adhere to, drawn up by external bodies, such as a professional association.
- Stakeholder organisations issue a certificate or label that demonstrates the quality of the work of the party involved. The party has an interest in working under the rules because if the certificate or label is withdrawn, the provider will no longer be perceived as reliable.

Facility management also knows commodities, such as components in cleaning, reception and catering services. The service providers are well-placed to check these components themselves, as the conditions mentioned above apply. But as soon as a result is not unambiguous and it depends on the interpretation of the auditor whether the result is ambiguous or not, independent and external control is required.

Facilities organisations without professional purchasing are characterised by a large number of suppliers. With many suppliers, it is hardly possible to assess the performance of each supplier annually. For professional supplier management, it is advisable to limit the number of suppliers.

A prerequisite for supplier assessment is adequate contract management (see also section 4.6). The programme of requirements is the foundation for supplier assessment later on. Overly general descriptions cannot be used as a basis for evaluation and easily provoke conflict instead of improvements.

5

In the case of decentralised purchasing, the results of supplier assessments must be available to all other organisational units that work with that supplier. This prevents situations in which underperforming suppliers are re-contracted by other organisational units.

5.3.3 Costs

Of course, it is essential that the services are carried out following the financial agreements, and for costs to fit within the budget. Table 5.1 shows the cost classification according to the EN 15221-4.

TABLE 5.1 Cost classification according to EN 15221-4

Cost of capital	The cost of borrowed capital; mainly the interest cost of bank loans
Enhancement of initial performance	Costs needed to meet any new requirements
Depreciation costs	Estimated or expected impairment of provisions
Total support costs	Combined costs for business support management (people and organisation) and real estate support management (space and infrastructure).
Material costs	Cost of goods (e.g. consumption, tools and parts)
Personnel costs	Salaries (gross annual salary, incl. social provisions, taxes, holiday allowance, profit-sharing, bonus) and other personnel costs
Facility management costs	Costs to organise facility management (overhead)
Revenues	Earnings from e.g. services to external companies, events
Taxes	For example, non-recoverable VAT, fees and other taxes

Total integral cost

A facility professional can make a fair comparison of quotations if s/he compares the total integral cost throughout the lifecycle. To the initial purchase price, all related costs up to and including the renouncement, are added. Examples are costs related to maintenance, repair, depreciation, insurance, taxes, funding, training and education, personnel, distribution and stock. In this way, the costs of doing business with a provider are systematically calculated and can be compared to alternatives. To obtain a complete picture, s/he also takes into account any savings generated by the product or service. Such a life cycle cost (LCC) analysis results in the total cost of ownership (TCO).

Total cost of ownership (TCO)

Fixed and variable costs

In addition, a distinction is made between fixed and variable costs. Fixed costs recur periodically and, in principle, do not vary with production or sales. Examples are subscription costs, coordination of services by external parties or services contracted at a fixed price. Variable costs depend on the time spent; the number of participants (in an event); or the volume, such as the number of lease cars. In practice, it is often hard to estimate how much time will be spent, how many people will attend an event, or what the purchase volume of supplies will be.

Controllable and non-controllable costs

Fixed costs themselves can be categorised as controllable and non-controllable costs. A controllable cost is a fixed cost that can be changed in the short term. For example, even though the janitorial cost has been budgeted as a constant cost, it may be possible to reduce the service and the cost at short notice. Non-controllable fixed costs are those costs that cannot be quickly adapted. The most common non-controllable fixed cost is

rent or lease payments and depreciation. In most basic calculations, the only real fixed costs are overhead costs. Overhead costs are ongoing costs required to operate the business, that cannot be allocated directly to the production or the service delivery, like costs of management, accouting fees, rent, and advertising.
Variable costs are directly related to production and sales. For example, the use of napkins or linen often varies due to an increase or decrease in events catering. Other variable costs include food, beverages, and some labour costs. Usually, in catering, the major variable cost is food and most of the labour costs. Variable costs are controllable; using less expensive ingredients, smaller portion sizes, and reduction of hours of temporary staff.

Actual and budgeted costs

Another breakdown of costs is possible, namely in actual and budgeted costs. The costs that an in-house facility management organisation expects to incur are calculated each year in advance and, after approval, laid down in an annual plan. FM-providers also work with budgets, based on their quotations. The actual costs are formed by what (the word says it all) is spent. Facility professionals strive to keep the actual costs equal to or lower than the budget or price agreed too. However, it is important to distinguish between influenceable versus non-influenceable costs. If the legislator imposes investments in improving the sustainability of real estate at short notice, the budget will most likely be exceeded, however, outside the sphere of influence.

Influenceable versus non-influenceable costs

Benchmarks

A benchmark is a standard, or set of standards, used as a point of reference for evaluating performance or quality levels. For benchmarks, costs are shown in key figures or ratios (see also section 5.5.4). Usually this is done per workstation or per employee. These two key ratios will lose their usefulness in the future. On the one hand, this is because companies increasingly also offer workplaces to external parties (such as independent professionals, consultants and trainees). On the other hand, the concept of the workplace is undergoing change, as many employees are working mobile. After all, what is a workplace today? As we explained in Chapter 3, the workplace is a rather broad concept. Some companies register the costs per person present, by recording how long someone (employees and visitors) is inside. They use a ratio of costs over total staff presence. Recording the costs per square meter of floor space is also quite reliable. Still, for benchmarks (comparisons with other buildings) one has to realise that these costs are profoundly affecting a buildings' efficiency.

Building related and user-related costs

When analysing costs, it is crucial to bear in mind that not all cost items are equally important. In this respect, the distinction between building related and user-related costs is relevant. Real estate costs - both rent and property - are by far the most important, but also hardly adjustable. They account for the most substantial proportion of total facility costs. Also, if property costs increase due to an increase in the amount of floor space, all related expenses increase proportionally. After all, the cleaning and energy costs are virtually proportional to the floor area. An example of a facility management service that - compared to property costs - is not as interesting, at least substantially so, is greenkeeping.

5.3.4 Quantifying indirect results

In Chapter 1, we discussed the added value facility management could have for an organisation. Added value is not limited to the actual service or product delivered (the result or the output). It relates to the indirect contribution that the product or service provided has on the achievement of organisational goals. This indirect effect is also referred to as the impact, the effect, or the outcome (see Figure 5.7).

Indirect contribution

FIGURE 5.7 FM's main process

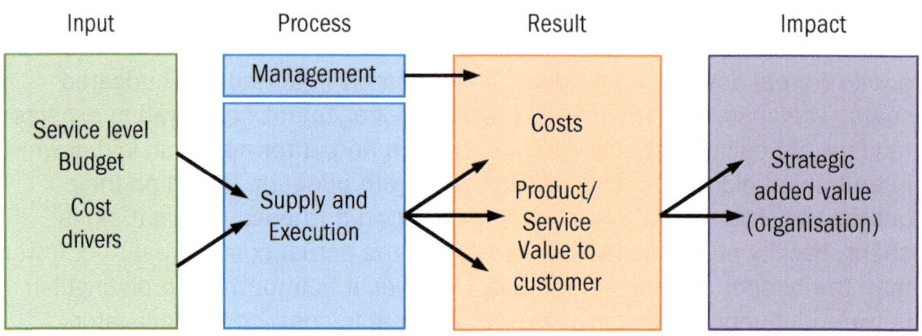

An example: the facility manager in a hospital is responsible for the patient's hospitality experience. The facility products and services offered, and *how* they are offered, contribute to a hospitable stay (result). As you have read in Chapter 3, this involves a combination of the product or service, a welcoming attitude on the part of employees and a comfortable physical environment (Figure 3.2). The effect of paying attention to hospitality leads to higher overall patient satisfaction (impact). Higher patient satisfaction is also not an end in itself, but a means to promote patient recovery. More well-being leads to stress-reduction, which, in turn, is beneficial for recovery. Patient satisfaction also contributes to maintaining or increasing market share (patient-loyalty and word of mouth), and to a better negotiating position with health insurers. The extent to which added value is achieved is assessed by the stakeholders: society, the external client, employees and owners.

Strategic contribution

You might imagine that measuring strategic contribution to the attainment of organisational goals is a lot more complicated than measuring cost, quality and customer satisfaction. The complexity results from the many other influencing factors involved, for example, the organisational culture, leadership style, terms of employment, quality of the IT-infrastructure and work pressure. Examples of external factors that have an influence are the market demand, the availability of resources, the economic climate and legislation and regulations (see Chapter 2 for a comprehensive DESTEP analysis). With so many influencing factors, there is no straightforward relationship between result and impact. Quantification of the exact contribution of facility management for these result domains is limited. However, this does not mean that there is *no* influence.

Influencing factors

In this section, we provide several substantiated assumptions based on the results of many scientific studies into the added value of facility management. We discuss the evidence-based contribution of facility management to productivity, image, sustainability and adaptability. Productivity is essential for every organisation because it affects its competitive position. The corporate image (see section 4.3) is relevant because it affects the organisation's profile in the market and the extent to which an organisation is perceived as an attractive employer (see section 2.3). Sustainability is also a crucial point of attention for every organisation, given the developments concerning the circular economy and the mandatory reduction of greenhouse gas emissions (see section 2.7). All events together determine the dynamics of the environment. As already mentioned, we live in a time of exponential changes; flexibility (adaptability) is the only answer to an unpredictable future.

Productivity

Productivity

Productivity can be measured as the ratio between the quantity of output (goods, products or services) and the input (labour, materials, machinery, knowledge) needed for it. There are three ways to increase productivity:
1 to produce *more* output with the *same* input;
2 delivering the *same* output with *less* input;
3 to increase the output *stronger* than the input increases.

Knowledge workers

Perceived productivity

That sounds pretty straightforward, but it's not. Because how do you measure input and output accurately? Especially when it comes to so-called knowledge workers. These are people who mostly think for a living and whose productivity cannot be measured in the number of pieces they have produced in a given time. Researchers have found something in response to this: they measure the extent to which respondents feel that they (can) be productive; we call this the perceived productivity. In addition, this type of survey consists of questions that measure satisfaction with aspects of the work environment that are known to affect productivity. Facility managers can also do this to measure the extent to which they optimally facilitate employee productivity.

Leesman Index

An example of a frequently used instrument is the Leesman Index. The Leesman Index indicates how well the work environment supports the employee in his activities. The Leesman Index consists of a digital survey with ninety questions about four main components:
1 the activities someone does at work (from concentration work to video conferencing and collaboration);
2 the physical amenities in the work environment (from desks and lighting to art on the wall);
3 the facility services offered (from coffee and cleaning to IT facilities);
4 the mobility of employees inside and outside the (office) building.

Two questions are asked for each item:
1 What do you think is essential for an effective work environment?
2 How well does this item support you in your work?

To determine the indirect effect, questions are asked about the impact of the work environment on pride, cohesion and productivity.

Figure 5.8 shows which factors influence the productivity of employees and what the interrelationships between these factors are. Reading from right to left, you will see that satisfaction with the work environment, satisfaction with the work itself, someone's motivation, and their knowledge and skills have a direct impact on productivity. Even more to the left, you will see the underlying factors: work environment (external conditions), physiology (gender, age, etcetera), psychology (personality traits, expectations, experiences), intrinsic rewards (pride, craftsmanship) and extrinsic rewards (salary, security, responsibility). The facilities organisation, therefore, has limited influence: via the work environment. Aspects of the work environment that have a proven effect on productivity are:

Work environment

- the physical conditions (temperature, humidity, light, sound, acoustics, etc.);
- space (layout, square metres, openness/seclusion, privacy);
- ergonomics and ambience (colour, aesthetics, quality of experience).

The workplace concept, the quality of the work environment and its look and feel, are determined by the organisation (strategy, culture, image).

FIGURE 5.8 Factors that impact performance and productivity and their interrelatedness

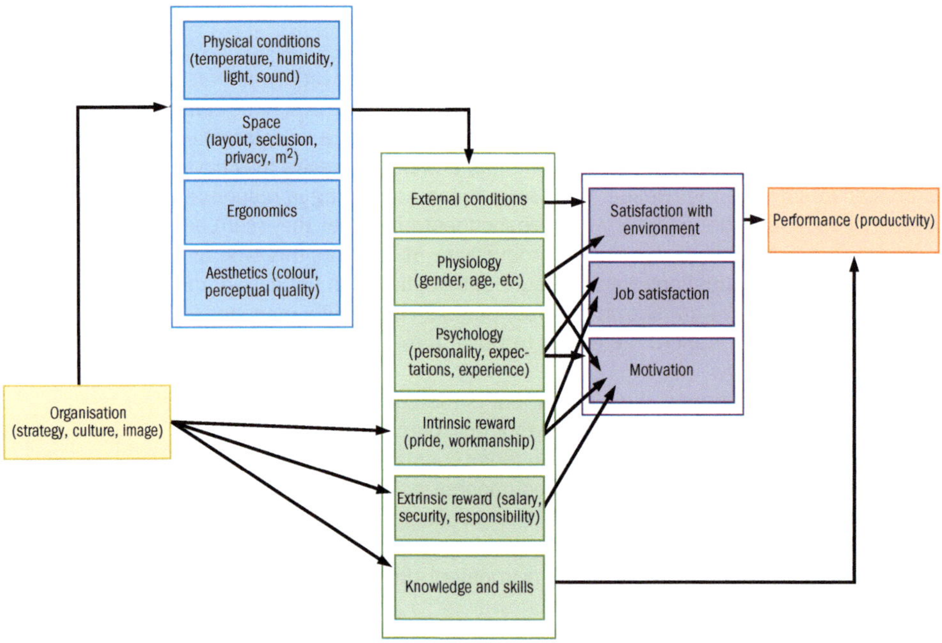

To give you an idea of the effect of the work environment on productivity, we provide some examples of research results (quoted in Jensen & Van der Voordt (eds.), 2017):

- Improving the work environment can lead to a 15% increase in productivity.
- Improving the indoor environment can result in a 3% increase in productivity, while productivity can deteriorate by as much as 6% when people exchange a naturally ventilated office for an air-conditioned one.
- The presence of plants in the work environment may positively affect productivity in two ways: by improving the indoor climate and by making people feel more comfortable in an environment with plants (psychological comfort).

5

‘It’s a misconception that accommodating many people on few square meters is cost-efficient.’

— Gideon van der Burg, managing director Benelux of Leesman, on www.facto.nl, 27 June 2017.

- Personal control over the work environment (temperature, light, air quality and noise level) has a positive effect on productivity. One study found a productivity improvement of 6 to 10% with complete personal control compared to a situation in which employees have no control at all over their workplace.
- Ergonomics is essential for employees’ health and comfort. Discomfort harms productivity. Flexible office concepts require simple but effective adjustability of furniture for ergonomic use by employees.

Healthy working environments

Because employee productivity is directly related to business performance, it is a crucial performance indicator for facility professionals. Also, the great attention to healthy working environments and vitality can be explained from its importance. Healthy work environments reduce the adverse effects of the environment on health. The underlying goal is to reduce productivity loss due to absenteeism. Related measures are usually aimed at improving the indoor climate (preventing sick building syndrome) and ergonomics (e.g. adjustable furniture). Many people are overweight and, in general, people spend too much time sitting down. Facility professionals can take health promotion measures for promoting healthy behaviour. Think of a healthy catering assortment, encouraging the use of stairs over lifts, stand-up meetings, and investing in an attractive environment to leave the office during breaks.

Healthy behaviour

Can you create a work environment in which everyone is optimally productive? That’s not as easy as it seems, because people and organisations differ from each other. Compare the way you and your fellow students learn for a test. Some people work in a structured way and start making summaries far in advance; others wait until the last moment and only read the summaries in the textbook. Some students prefer to learn from written material; others remember information the easiest when explained in class. Also, the impact of the environment is not always unequivocal: what one person likes (inspiring crowds), the other finds disturbing (distraction from work). Organisations also differ from each other in culture, size, type of activities, goals and the risks they accept. An optimal workplace concept is, therefore, always a matter of customisation. However, there are a few starting points that must be taken into account. According to several extensive studies, the most important predictive factors for the productivity of knowledge workers are:

Predictive factors for the productivity of knowledge workers are

- the ability to work undisturbed;
- the possibilities for (mainly informal) interaction with colleagues;
- freedom of choice.

A work environment that offers sufficient diversity of workplaces, so that people can choose the setting they need at that specific moment in time, seems to be the best solution for now.

The WELL Building standard

WELL Building Standard

In 2014 the WELL Building standard was launched. WELL is a performance-based system for measuring, certifying, and monitoring features of the built environment that impact human health and well-being. Ten aspects of the built environment are included in its scope: air, water, nourishment, light, fitness, comfort and mind. WELL is a global tool that is utilized in more than 50 countries.

WELL was developed by integrating scientific and medical research and literature on environmental health, behavioural factors, health outcomes and demographic risk factors that affect health with leading practices in building design, construction and management.

WELLv2 focusses on ten concepts that drive the quality of the built environment from a user and sustainable perspective. Each concept is comprised of features with distinct health intents. Features are either preconditions or optimizations.

1 *Air.* The WELL Air concept aims to ensure high levels of indoor air quality across a building's lifetime.
2 *Water.* The WELL Water concept covers aspects of the quality, distribution and control of water in a building. It includes features that address the availability and quality of drinking water, as well as measures to avoid damage to building materials and environmental conditions.
3 *Nourishment.* The WELL Nourishment concept encourages the design of food environments in which the healthiest choice is the most natural choice.
4 *Light.* The WELL Light concept aims to create environments that are optimal for visual, mental and biological health.
5 *Movement.* The WELL Movement concept promotes physical activity and active living. It discourages sedentary behaviours through environmental design strategies, programs and policies.
6 *Thermal Comfort.* The WELL Thermal Comfort concept aims to promote human productivity through improved HVAC system design and control and by meeting individual thermal preferences.
7 *Sound.* The WELL Sound concept aims to bolster occupant health and well-being through acoustical comfort.
8 *Materials.* The WELL Materials concept aims to reduce human exposure to hazardous building material ingredients.
9 *Mind.* The WELL Mind concept promotes mental health through policy, program and design strategies that address factors that influence cognitive and emotional well-being.
10 *Community.* The WELL Community concept aims to establish an inclusive, integrated community through social equity, civic engagement and accessible design.

WELL operates on a points-based system. Its potential for impact determines the point-value of a feature. This is defined as the extent to which a feature addresses a specific health and wellness concern or opportunity for health promotion, and the potential impact of effective intervention. Projects can be awarded a Silver, Gold or Platinum certificate. Building certifications range from Bronze to Platinum.

Image

In Chapter 4.3, we examined the image and identity of the facilities organisation. However, facility management also influences the image and identity of the (parent)organisation, mainly through property management. Real estate is a communication channel through which the identity and organisational strategy, vision and culture can be expressed.
Table 5.2 gives examples of location and building aspects that contribute to branding.

TABLE 5.2 Building aspects that contribute to corporate branding (source: Jensen and Van der Voordt (eds.), 2017)

Location aspects	Building aspects
Type of location	Facade
Reputation	Building dimensions (height and volume)
Landscaping	Shapes, use of colour and materials (interior and exterior)
Amenities	Entrance area
Visibility	Architecture
Accessibility by car	Horizontal and vertical zoning
Accessibility by public transport	Accessibility for disabled people
	Functionality
	Thermal comfort
	Lighting
	Security
	Catering services
	Visibility of sustainability measures

5

Building performance evaluation (BPE) standards - like the Real Estate Norm (REN) or the Dutch 8021 - can be used to evaluate the performance of buildings in use, compared to the desired situation or ambitions. To measure the stakeholders' impression of the location and building (facilities) respondents can be asked to what extent they associate these with certain values, or respondents or the associations the building and the location evoke. A more indirect way to get a first-hand idea of an organisation's image is to analyse social media expressions about the organisation.

Sustainability

Throughout this book, the message 'pay attention to sustainability' is repeated. Organisations that want to make their business operations more sustainable are in the best position to do so if they are building owners. A tenant has less say because the owner is generally the party that has to invest. Especially in the case of sustainability investments, the motivation to invest is higher if the benefits (e.g. energy savings) accrue to the same party.
In addition to a positive effect on the ecological footprint of organisations, research shows that sustainable accommodation can lead to:

- increased productivity: healthier buildings and a higher quality working environment lead to greater job satisfaction, with the indirect effect of increasing productivity (see also Figure 5.8);
- lower operating and maintenance costs: reduced use of resources and less waste lead to lower costs;

- competitive advantage: a sustainable image can lead to more turnover and customer appreciation;
- sustainable image: this can increase the attractiveness of an organisation as an employer.

An analysis of the EN 15221 standard shows that only twelve of the more than one hundred facility products and services described, explicitly mention sustainability. Sustainability is discussed almost exclusively in descriptions of facility products and services in the area of 'space and infrastructure'. This is logical because of the strong intertwinement with the built environment. Examples of ways to measure the sustainability performance of buildings are the Building Research Establishment's Assessment Method (BREEAM) and Leadership in Energy and Environmental Design (LEED). In these systems, the sustainability performance is determined for underlying indicators, and, subsequently, the weighted total score is calculated. Based on the overall rating, a building is classified (see section 4.8.1).
Examples of ecological indicators are:

- energy consumption;
- water consumption;
- use of materials;
- waste;
- CO2 emissions;
- land use and the ecological value of the site.

Examples of economic indicators are:

- the adaptability of buildings;
- ease of maintenance;
- the total cost of ownership, see section 5.3.3;
- process quality (planning and preparation);
- innovation.

Examples of social indicators are:

- indoor climate quality;
- accessibility by public transport;
- comfort (visual, thermal and acoustic);
- accessibility for the disabled;
- safety and security.

It is not necessary to certify a building. A facility manager may also select the most critical indicators and use these for his/her balanced scorecard (see section 5.4.3). As explained in section 4.8.3, ISO 26000 can also be used to provide insight into the contribution of facility management to sustainability.

Adaptability

Chapter 2 deals with the exponential times in which we live. Changing circumstances can lead to a changing demand from society (manifested, for example, in legislation and regulations), from owners (for example, in terms of risk profile, return or contract duration) and from users (functionality). As a result, organisations continually have to adapt to new circumstances. To be able to do so, they must be agile. We distinguish between the adaptability of hard services (the built environment) and the flexibility of soft services.

Hard services
Soft services

Not all changes in organisations have a direct impact on the built environment. Many changes primarily involve modifications to the organisational structure or processes. These types of changes can have consequences for soft facility services.
An example: a production company receives an unexpectedly large order and decides to work in shifts to meet the need for extra capacity. This decision has consequences for the opening hours of the production location and therefore for the deployment of security, catering (provisioning times and supply), cleaning (more intensive use, less time to clean outside production time). Public transport may not be available to transport night shift workers to the location so that the deployment of shuttle buses is necessary. The flexibility of services is largely determined by contract management. If facility services are outsourced, the question is whether the contracts are flexible enough to accommodate a (temporary) expansion of demand within existing settlements. The same applies if an organisation faces decline. In that case, the question is whether the contracts of facility suppliers allow the purchase of services to be adjusted downwards. Another form of flexibility has to do with the channels that a facilities organisation uses for communication with customers. It goes without saying that digital self-service increases the accessibility of services and thus, the flexibility offered to customers.
The same example can also lead to a request for building adaptation: an extension or reduction of the number of square meters or a change in functionality.

Four factors determine the flexibility of buildings: the extent to which it is possible (1) to make adjustments, (2) to extend, (3) to split (or divest) and (4) the degree of multifunctionality. The degree of flexibility of a building can be measured, for example, with the Flex 2.0 or the Flex 2.0 Light (both by Geraedts, 2015). The Flex 2.0 is a measuring instrument with 38 performance indicators that measure the degree of adaptability of buildings based on the construction and spatial and installation characteristics of a building. The Flex 2.0 Light comprises a selection of the seventeen most decisive performance indicators. The scores provide insight into the possibilities and limitations for adapting property at different levels: the location, the construction, the facade, the technical installations, the layout plan and the furniture. You can imagine that a building that is only adaptable at the level of the furnishing plan and fittings quickly becomes a hindrance when more fundamental adjustments are needed. Facility professionals who see flexibility as a significant added value of facility management will opt for solutions that offer maximum adaptability at all levels when planning for accommodation adjustments.

Flex 2.0

5.4 Performance management

An organisation is a coherent system of processes and activities, which must be aligned to achieve the organisation's goals. One cannot achieve goals without steering. Performance management is to steer the organisation by (1) systematically defining an organisation's mission, strategy and objectives, (2) translating these into all parts of the organisation in a structured way, and (3) making them measurable using critical success factors and performance indicators. Performance management is closely related to the business model of organisations. That is why this section begins with a brief explanation of this.

Performance management

5.4.1 Business model

Business model

Building blocks

A business model is a blueprint for the implementation of a strategy in the company's organisational structures, systems and processes. The business model canvas (BMC) of Osterwalder and Pigneur (2010) is a useful tool to map the business model of an organisation or organisational unit. It consists of nine building blocks (see Figure 5.9).

FIGURE 5.9 Business Model Canvas

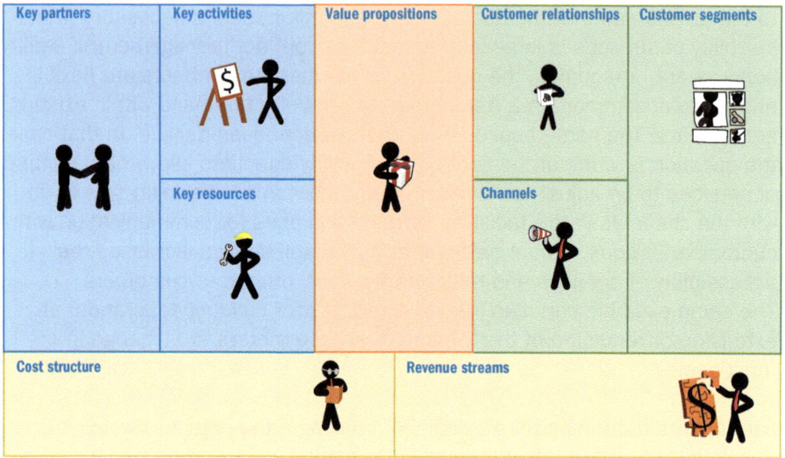

The building blocks of the model provide insight into all aspects of an organisation. In the centre of the canvas model is the value proposition: that which distinguishes an organisation (see also section 1.5) and through which money can be earned. The value proposition is the starting point, the *why*. On the left, the facility manager maps out how the organisation creates such value: with which partners, assets and activities. The right side is about bringing the value proposition to the market. Who are the customers and how does the organisation reach them? At the bottom of the canvas, the costs and revenues are mapped. These are, in fact, four main focus areas: customers *(for whom)*, supply *(what)*, financial viability *(costs and revenues)* and infrastructure *(how).*

For each of the components of the business model, targets are set, which are derived from the central value proposition. The business model can be elaborated on different levels in the organisation. If you make a BMC for an organisation, facility management is a business asset. If you make a business model for an in-house facility management organisation, then you take the strategic added value of facility management for the organisation as a starting point.
The same questions apply: which customer groups can you distinguish? What are the needs and wishes of customers? Which channels do you use to reach the customer? On the left side, the building blocks relate to how: which partners are needed to create added value? Internal partners are, for example, the departments Property Management, ICT, HRM or Marketing and Communication and external partners are suppliers, but can also be licensing

authorities? Which activities are required, and what is the desired service level? Which company resources does the facilities organisation deploy? What costs and revenues are involved? For an in-house facilities organisation, these revenues are sometimes challenging to determine, as internal support does not make a profit. Nevertheless, it does make a positive impact, such as an increase in customer satisfaction, a more productive working environment, cost savings (without loss of quality), greater flexibility, lower CO2 emissions and less energy consumption. Providing insight into the revenues is a prerequisite for giving insight into the added value of facility management. Suppliers also have a business model. The value proposition is how they make money and distinguish themselves from competitors in the market. For example, hospitality services with 5-star service, sustainable maintenance, innovative workplace concepts or carefree service through integral facility management. A supplier carefully balances costs and revenues to safeguard the financial health of the business. Like any entrepreneur, the supplier has partners and suppliers. Partners can be, for example, entrepreneurs who work together to provide a complete range of services to the customer, suppliers with whom products are developed or customers with whom they collaborate. The customer is the contracting organisation (see also the 9-grit in Figure 4.3).

5.4.2 Plan, check, act

Performance managers are, in fact, engaged in three activities. To use the terms of the Deming-cycle: plan, check and act. Do represents the implementation or execution (see also Figure 4.13).

Deming-cycle

Plan means preparing the service in such a way that it is clear what will be checked, who will do so, and what the consequences of the check will be before the order is given. In this phase, everything that needs to be measured to assess the service and adjust its performance is recorded. Compliance with the SMARTI-criteria (see introduction) is crucial, as it prevents frustrating discussions and differences of opinion about the interpretation of terms and specifications. Equally important is to ensure that all stakeholders are aware of these. Everything is laid down in a contract with a service level or programme of requirements, as described in previous sections. The performance manager will always make sure that the supplier is precisely aware of the content of the listed documents when signing the contract. If a service is outsourced, these documents must be ready before the tender is launched.

The quality of the facility services must be *checked* regularly. Not only after delivery but also during execution. In the eyes of the performance manager, the do- and check phase take place simultaneously. After all, most facility services are performed in the form of a continual process. Of course, this is different for projects such as a renovation or relocation project. At a certain point, these projects are finished, and the result can be checked against the requirements. However, also in these types of projects, it is advisable to check during implementation; otherwise, it may be too late (or too costly) to make adjustments. This checking or measuring is also called auditing or monitoring.

Auditing

Similarly, the performance manager sees *act* (adjustment) as a continuous process. Ideally, when undesirable deviations from the agreed performance are detected, the supplier or service provider should be informed directly to discuss how corrections can be made. Figure 5.10 visualises the management of organisations.

FIGURE 5.10 Governance of organisations

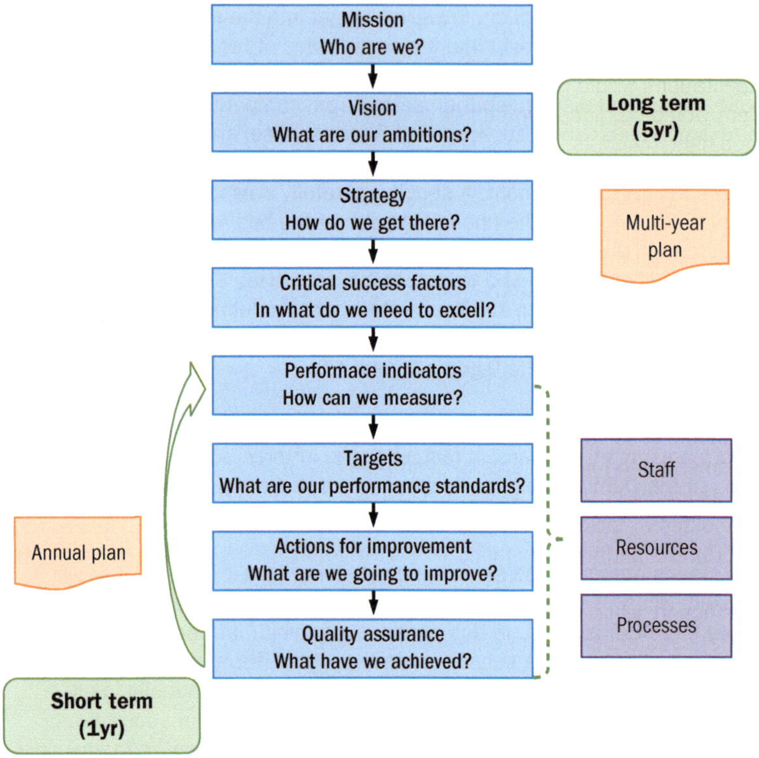

5.4.3 Balanced scorecard, KPI and CSF

Balanced scorecard (BSC)

One of the methods used to structure performance management in an organisation is the balanced scorecard (BSC) of Kaplan and Norton (1996). The BSC is a tool for monitoring the strategic decisions taken by the company based on indicators previously established. These indicators should be thoroughly assessed at least four aspects: financial, customer, internal processes and learning & growth. Applied to an in-house or external FM-supplier:

1 *Financial perspective*: What financial performance does the board expect from the in-house facilities organisation? Or what financial performance do the shareholders expect from the commercial provider?
2 *Customer perspective*: How do (internal or external) customers evaluate the organisation and what customer requirements do the internal or external FM-supplier have to meet?
3 *Internal processes perspective*: In which processes must the (internal or external) FM-supplier excel to keep its customer satisfied?
4 *Learning & growth perspective*: To what extent is the internal or external facilities supplier able to learn, adapt, and improve itself?

As you can see from the four perspectives mentioned above, the balanced scorecard is not only focused on financial results. Setting performance indicators for each of the four perspectives enables a holistic view of the organisation's performance. *Balanced* indicates, that to achieve success and continuity in the short-term and long-term, all four perspectives must be taken into account. The leading perspective depends on the type of organisation, but for successful business operations, all perspectives deserve attention. Objectives that detriment other goals must be avoided.

TAPPING INTO A NEW MARKET

We give an example of the perspectives in tapping into a new market. Imagine a national contract caterer entering a new market by launching a home delivery service for private customers. This would involve financial targets relating to investment costs, payback-time and turnover. From a customer perspective, one can think of marketing objectives that are necessary to work the private market and collaboration with home care. From the internal process perspective, goals are needed that relate to the work processes in the organisation. Do they need to be adapted to enter the new market? In this example, the delivery process will have to be restructured. From an innovation and learning perspective, think of training programmes and recruiting new employees who have the necessary knowledge and experience in working in the private market of meal facilities.

Each general objective must, therefore, be worked out into specific goals for each perspective. This applies to the entire organisation, including facility management support.

Next, critical success factors (CSFs) are determined for each objective. A number of the critical success factors are derived from the criteria used by the client when assessing the (facility) services. To a certain extent, they form a description of the customer's wishes. By monitoring these factors, you can determine whether you are doing the right things from the customer's point of view. Critical success factors also relate to the financial, internal processes and innovation and learning perspective (see Table 5.3 for the example).

Critical success factors (CSFs)

TABLE 5.3 Example of critical success factors from a customer perspective

BSC-perspective: customer	Financial	Customer	Internal processes	Learning & growth
What does the customer expect from the facilities supplier?	Market conformity	Customer satisfaction	Prompt service	Taste and consistency of meals tailored to seniors. Knowledge of private market.

Critical success factors must be specific. Table 5.3 contains fairly general concepts, such as prompt service and knowledge of the private market. To continually monitor whether the organisation is achieving its goals, all parts of the organisation must provide insight into their performance. To measure this performance, performance indicators are also determined for the

Performance indicators

critical success factors (Table 5.4). Performance indicators differ from critical success factors. CSF describe the condition and help determine whether the organisation is doing the right things (effectiveness). A KPI is a critical (key) unit of measurement with which the organisation can measure unambiguously whether that condition is being met. To give a simple example from everyday life: arriving somewhere on time is a CSF, the number of kilometres per hour you drive and the expected time of arrival on the navigation system are the KPIs. Determining the unit of measurement naturally also involves setting a target value or standard, i.e. 100 or 130 kilometres per hour on the motorway, 30 or 50 kilometres per hour in urban areas.

When it comes to lead time, it should be described how much time may elapse between order and delivery. Suppose private customers are mainly single older people who are still living independently. These people generally have a small refrigerator. Once-a-week delivery is unattractive for the customer, as the entire refrigerator will become filled with meals that have a limited shelf life. Maybe delivering once every three days is an option, preferably on scheduled days. In that case, it is useful if the meals are delivered by 5 p.m. at the latest so that customers can put a fresh meal on the table that same day. Insights about customer segments are required to adapt processes to customer's needs. For example, it is well-known that the taste buds of seniors are less sensitive and have a reduced appetite. The supplier will have to take this into account in the recipe and portion size.

TABLE 5.4 Example of CSF and associated performance indicators

		Perspective		
	Financial	**Customer**	**Internal processes**	**Learning & growth**
CSF	Market conformity	Customer satisfaction	Prompt service	Taste and consistency of meals tailored to seniors Knowledge of private market
Performance indicator	Sales price of meal equal to that of the reference group	Score of 8 for customer satisfaction	95% of deliveries within agreed delivery time	Test panel scores a minimum of 7 for recipes and portion size

It should be feasible to generate management reports based on performance indicators. By determining the critical success factors and the associated performance indicators, quick insight can be gained into the performance of the organisation. Is the organisation performing well in every aspect? Are interventions needed? And what is the effect of previous interventions on the current performance?

5.4.4 Methods for embedding performance improvements

There are several ways to anchor the pursuit of performance improvement in an organisation. We describe three conventional methods: Total Quality Management, the Facility Excellence Model and Lean Six Sigma.

Total Quality Management

Total Quality Management (TQM)

Total Quality Management (TQM) is an extensive and structured management approach for continuous quality improvement of products and services. Regardless of the size of the organisation, or whether it is a public, private or third sector, the following guiding principles apply:
- the customer comes first;
- the need to take a long term, stakeholder centric view;
- understanding the cause and effect linkages between why an organisation does something, how it does it, and what it achieves as a consequence of its actions.

The European Foundation for Quality Management (EFQM) has developed a new management model to support the implementation of TQM (figure 5.11). This model is a framework for testing the coherence and alignment of an organisation's ambitions for the future, referenced against its current ways of working and its responses to challenges and pain-points. The EFQM-model structure is based on the simple but powerful logic of asking three questions:
1 *Why* does this organisation exist? What purpose does it fulfil? Why this particular strategy? (Direction).
2 *How* does it intend to deliver on its purpose and its strategy? (Execution).
3 *What* has it achieved to date? *What* does it intend to achieve tomorrow? (Results).

Central to the rationale of the EFQM-model, is the connection between the purpose and strategy of an organisation and how this delivers outstanding results and sustainable value for its most important stakeholders.

FIGUUR 5.11 EFQM-model

5

EFQM has developed a diagnostic tool called RADAR for organisations to diagnose their current strengths and opportunities for improvement. The acronym describes the logic behind the tool. In essence, a high-performance organisation should: (1) determine the *Results* it is aiming to achieve as part of its strategy; (2) have in place several *Approaches* that will deliver the required results, both now and in the future; (3) *Deploy* these approaches appropriately; (4 & 5) *Assess and Refine* the deployed approaches to learn and improve.
These RADAR elements are broken down into a number of attributes. Each attribute is elaborated into what is meant, and what the organisation should be able to demonstrate clearly. For each attribute, a score is given to indicate the level of mastery. The scores can be used for self-assessment (as a tool for organisational development) or for external review (to demonstrate quality control to clients and business partners).

The effectiveness of TQM is evidenced by a large-scale survey of the financial performance of 120 EFQM winners (European TQM Awards). These organisations have been followed over a period of eleven years. They were compared with organisations of approximately the same size and within the same industries. When the first award was presented, the winners usually managed to reduce costs with TQM. Three years later, their share value was on average 36% higher, while their turnover was 17% higher. In the last year of the survey, the performance of the winners turned out to be better again: on average, 77% more turnover and 18% more operating income.

Facility Excellence model

Facility Excellence model

Consultancy firm Berenschot has developed the Facility Excellence model for facility organisations. It is a useful tool for mapping the development phase of the organisation in a relatively short time. The FE-model is based on four different development phases (see Figure 5.12).

FIGURE 5.12 The four development phases according to the Facility Excellence model

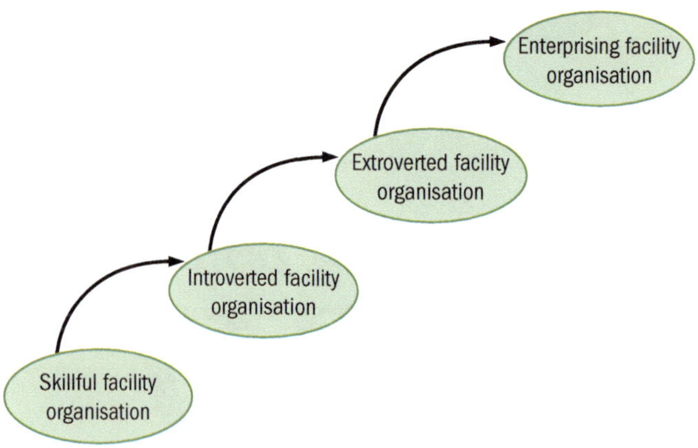

Westerveld, De Zeeuw and Bakker (2000) explain the four phases of development according to the Facility Excellence Model:
Phase I: Skillful facility organisation. The skilled facilities organisation is execution-oriented, versatile in providing services and helpful. The parent company experiences the facility services as self-evident. The facilities department is positioned 'invisibly' in the parent body.
Phase II: Introverted facility organisation. The facilities organisation is focused on control and is supply-driven. It is a well-structured provider of facility services within the parent organisation.
Phase III: Extroverted facility organisation. The facilities organisation is loyal to the parent organisation. The facilities organisation is flexible, which means that it is geared to changes and, in line with developments at the internal client, looks for ways to optimise facilities services.
Phase IV: Enterprising facility organisation. The facilities organisation operates independently, is competative and market-driven. The facilities organisation acts as a partner to its internal and external customers.

Development phase

To determine which development phase the facility company is in, a self-assessment is carried out based on the current situation. This self-evaluation consists of a questionnaire for the facility manager, employees of the facility organisation and the internal client (parent company). Each question has four possible answers, corresponding to the four different phases. The questions are clustered into seven categories:

1 position and role of the facility organisation: the relationship between the facility management and the parent company;
2 management and organisation: how the facilities organisation is structured;
3 financial management: among other things, the budgeting method;
4 personnel management: personnel policy, competencies and training;
5 process management: how processes are managed and improved;
6 customers: contact, relationship and customer approach of the internal and external customers;
7 suppliers: cooperation with suppliers and forms of contract.

An average score can be calculated for each category. The ratings are plotted in a cobweb, showing at a glance how the facilities organisation is structured. Figure 5.13 depicts an example of such a cobweb.

The (inner) blue cobweb represents the current situation, which is how the facilities organisation perceives itself to function. Because the facilities strategy is derived from the organisational strategy, the cobweb needs to fit the ambitions of the parent organisation. In Figure 5.13, the average desired position would, therefore, be phase 2 (red spider's web).
To determine the ambitions of the facilities company for the longer term, the same group of employees answers the questions again with the following question in mind: what are our parent company's expectations for the future? This results in the green spider web. In this example, the facilities company is developing from an introverted to an extroverted facilities organisation. The strengths of the facilities organisation are those that correspond to the desired situation. The aspects where growth is still needed are areas for improvement. In this example, the major differences between the desired and current position are in the field of personnel, processes and customers.

FIGURE 5.13 Example of cobweb representation of average scores

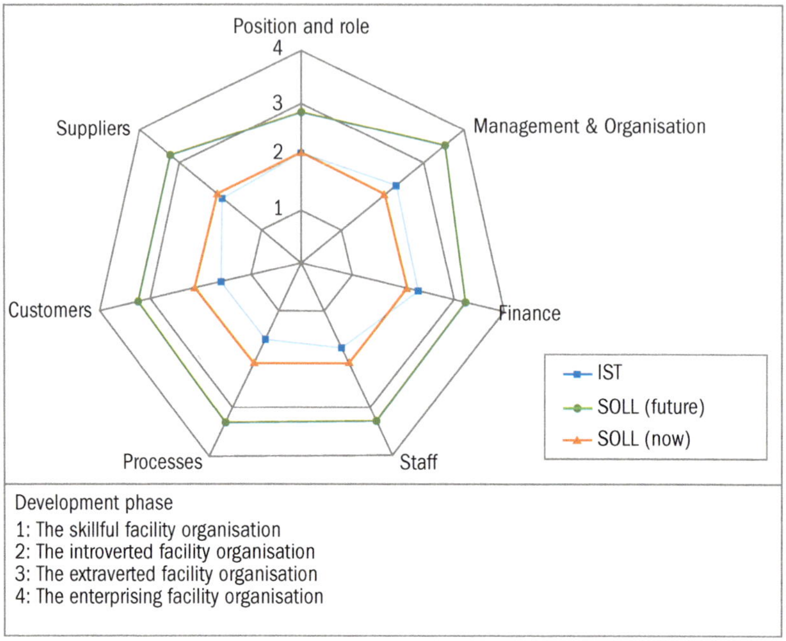

The Facility Excellence model in Figure 5.12 shows a different classification of the development of facilities organisations than Figure 1.9. Yet, the diagrams do not contradict each other at all.

Lean Six Sigma

Lean Six Sigma

Lean Six Sigma is an improvement methodology. It aims to design processes in such a way that the organisation only undertakes those things that customers value (effectiveness), at the lowest possible process costs (efficiency). The difference between Lean Six Sigma and Total Quality Management is that TQM has a theoretical foundation, and Lean Six Sigma originated from practice.

Lean Six Sigma focuses on increasing customer satisfaction and achieving (financial) results for the organisation. Quality is not the end, but a means to improve the operating result. By focusing on what is really important to the customer, and on reducing the number of errors in the execution, process steps are reduced (Lean), and the outcome of processes becomes predictable (Six Sigma). Predictable processes have low variation in process time, and quality delivered. Operations with a considerable variety in time and quality are unstable: things go wrong, and the result may not be what the customer requires.

Predictable processes

Although a process can be carried out properly from a technical point of view, the customer may not be satisfied with the result achieved. An example: An SLA states that a malfunction report to the climate installation must be resolved within 24 hours. If the malfunction is remedied within one day, while it is freezing outside and the customer is stuck in a cold room, the customer will still not be satisfied. Although all parties in the process have done an excellent job, the ultimate goal has not been achieved. The customer is dissatisfied because the starting point was wrong. If the customer's point of view was taken into account, first, the room would be brought back up to temperature by installing radiant heaters or the customer would be offered a similar, but heated, workspace.

The first step in the Lean Six Sigma methodology is to map the present situation, in consultation with the employees involved, as they are often best placed to identify weaknesses and opportunities for improvement in a process. In addition, all actors in this value chain need to be involved. The value chain is an overview of all the steps that are taken, from receiving fault reports to receiving invoices. Particularly at the points where processes transcend departmental boundaries, a considerable improvement can be achieved.

Value chain

In America, Lean Six Sigma is an integral part of the business strategy of almost all top 500 companies. The methodology has developed over the years and is generically applicable to all conceivable processes and in every imaginable type of organisation, albeit it most appeals to production companies. Service organisations more often opt exclusively for a lean-approach. They focus on minimising waste and maximising customer value, but to optimise customer value service processes require a certain degree of flexibility at the expense of predictability. Just like service management, Lean focuses on 'delivering customer value'. Especially organisations with 'operational excellence' as a value proposition, see opportunities for applying lean principles in service management.

There are five starting points for the lean organisation of processes: *value, value stream, flow, pull* and *perfection* (see Figure 5.14).

FIGURE 5.14 The 5 key principles of lean processes

Next, we discuss these five starting points:

Identify value

1 *Identify value*. The value refers to a product or service (or parts thereof) that the customer is willing to pay for. To do this, the organisation first needs to know who the customer is (this can also be internal departments or colleagues) and where this customer derives value from. This sounds logical, but often things are done 'because they have always been like this', 'because the boss wants it', or 'because we think that this is what the customer wants.' Customer value is not exactly the same as customer demand. It's about understanding the underlying need. Successful companies understand their customers' (potential) value so well that they surprise them with a product they hadn't asked for yet, but which turns out to be of great value.

Map the value stream

2 *Map the value stream*. If you know what is of value for the customer, it is important to distinguish which activities within the service process add value to the customer and which do not. All activities (process steps) that do not add value are called waste within Lean. Examples are waiting times, unnecessary administrative actions, sending emails in cc to people who do not need the information, and over-performing. The latter seems to clash with the idea that the organisation should surprise customers with extras so that they become even more satisfied and loyal. However, this is not necessarily so. If the customer values these things (i.e. wants to wait or pay for them), it is not wasteful. A handy tool for gaining insight into this is the Value Stream Map, in which all process steps from triggering the customer to delivering the service or product are mapped out in detail. Subsequently, for each activity, it is determined whether or not it adds value for the customer. Activities needed to keep the business running should be kept to a minimum. Activities that are neither necessary for the customer nor for the functioning of the organisation must be eliminated.

Flow

3 *Create flow*. The purpose of 'flow' is to let the service or product 'flow' to the customer without loss of time. This is only possible if processes are considered in full, instead of as functional (separate) boxes.

'Pull'

4 *Establish a pull*. Pull stands for: do what is needed, when it is needed. It is only useful actually to produce when called upon by the customer. This principle is called 'pull'. The outflow of products is the trigger for the organisation to ensure new inflow, to avoid unnecessary stocks. In service processes, this seems almost self-evident, because no stocks of produced services can be kept. But it is also about continuing to 'produce' without customer demand: for example, paying more attention to a request or preperations than necessary.

Seek perfection

Incremental changes

5 *Seek perfection*. Lean organisations are learning organisations. It aims to perform a little better every day instead of setting up a large project now and then. We call this incremental changes: a small change or addition to what is already there. A useful tool is a Week Start, which makes continual improvement part of the daily routine.

5.5 Financial management

Facility budgets

This section discusses the establishment of the facilities budget, budget control and benchmarking. Facility budgets cannot be isolated from service processes. In section 5.4, we have already discussed the importance of performance management and tools to optimise the quality of operations.

5.5.1 Relation with process management

The optimisation of processes and the deployment of people and resources has consequences for the financial results of the facilities organisation. To be able to carry out a process, the deployment of people and resources is necessary. The design of a process determines the need for people and resources, and thus the cost price of the service or product. Based on this cost price, the selling price can be determined by adding a surcharge for risk and overhead costs. Sometimes, a maximum price is agreed for the service to be performed or the product to be delivered. In such cases, the deployment of people and resources can be determined on the basis of the price-based approach.

The difference between cost price and selling price is the operating result. A facilities provider may be required to achieve a certain return. In practice, this is only the case with commercial providers of facility services and products. In such cases, the yield target is the starting point.

Operating result

Every FM-organisation, whether or not it operates commercially, has to deal with external influences that affect the process itself. For example, a new demand arises in the market or at the internal customer, or the need for a particular service or product disappears. External influences that affect people and resources include, for example, an ageing population and technological developments. Examples of external forces on the cost price are soaring oil prices and falling rents. External influences on the selling price are increased price competition as a result of a surplus of suppliers (see also Chapter 2).

5.5.2 The facilities budget

Budgeting is a means of controlling operational costs. Budgets have to be continuously tested against actuals and adapted to changing circumstances. The disadvantage of budgeting is that it provides (almost) only financial information, which is usually reported afterwards. Depending on the industry, the relative size of the facilities budget differs with the budget for the primary process. In most cases, the facility manager's budget is substantial. In office-based organisations, the facility budget accounts for approximately 25% of all costs incurred by the organisation - the most significant cost item being the salaries of the employees. Hence, it is logical to take a critical look if savings are required.

Budgeting

However, a large part of the facilities budget is spent on long-term (outsourcing) engagements and accommodation costs. Property owned is shown as fixed assets on the balance sheet. The degree of influence ('ability to reduce costs') is, therefore, often overestimated. Saving floor space does not always yield money in an existing property: vacancy cannot simply be solved. The facility manager also has limited influence on the decision to dispose of services; he has the power to make proposals but needs approval from the board. And because services are labour-intensive, saving on labour costs often means that the same work has to be done with fewer people. Eventually, however, this will reach its limits, and there is a danger that facilities services will fall into a downward spiral. It is, therefore, necessary to include the effects of cost savings on the attainment of added value when setting priorities.

The establishment of the budget

Control principle

How the budget of the in-house facility management supplier is established depends on the control principle. This concerns how responsibilities and powers are delegated within organisations. A facility management division may have a profit focus, a not-for-profit focus (profit is allowed, but is not a requirement), or be seen as a cost centre.

Cost centre

If the in-house facility management supplier is considered to be a cost centre, the main job of the facility manager is to provide services at the lowest possible costs. The in-house facility supplier must account for the costs of service provision, but of course, in relation to the requested and delivered quality. At management level budgets are set, usually by reducing last year's budget by a percentage (savings) or increasing it to compensate for inflation or increased demand. Within the financial framework set by the board, the facilities organisation will provide its services to the best of its ability. Still, it will struggle to meet any additional requirements of its internal customers.

Not-for-profit

An in-house supplier of the not-for-profit type charges its costs to the internal customer. However, in general, the internal customer does *not* have the freedom to purchase services elsewhere. Therefore, there are few direct incentives for not-for-profit facility departments to optimise matching the supply to the demand for services.

Profit centre

A supplier of the profit centre type is judged on the cost, quality and the results achieved. A workplace can, for example, be fitted out more luxuriously if this has a proven positive effect on the productivity of employees. Business cases include (in)direct contributions to an organisation's performance as revenues.

Not infrequently, the principle of free market is applied in order to keep the in-house FM-supplier on its toes. As the internal client is not obliged to do business with the in-house supplier, it will try to get as much value for money as possible. If the in-house supplier overcharges its internal client, the client will turn to external competitors. This situation is an enormous incentive for the in-house facility management department to operate market-conformally — also, for professionalising the client-supplier relationship.

If internal customers are (partially) free to source services from suppliers other than their own facility management organisation, the facility budget is often determined in consultation between facility management and the other organisational units. Although there is still a facility budget, this falls under the joint responsibility of the organisational unit and the facilities management department.

Zero-base budgeting

Zero-base budgeting

A rather thorough way of establishing budgets is zero-base budgeting. The name says it all, zero-base budgeting indicates that every year the budget is built up from scratch based on agreed services provisions for the coming period. Zero-base budgeting is based on the principle that your budget is essentially zero, but you have to prove that you need a certain amount of money. As demand may fluctuate, facility departments that work in this way must be capable of absorbing changes in demand at an adequate pace. This can be achieved by contracting temporary staff or by outsourcing services. In addition, it is important to ensure that the organisation builds up sufficient financial reserves to be able to temporarily absorb declining revenues. In rapidly changing circumstances zero-base budgeting is preferable, in more stable situations zero-base budgeting can be considered once every five years. In theory, this type of facility management organisation could develop into independent commercial providers, but in reality, this is uncommon.

Interdepartmental charging

Facility departments of the not-for-profit and profit centre type act as in-house suppliers and accordingly charge their clients for the services they provide. There are three ways to charge facility costs to users:

1 based on a fixed rate (e.g. per square metre, employee, patient or workplace);
2 based on actual costs (the actual unit costs are often difficult to determine, and this method requires a considerable administrative effort);
3 based on a fixed rate with variable surcharges (a combination of 1 and 2).

With the increasing influence of internal customers on the range of facility products and services, the need arises for an interdepartmental charging method that takes account of the differences between internal customers. In a situation where some departments use resources very economically and others opt for luxury, it is not considered fair to charge services without distinction. Still, the organisation needs to maintain a certain basic level of quality of building services, such as security and building maintenance. Furthermore, the physical environment is the organisation's calling card and it is therefore undesirable for the image if departments differ from one another too much.

Interdepartmental charging

The combination of a fixed rate for basic facilities such as maintenance, cleaning, reception and security and a surcharge for additional services or a higher service level is, therefore, a commonly used method. Costs that are beyond the control of an internal customer include, for example, the cleaning of central facilities (reception hall and circulation areas) and the security of the building and grounds. These costs are distributed according to a fixed basic rate, in proportion to the number of square metres used or the number of FTEs employed by a department. Surcharges are applied for costs the customer can control through the choices he makes. These variable surcharges correspond (where possible) to the actual costs.

Basic facilities

Additional services

A simple example of this is vending machine catering. A standard service level is a coffee machine with a basic range of coffee, cappuccino and tea. For this provision, each internal customer pays the standard rate. However, it is possible to enjoy a more luxurious selection by adding Wiener melange, chocolate milk and soup for a surcharge on the standard rate. These options are listed in the products and services catalogue. Agreements on additional services are included in SLAs, with variable surcharges.

Project budgets

The dynamic changes discussed in Chapter 2 also affect the dynamics of a project manager's work. In response to this, forms of agile working are prevalent. Agile is a collective term for iterative ways of working that offer

Agile working

maximum scope for adaptation based on progressive insight. The starting point of agile working is that dynamics and change are part of the project and should be managed within the budgetary frameworks; in contrast to avoiding changes through detailed prescriptions.

Scrum

A specific form of agile project management is Scrum. This is a dynamic project approach in which customer and supplier collaborate in teams, delineating time and budget. A fundamental principle of Scrum is the dual recognition that customers will change their minds about what they want or need and that there will be unpredictable challenges. In this context, a predictive or planned approach is unsuitable. As such, the Scrum method focuses on how to maximise the team's ability to deliver quickly, to respond to emerging requirements, and to adapt to evolving technologies and changes in market conditions. In this type of project, sprints of two to three weeks are typically used. The capacity for each sprint is determined in hours (this is the budget). The tasks within the sprint are classified by task severity and are prioritised by the team. Scrum aims to increase productivity within the available budget of hours by becoming more efficient as a team. There is plenty of room for progressive insight, which should lead to more quality.

Suppliers' budget

Commercial providers

The above is described from the perspective of an in-house facilities department. Now let's look at the perspective of commercial providers and consultants.

For the delivery of their services, providers and consultants draw up a quotation in which they translate the expected deployment of people and resources to the price for which they offer their services. The hourly rates with which they charge include direct and indirect labour costs plus a surcharge for risk and profit margin. With stiff competition and price as a decisive factor, providers will take into account what competitors do. The strategic importance of obtaining a contract also plays a role in the final quotation. If the market is weak, providers will accept a lower profit margin if this gives them the security of work. If a provider has more orders than capacity, it will set a higher profit margin. Some projects are attractive because they can be used for marketing and branding purposes; these are clients of name and impressive projects. Winning such a project can have side effects that make a smaller margin acceptable.

Surcharge

Fixed price

If a contract is awarded for a fixed price, the provider or consultant enters into an obligation to achieve a specific result. If more human resources or material is required to reach the agreed result than foreseen, the additional costs are for the expense of the provider or consultant (hence the risk premium).

However, not all consumption can be accurately predicted upfront. In these situations, providers will only agree to a fixed price if it includes sufficient coverage for the commercial risk they take. This is not attractive for clients, as it brings the risk of being overcharged. Instead, an estimated calculation is based on agreed-upon prices per unit, such as the cost of changing towel rolls per unit, façade cleaning per square metre and the hourly rate for hiring services (e.g. consultancy). Periodically, the actual consumption is invoiced. This is a contract based on recalculation of costs. The risk of exceeding budgets then lies with the client.

Recalculation of costs

5.5.3 Planning and control

The planning-and-control-cycle consists of two elements, where *planning* stands for preparation, coordination and process management, and *control* for auditing and adjustment.

Planning-and-control-cycle

Planning is about making and establishing agreements about how, when, and by whom activities will be carried out; this is where the implementation of the policy takes shape. Planning also includes coordination, in terms of time and the coordination of activities. Planning is necessary at all levels of the organisation. Strategic planning has a long-term scope, tactical planning a medium-term focus, and operational planning focusses on daily operations, the short term.

Planning is the first step towards controlling business processes. Management control is a collection of procedures and techniques with which the organisation tries to achieve an effective and efficient deployment of people and resources. The execution of the activities must continuously be reviewed against the budgeted planning and objectives. If everything goes according to plan, no intervention is required. However, if a difference between intention and practice occurs, the management must take action.

5

If management solely relies on financial data, action may be taken too late. After all, financial results follow from underlying outcomes, such as customer satisfaction, occupancy rate and the number of complaints. For this reason, attention should also be paid to operational performance and results that cannot immediately be translated financially but are nevertheless essential (see section 5.3).
These performances are expressed in:
- cost compared to the budget;
- customer satisfaction;
- contract compliance by suppliers and sub-departments of the facilities division.

Figure 5.15 shows the planning and control steps in the PDCA-cycle for all three organisational levels (blue). Here you can see that the planning (green) at a tactical and operational level is derived from the strategic framework (top-down). The control cycle (red) works bottom-up; from the lowest to the highest level information and figures are increasingly aggregated and combined (see Figure 5.2).

PDCA-cycle

In general, the planning-and-control-cycle consists of at least the following activities:
- drawing up an annual facility plan with the corresponding budget;
- the preparation of monthly or quarterly reports with an explanation of differences between the result and the standard or prognoses;
- preparation of an annual report that includes an account of the result for the past financial year.

For the facilities policy (and therefore budget) to fit in with the plans for the primary process, these must be known. That is why the facilities policy plan is often only approved after the policy plans and budget requests from the primary process have been determined.

Budget control

Although budget control is a financial-economic function, it is briefly described here since it is important to have insight into how the supervision is arranged. The General Board appoints a facilities director. He or she is ultimately responsible for the facility services, has the authority to take measures and to make investments for the benefit of the intended objectives (within the budgetary framework). He or she must have regular meetings - for example, once a month - with the General Board to account for the results of the facilities management division. A person with ultimate responsibility should not be allowed to monitor himself, to prevent misuse of powers. Controllers, therefore, supervise the financial management of the facilities organisation. A controller is a financial expert. S/he ensures that all (mainly financial) information requested by the general manager, is compiled. This information mainly provides insight into the extent to which the facilities organisation meets the agreed performance.

Controller

FIGURE 5.15 Planning and control in the PDCA-cycle

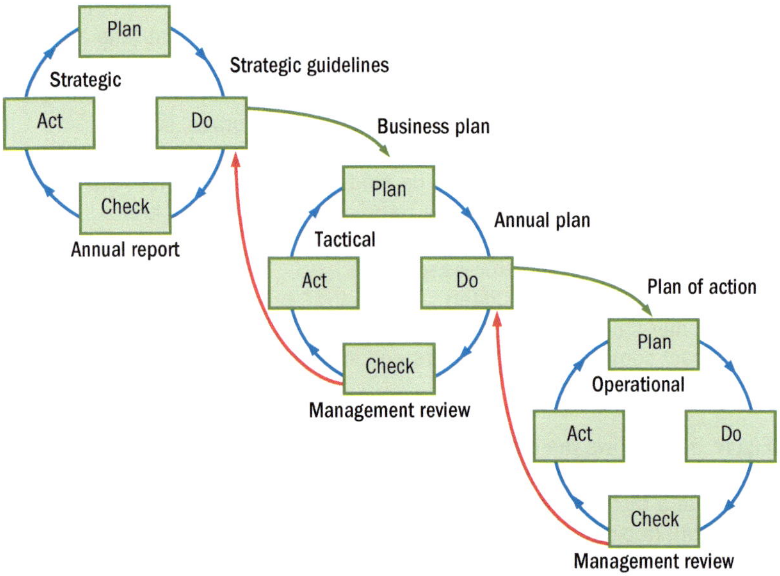

5.5.4 Benchmarking

As will have become clear by now, the facilities field is extensive. As a result, facility costs are also very diverse in nature. In part, these costs are difficult to adjust because they are linked to investment decisions made in the past. Or because they are the result of a long-term lease or rental contract (constant costs). However, a significant part of the facility costs can be influenced (variable costs). It is essential to know what the savings potential is because having fewer costs can contribute to a larger profit or at least to the financial health of an organisation.

Savings potential

If the board or the facility manager wants to determine how well the facility organisation is doing in terms of financial performance, it is not enough merely to compare the actual costs incurred to the budget. This comparison does not give any insight into the competitive position of the in-house facility management organisation. Comparing one's performance with that of similar organisations provides insight into the potential for optimisation.

Competitive position

Compare it with the grade you got for an exam. Suppose you score a grade six (out of 10). One could say that you stayed 'within budget'; after all, you passed the exam. But how well did you do? If you belong to the 10% of students who passed the exam at the first attempt, one can conclude that you are a good student. However, if 80% of your fellow students achieved a higher grade, the outcome is the opposite. And should you want to improve your performance next time, it is better to ask advice from a more qualified student than from a student with lower grades than yourself. However, this does not mean that you should necessarily measure yourself against only the *best* student in the class, because if this aim is out of reach, it may result in demotivation.

The same goes for organisations. This mutual comparison is referred to as benchmarking and is aimed at gaining insight into one's performance. Benchmarking is a systematic process to optimise performance by comparing one's business functions, procedures or working methods with those of comparable companies. A benchmark may reveal that an organisation performs in line with the market, and the quality is optimal given the costs. In this context, benchmarking is the comparison of facility services with the aim of gaining insight into the quality of one's own service.

Benchmarking

- Benchmarking is all about *comparison*. It shows differences with similar organisations in the same industry.
- Benchmarking is about *insight*. It raises questions. For example, why a process has been carried out in a certain way for ten years, after which it turns out that it can be done much more efficiently with new techniques.
- Benchmarking is all about *learning and growth*. It stimulates improved performances, as the comparison often evokes a healthy competitive atmosphere.

Benchmarking only makes sense when the comparison is fair. There is a tendency to only compare on costs, but that is a limited approach. Suppose two friends have each bought a used car and want to assess who made the best purchase. If they only include the purchase prices in their comparison, this can never lead to a proper conclusion. If one paid 10,000 euros and the other 20,000 euros, they only know that the first car was cheaper than the second - nothing else. To determine who made the best purchase, they also have to take into account the age of the vehicles, the make, the state of maintenance, fuel consumption and market value.

It is possible to compare more aspects than just costs, e.g. service level and satisfaction. However, these must then be clearly defined for a fair comparison. Non-measurable or quantifiable quantities cannot be objectively compared. The NEN-EN 15221-7 lists different types of benchmarking, see Figure 5.16.

FIGURE 5.16 Classification of benchmarking according to the EN 15221-7

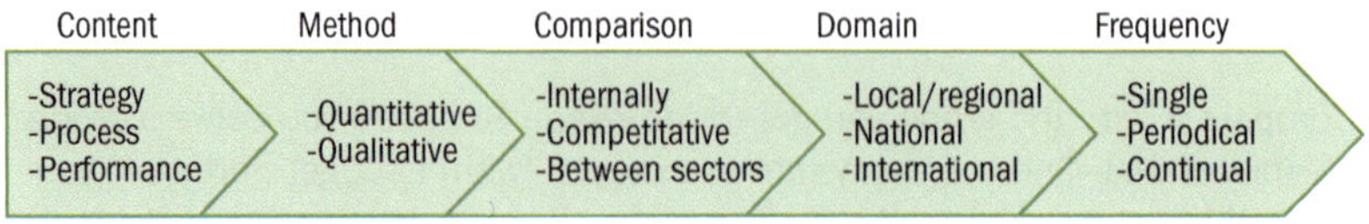

Factors

For effective benchmarking of facility management, the following factors should be included in the analysis: industry or branch; cost influencing factors; customer satisfaction; service level; costs.

Using the diagrams in Figures 5.17 up to and including 5.20, we explain hereinafter why these aspects are so important when benchmarking.

Figure 5.17 shows the operating costs of five buildings (A up to and including E). You immediately see that building A is the most expensive and building E the cheapest. So, one could say that building E 'performs' best. But does it?

FIGURE 5.17 Benchmarking on costs

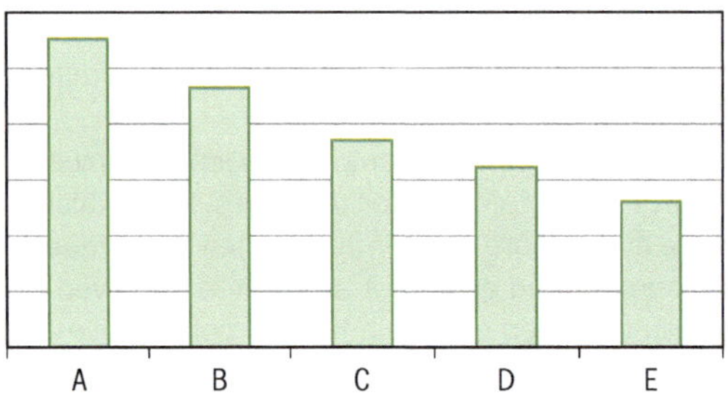

Figure 5.18 provides additional information. It includes the average values (in this case, costs) of all relevant comparisons. Bars indicate the bandwidth of buildings with the same cost-influencing factors. This bandwidth has been established based on the two middle quartiles (a statistical term). A reliable database with a sufficient number of buildings is needed to obtain reliable results. The operating costs of all buildings have been entered into the database. Then the average is calculated, without taking into account the most expensive 25% and the cheapest 25% of the buildings. This means that an average is calculated over 50% of the buildings entered. Of the remaining buildings, the bars indicate the most expensive (max), the average (av) and the cheapest (min).

FIGURE 5.18 Benchmarking on costs and cost drivers

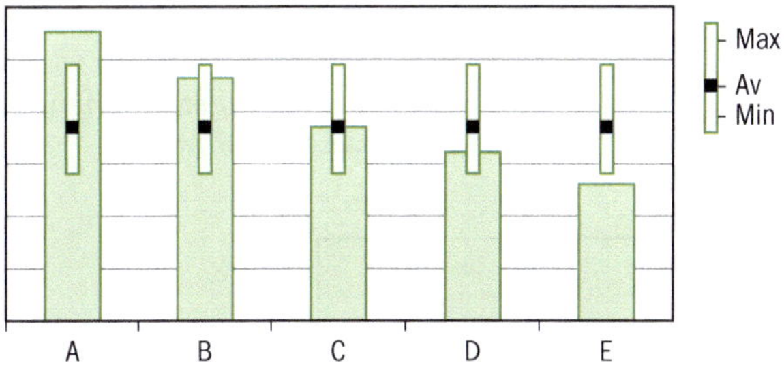

The advantage of this statistical calculation is that it provides more information at a glance. First of all, it is striking that A is among the most expensive 25% of buildings; in other words, it has far too high operating costs. Such an outlier requires an explanation. But there is also something special about building E because it is among the cheapest 25% of buildings which also raises questions. Are some costs excluded from the operational costs? Or is it a miscalculation? Figure 5.18 shows that building D appears to perform the best because it has the lowest costs within the bandwidth. But buildings B and C are not doing badly either.

However, to get a more thorough understanding, we need to make things more sophisticated. Figure 5.19 shows the catering costs for buildings A up to and including E. Buildings A up to and including E appear to have three different service levels for catering. In the lowest level, only lunch is provided, at the medium level, lunch and breakfast are provided, and the highest quality level includes breakfast, lunch, and dinner. The catering costs of buildings B and D appear to be far too high. Although A has the highest costs and E is the cheapest, they nonetheless perform equally well.

FIGURE 5.19 Multi-bandwidth benchmarking

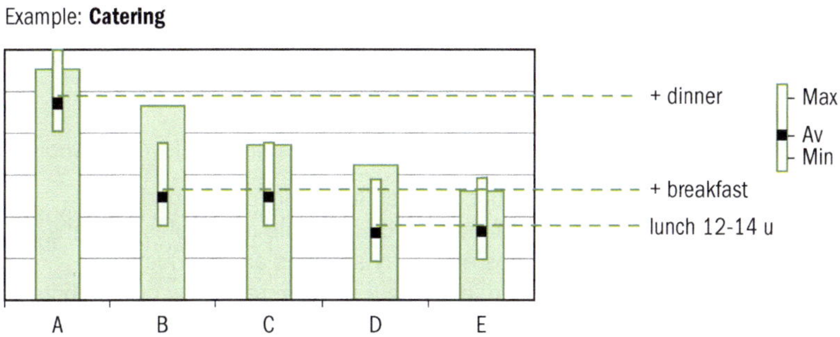

But we're not finished yet. After all, we have seen that a satisfied customer is essential for business success. That is why customer satisfaction is added to Figure 5.20. In building B, we find the highest customer satisfaction and the costs are still within the acceptable bandwidth. In building E, the costs also fall within the bandwidth, but customer satisfaction is lowest. Building A also scores well, and building D has the worst performance.

FIGURE 5.20 Benchmarking on costs, cost drivers and expected satisfaction

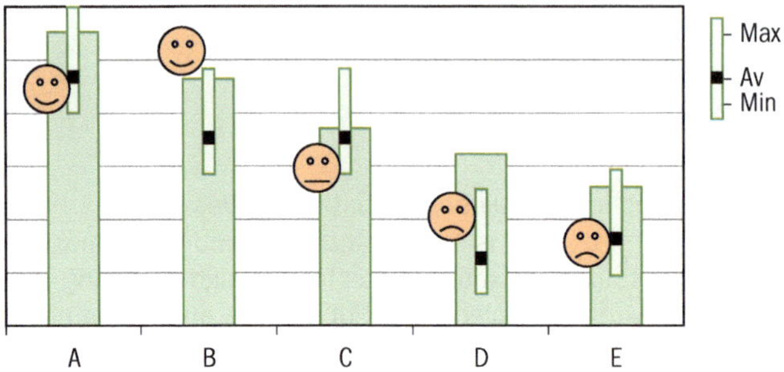

External benchmarking

Benchmarking can take place both externally and internally. External benchmarking examines best practices outside the organisation to learn from peers.

Internal benchmarking

Internal benchmarking is about learning from business units or departments within an organisation. If an organisation manages multiple premises, the energy costs of those premises can be compared, for example, or the maintenance costs, the occupancy rate of the company restaurant or customer satisfaction with the reception. It is also possible within a department to compare performance over subsequent years (trend). For example, you can look at the development of facility costs per workplace over the years.

Facility metrics

Examples of frequently used benchmarking facility metrics are:
- total facility costs per employee;
- accommodation costs per square meter GLA;
- number of square meters GLA per workplace;
- cleaning costs per square meter GLA;
- energy consumption per square metre GLA;
- catering costs per employee.

A facility index becomes more useful when it is placed in a model or system, ensuring that the information is comparable (no apples to oranges) and robust. EN 15221-4 defines all facility services and products. It standardises their classification, allowing a uniform recording of the costs of facilities and activities. For benchmarking purposes, it is useful to make a comparison with companies in the same industry. Many circumstances that affect operations are likely to be on par, e.g. operating and response times, the functionality of premises and characteristics of the location.

Participating in a benchmark over several years with the same organisations provides the opportunity to improve facility performance jointly. An international example of a long-term (facilities) benchmark is the Leesman Index. This is a worldwide benchmark that measures the design impact of workplaces on an employee's overall experience; from their perception of personal productivity and level of enjoyment to their pride in their workspace. As the largest independent research project of its kind, the database now encompasses findings from more than 600,000 employees worldwide, occupying 4,263 buildings in 93 countries. A European FM-benchmark is the Occupier Cost Index (OCI) published by Colliers annually. It compares the workplace costs per FTE of most European countries based on EN 15221-4. The Colliers International benchmark in 2018 included 3,900 buildings from 28 countries, with a total floor area of 25.2 million square metres. IFMA's Research and Benchmarking Institute (RBI) annually performs the Operations and Maintenance Benchmarking Survey. The 2018 report is based on the responses of 2,000 facility managers located throughout the United States and Canada, representing 98,000 buildings in 35 industries.
Many consultancy firms provide research and benchmarking services, such as Gartner, Verdantix, Measuremen, the FM Research and Benchmarking Institute, to name a few.

When benchmarking, one must avoid the following pitfalls:

- *Garbage in = garbage out.* If poorly classified or incomplete data is provided, a benchmark fails to give a reliable result. Careful handling of the data collection by all companies under comparison, however, ensures that the result is valid. By applying a standard such as EN 15221-4, consensus can be reached on definitions and classification. However, this is still no guarantee of quality because the costs incurred by an organisation are not accounted for according to the EN classification. In the conversion of financial data from the general ledgers mistakes and errors may occur;
- *Without a performance level, comparisons are false.* A cost figure without a notion of the service level has little value (see Figures 5.17 up to and until 5.20);
- *Uncertainty about the right peer group.* A correct peer group (e.g. from the same industry) has characteristics that match the organisation, the nature of the process, the type of accommodation and the quality of the services provided. This exceeds the total surface area and the number of workplaces. For example, steep security costs in the banking industry are not a topic of discussion, because security is paramount;
- *Jumping to conclusions.* A benchmark has a signalling function. Outcomes of benchmarks should, therefore, be treated as a signal. Further research is always needed to arrive at appropriate conclusions.

Young Professionals in Facility Management

Brigitte van Gompel, Tender Manager at Heijmans Utility, Rosmalen, The Netherlands - graduated bachelor International Real Estate and Facilities Management at BUAS in 2016, and graduated Master of Science in Facility and Real Estate Management at Saxion University of Applied Sciences/ University of Greenwich in 2017.

It's not the beer, but the overall experience that counts

As the daughter of a hospitality entrepreneur, hospitality is in the blood. As a child, I knew for sure: when I grew up, I would make people happy with snacks and drinks. But instead of studying Hotel Management, I chose Facility Management. What appealed to me was that in this field, the focus is not only on hospitality but on the quest to understand the factors that drive customer satisfaction. During my training at the BUAS, I studied how the environment can influence customers, end-users or visitors. As I also experienced during my traineeship in Australia, 'hard FM' cannot do without the human touch. During my master's I further studied the interplay of building, design, organisation and people; which results in the added value of facility management!

It's not so much the daily organisation of facility services that appeal to me, but the improvement and devising of a customised concept. This came together in the traineeship Heijmans offered me. Heijmans is not a traditional facilities provider but a construction and installation company. In 2018 I started in the role of tender manager. In this role, I work with a team to provide the best possible response to tenders for large and highly complicated maintenance contracts. The technology and the structural condition of the building have a significant influence on the functioning and experience of users. For example, the climate in which a student can best concentrate, a professor can best do research, or that helps you relax at an airport.

Technology is the basis of all our tenders. My role is to make a difference by the added value Heijmans provides in addition to technology. I devote a large part of my work to customer analyses, industry-analyses, site visits and dialogues, for identifying the underlying issues.

The winning answer always is a combination of excellent knowledge and management of technology, a well-designed (user-friendly) building, the right team of mechanics and managers, attention to pleasant cooperation, a surprise effect, and an eye for the client's ambitions. This is how you make the difference as a tender manager with a facilities background!

Summary

- A business model provides a blueprint for the implementation of the strategy in the company's organisational structures, systems and processes.

- Performance management is the governance of the organisation by systematically defining an organisation's mission, strategy and objectives; implementing these in a structured manner; and making them measurable using critical success factors and performance indicators.

- Measuring facility performance is important. The performance of the facility organisation can only be assessed if it is measurable. By measuring the performance, insight is gained into the strengths and weaknesses. A facility manager can thereby demonstrate that s/he adds value to the organisation and is in control.

- Management information is required to steer processes. Performance indicators (key performance indicators) are indicators through which it can be monitored if service delivery is following the terms of the agreements.

- In day-to-day practice, the following leading dashboard indicators are used: customer satisfaction, service level, and costs. Information about customer satisfaction can be obtained from customer satisfaction surveys or customer panels, but also the complaint registration. Information about the service level can be obtained by a facility manager by carrying out a supplier assessment (in the case of outsourced services), by using a mystery guest and by recording deviations from established standards such as lead times. The facility manager can obtain information about costs from the financial reports.

- It is necessary to express the added value of facility management in performance indicators. A multitude of indicators is possible. The choice depends on relevance and alignment to organisational objectives. For (almost) every facility manager, productivity, image, sustainability and agility are essential performance areas.

- Frequently used methods to anchor the striving for performance improvement in an organisation are: Total Quality Management (EFQM model), Facility Excellence Model and Lean Six Sigma.

- The optimisation of processes has consequences for the financial results of the facilities organisation through the deployment of people and resources. Financials cannot be viewed separately from service processes.

- The facility budget can be established in various ways, depending on the governance principle. It can be set by management (cost centre type) or (partially) by negotiations (not-for-profit and profit centre).

- Management control is a collection of procedures and techniques with which the organisation attempts to achieve effective and efficient deployment of people and resources. Budgeting is a means of exercising control over activities.

- In general, the planning and control cycle consists of preparing a facilities policy plan with the corresponding budget proposal, monthly reports, quarterly reports and an annual review.

- Benchmarking is all about comparison, understanding and learning and growth. It does not make sense to look only at costs. Cost influencing factors, service level and customer satisfaction, must also be taken into account for a meaningful comparison. Benchmarking has a signal function: the underlying causes of deviations from the peer group always need to be investigated.

Questions and Assignments

5.1 Look up the Occupier Cost Index on the site www2.colliers.com, choose the most recent year and describe in your own words what facility costs this index covers.

5.2 "To measure is to know" is a well-known expression within the FM-industry. What is meant by this, and how can you best do this?

5.3 Regular measurement of customer satisfaction is essential. However, there are several ways to do this. For example, through a survey, customer panel or by inviting users to post a review. Which method do you prefer and why?

5.4 Facility managers can also take health-promoting measures. In healthcare, this is often referred to as a "healing environment". Compare an academic hospital with a local hospital: which hospital do you think applies the principles of "healing environment" best and why?

5.5 A director of a large transport company approaches his facility manager with the following story. During a party, he had a conversation with an acquaintance who owns a production company. While they were talking, they came up with the costs of security. The transport company pays twice as much for security as the production company. Both are located in an industrial park. The director asks the facility manager to explain the difference — mention four possible causes.

5.6 As of January 1st, 2023, every office in the Netherlands with a surface area of more than 100 m2 must have a C energy label as a minimum. This implies an energy index of 1.3 or better. If a building does not meet the standard, it may no longer be used as an office. Mention five measures that you, as a facility manager, can take to comply with this requirement.

5.7 As a student, you are a beneficiary client of your school's facilities department. What are the success factors for student satisfaction with classroom furniture, the caterers' sandwiches, the furnishing of the library, the WIFI network and the bicycle shed?

5.8 You are a catering manager at Sodexo and take care of lunch and banqueting at a large insurance company. Over the last six months, you have noticed that the number of banqueting activities has decreased considerably. Still, the rinsing kitchen is busy as ever: glassware, cake plates and cutlery are often borrowed and returned for cleaning. You find out that the facility manager has decided to abolish the compulsory shopping system, causing many departments to use websites like ubereats.com to deliver their drinks or lunch. You request a meeting with the facility manager: what is your first question?

5.9 Explain why 'interdepartmental charging of facility costs' often leads to heated discussion.

5.10 For each phase of the Deming-cycle (PDCA), describe two activities with concerning your studies for this academic year in terms of the PDCA-cycle.

6 The future of the Facility Management profession

In this chapter, we explore the future of the facility management profession. The main questions that are being addressed are:

- What are the effects of social developments on how companies operate and how people work?
- What opportunities does the networked economy offer for facility professionals?
- What do we currently see in the field?
- What challenges will the future bring?
- What is the influence of the type of network on the role of facility management?
- What should one be capable of in the role of area manager, community manager and community builder?
- What are the similarities and differences between these roles?
- What are the changes every facility professional will have to deal with?
- How useful is scenario planning?

◀ Through the eyes of the facility manager:
'Working outside the accommodation must also be facilitated. In this case, through information technology.'

Expert Column

6

The future of work, the workplace and the work experience: options, options and more options

What might the future of work be? Considering the ongoing evolution and growing impact of technology (ML, AI, IoT, big data), the acceleration of the pace of change, growing uncertainty all over the world, the intensifying battle for top talent and other factors, the future of work will very likely be different from what it is today and, for sure, much hard(er) to predict. And where and when might that work be conducted in the future? The number of workplace options will most likely increase. From the traditional office or for some the home office to now also co-working spaces, food and beverage venues, public places to literally wherever and whenever in the future as technology will enable us. Combine this with a rapidly diversifying global workforce with different needs and wants, and you can see that the regular nine to five Monday - Friday workdays in a traditional office are numbered.

So, what to prepare for? Flexibility and agility to have the ability to offer customizable outcome-enabling workplace services. A key responsibility for facility managers in the future will be to optimise and constantly rebalance a flexible portfolio of agile workspaces and options, with a customised offering of workplace services and experiences that enable individuals and teams to be successful under ever-changing circumstances. It will be about the experience, a sense of belonging, flexibility and personalisation as employees will want it their way, right then and there.

And with employees getting the opportunity to work much more flexibly (where- and whenever), what might bring them together? What will be pulling them into an 'office' in the future? Enhanced workplace amenities such as exciting food at work, onsite sport and fitness facilities or event venues will become much more important. The human-longing for authentic, meaningful personal connections in an ever-connected world will only increase. It will be the role of the facility manager in the future to use all facilities and services to bring employees together and enable them to thrive, where and whenever they will want to work. Exciting times are ahead of us.

Michiel Bakker
Director, Global Workplace Programs at Google, Mountain View, California, USA

6.1 Introduction

In Chapter 2, we outlined the importance of always being well-informed about social and technological developments, as these can have a significant impact on the business context and on the facilities discipline itself. These developments occur at an enormous pace, while the working environment can be considered as a rather static whole. Aligning these diverse dynamics is an exciting challenge for facility professionals!

As the future starts today, it is essential for you as a student to develop an antenna for upcoming developments that will determine the facilities playing field in future. In this chapter, we discuss the progress of facility management and possible scenarios for how the facility profession can develop.

The structure of this chapter is based on the scenario planning method (see Chapter 2). First, we will select two driving forces to be positioned on two axes (section 6.2). The four quadrants around these axes represent four scenarios, each of which we discuss in detail (section 6.3). It is not so much the question which scenario is most likely. All four scenarios can occur simultaneously, depending on the industry, the nature of an organisation, and even within industries. The relevant question is what opportunities these scenarios may entail for the facility management industry. That is why we explore each of the four scenarios in section 6.4 et seq., each time addressing the following questions:

- What do we currently see in the field?
- What is the role of facility professionals in this scenario?
- What qualities should a facility professional have in this role?
- What are the crucial issues for the future?

In the last section, we will consider the general outlook for the future of in-house facilities management.

6.2 The selection of the driving forces

As explained in Chapter 1, the structure of this book is based on a cone-shaped model of which the basis is formed by the organisation, with people on top (see Figure 1.14a). The relationship between people and organisation, with facility management as the connecting factor, is what this book is all about.

In our time, networks are what factories were to the industrial revolution. Every working person has a relationship with one or more organisations - whether that organisation is small or large. Networks provide access to resources, such as knowledge, new assignments, new contacts, money or (shared) equipment. Without these resources, success is out of the question.

At the time of the industrial revolution, people started to work in ever more extensive, stable relationships. The small-scale activity was increasingly being replaced by factories and large-scale organisations in large buildings. In the network economy, we see the opposite direction: large organisations withdraw to the core qualities and become smaller. At the same time, highly flexible networks of small, dispersed units are emerging: the independent professionals. The nature of the relationship between people, between people and organisations and between organisations, determines the future roles of facility professionals.

Network

The connections between them form a network. A network can have an open, partially open or closed character. *Open* means that the network (of people or organisations) is accessible to everyone willing to participate. An example of an open network is a community centre. Anyone in need of contact or information can drop in for a cup of coffee and a chat, even if they live in a different neighbourhood. An *exclusive* network is a network of which someone has to be a member to participate. If you work for an organisation, you are a member of the organisation, and you collaboratively work with other members (colleagues) to achieve organisational goals. Organisations can also be part of a network; for example, through the Chamber of Commerce (open network, because accessible for every company) or in partnerships (private, exclusive relationship with mutual goals and agreements laid down in a contract).
For our first driving force, we select *the relationship between people and organisations - from open to exclusive networks.*

In considering the future - apart from the characteristics of networks – we want to include the nature of organisations. Many factors characterise an organisation, such as culture, market orientation (business-to-business, business-to-consumer), profit motive, social relevance and introverted or extroverted. Another characteristic is the degree to which people within the organisation collaborate. Large organisations have the disadvantage that they are rather cumbersome. However, in these changing times, agility is a prerequisite for businesses to survive. One way in which companies try to increase their agility is by contracting fewer permanent staff. This explains the demand for temporary staff, which can be hired on a project basis, for example. One consequence of this is the emergence of independent professionals (see section 2.5), at the cost of loss of loyalty for organisations and income uncertainty for self-employed workers.

Collaborate

The rapid increase in the number of independent professionals and the emergence of the networked economy is turning the character of collaboration from predominantly permanent and long-term to a more flexible and temporary one. That is why we take as our second driving force: *the nature of cooperation, the scale of which can vary from a collaboration between individuals to collaboration in large-scale collectives.*

Figure 6.1 depicts the matrix thus created. On the vertical axis is the relationship - with collaborative organisations above and confined organisations below. The horizontal axis represents collaboration - within the collective on the left and among individuals on the right.

In short, in the matrix (from top left in a clockwise direction) we distinguish between the collaborative organisation; the open community; the private community and the confined organisation. We note that this matrix always refers to a physical business environment. This physical environment may be a site, building or even a space in a building. Networks of independent professionals are often called communities. For the sake of clarity, we use the term communities when referring to networks of individuals (right side of the matrix), and the term networks when relating to the collaboration of collectives (left side of the model).

FIGURE 6.1 Scenario matrix for the future of facility management

6.3 The four scenarios

In this way, four scenarios emerge, which represent different forms of cooperation. By naming the four scenarios, one gains a better understanding of what is intended and can communicate about it more easily with others. It's even more effective if those names are derived from metaphors because that promotes a more creative way of thinking without biases. In this outlook, we choose the sports world as metaphor; it enables us to make an appealing comparison between the role of facility professionals and professionals in sports. The upper two quadrants of the matrix are mainly about meeting places. This is where organisations (left) and individuals (right) come together to interact, motivate and collaborate. Both righthand quadrants are about individual sports (independent professionals), while both lefthand quadrants are about team sports. And in the two lower quadrants you will find the private sports associations, whether or not open to other athletes.

Scenarios

Metaphor

Clockwise, starting in the upper left quadrant, we will name it:
- sports complex - open accommodation for various team sports and supporters;
- fitness centre - accessible accommodation for individual athletes;
- golf club - private accommodation for individuals, where others may play golf by invitation;
- football club - restricted accommodation only for members and visiting clubs in organised competitions.

A sports complex is at the disposal of several sports clubs, often of various sports disciplines, each of which has its autonomy. One can identify members of each collective by the club clothes they wear. A fitness centre

Sports complex

Fitness centre

is easily accessible to individuals. Someone has to be a member, but there are no specific admission requirements, and often it is also possible to acquire a day-pass. A golf club has permanent members; other golfers are welcome, provided they are registered first and that they have required qualifications. And finally, the football club, hosting teams and their supporters (as guests), but where, apart from their members and supporters, no other athletes are expected. Here, too, members of a club can be recognised through the outfit worn.

Golf club

Football club

Multiple quadrants

Incidentally, organisations may belong to multiple quadrants at the same time. Next, an example of such a mixed form is described.

On the campus of Wageningen University, one will find collaborative organisations *(the campus)*, private communities *(incubators and business centre situated on the campus)* and open communities *(student challenges)*. The university in itself is a mix of a confined organisation *(the university staff)*, a private community *(students)*, an open community *(visitors and temporary individual users)* and an open organisation *(located on campus, as a collaborative partner for related businesses)*.

One of the buildings on campus is the Plus Ultra building developed and operated by Kadans Science Partner. Plus Ultra offers space to starting, growing and established companies in the world of agro, food, bio-based, healthy lifestyle and living environment (business centre - example of a private community). These are the research areas of Wageningen University *(a confined organisation with characteristics of all other manifestations)*. With a diverse range of offices, laboratories, experiment halls, meeting rooms and meeting places, Plus Ultra facilitates the establishment of a community between all users of Wageningen Campus *(example of* a collaborative *organisation)*.

The need for extra space soon arose because many companies wanted to be situated on campus. Also, Wageningen University required a permanent space for the student incubator *(example of private community - incubator)*, and for the hosting of the future student challenges *(example open*

6

community - highly variable group of participants) in and around the Plus Ultra community. To accommodate all these start-ups and challenges a long-term lease has been signed with the university.
Kadans Science Partner then took the initiative to develop a new business centre, Plus Ultra II, next to and connected to the Plus Ultra building. This new building, with a surface area of more than 10,500 square metres, provides space for both starting, growing and established companies, with each party contributing and helping to build the community *(example of a mixed-use building)*.

In the following sections, we will discuss each quadrant of the matrix in more depth, starting with organisations that are comparable to a sports complex.

6.4 Collaborative organisations (comparable to a sports complex)

Collaborative organisaties

Some companies and institutions collaborate intensively with other companies and institutions in running their primary process. Truck factory Scania, for example, works closely with its suppliers for just-in-time delivery of parts and logistics, and with subcontractors for maintenance of the production line. However, these organisations do not share physical space. Scania's facility manager, therefore, only deals with his/her site. To fit the sports complex metaphor organisations must be located in a defined area with -to some extent- shared facilities. Like a sports complex accommodates various clubs and their private facilities such as a clubhouse, dressing rooms and sports fields - but also offers shared facilities for parking, signposting and, for example, green keeping. They also join forces in negotiating with the owner of the complex (usually the municipality).

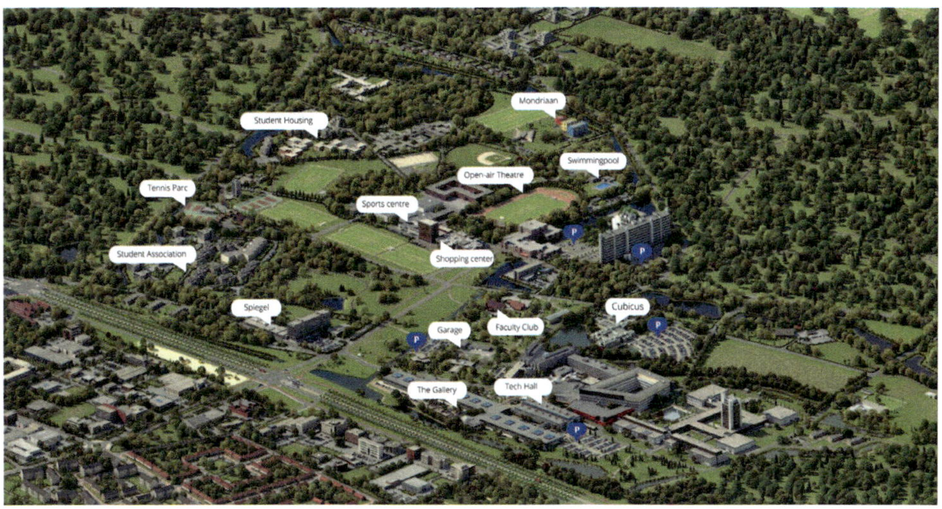

Twente University Sports Complex

6.4.1 What do we currently see in practice?

University campus

An example that fits the sports complex metaphor is a university campus. A whole range of innovative companies (start-ups and scale-ups) typically cluster around the university, drawing on the same kind of knowledge as the university does. The university benefits from connecting with these companies; for research and for being able to introduce the latest developments in education. In Eindhoven, for example, the High-Tech Campus is home to innovative technology companies. These companies successfully combine their strengths in terms of synergy and mutual cooperation. The promising motto of this campus is 'Turning technology into business', with the subtitle 'The smartest square km in Europe!' The university campus is also home to all kinds of service companies, restaurants and providers of leisure (e.g. fitness) and (temporary) living space. This mix results in a lively and inspiring study, work and living environment where students, university staff and companies enjoy staying. The collective knowledge accumulated on the campus thus creates a self-reinforcing effect with added value for all participants. A campus can also be established around a successful company. An example of such a business campus is the Brightlands Chemelot Campus in Geleen, the Netherlands, where innovative companies in chemistry and biotechnology gather around chemical giant DSM.

Campus

Shared area

Business zones

The campus examples focus on knowledge and innovation, in which a shared area is an enabling factor. However, companies may be located in the same area, without common ground in terms of the nature of the work or knowledge they depend on. These are, for example, business zones, which can be found on the outskirts of any town and along any motorway. Companies that share a business park have partly common interests. To give a few examples: a clean and safe environment, excellent infrastructure, well-maintained green space, accessibility for employees, area branding and sometimes also catering facilities or events. In situations where the entire site is under the management of users (companies), one speaks of park management. Companies are jointly responsible for an adequate level of facilities and often want to invest in quality improvement - provided that everyone contributes. By establishing a so-called Business Improvement District (BID), it becomes possible to oblige all entrepreneurs and owners to make a financial contribution. Business Improvement Districts can also be established to manage shopping areas, for example, a BID for an inner-city. An example of an award-winning concept for a business park is Industrial Park 'de Kleefse Waard'. The example shows how this industrial park profiles itself on its website. What stands out are the shared facilities that add to the quality of the business park (see screenshot).

Business Improvement District (BID)

● www.ipkw.nl

About Industrial Park Kleefse Waard

The Industrial Park Kleefse Waard (IPKW) is an award-winning business park that accommodates, facilitates and connects sustainable energy-related entrepreneurs. In an environment that is interesting both from a business economic perspective as well as from a social entrepreneurship perspective, cooperation with innovative companies is vital.
Sustainability has always been in our heart. As a former site of the AKU (Algemene Kunstzijde Unie) and later AkzoNobel and now owned by Schipper Bosch, we propagate sustainability in various ways. With the kind of companies we help excel, we give substance to our ambition to be the most sustainable business park in the Netherlands by 2025.
The themes that lie at the heart of this are not only relevant within the business community and public authorities but also appeal to educational institutions. In this way, we give substance to the modern campus concept and ensure an optimal business climate.
Are you working on tomorrows world and do you want to excel in your market? Then join other companies with the same focus, at a business park that stimulates and facilitates innovation.

6

In very different environments, too, one can find collaborative organisations that jointly design and share an area, such as at train stations and airports. Here, shops, food services and providers of all kinds of services and transport are part of a carefully designed retail concept that together shape the passenger experience. A well-thought-out mix of shops and services increases the attractiveness of the area, which in turn leads to increased passenger volumes and thus to improved financial results for the companies involved. As this requires different knowledge and skills than the logistical operations, the real estate management of stations and airports is often accommodated in a dedicated operating and management organisation. **Attractiveness**

A socially driven form of open organisations can be found in neighbourhoods. Neighbourhood support often takes place in buildings that are qualified as social real estate. Social real estate is real estate with added value for society, such as community centres, cultural centres, shelters, health centres and sports facilities. These days, social functions in **Neighbourhoods**

neighbourhoods and districts are hardly ever developed in isolation. Instead, a combination of features is used as a basis for more efficient use of space and more effective service provision to citizens. Examples of functions that can be combined are a well-baby clinic; a medical aid depot; general practitioners, dentists and therapists; a gym for surrounding schools and local sports clubs; a Municipal counter; and a communal meeting place for playing billiards and cards, sharing meals, but also for language classes.

6.4.2 What is the role of facility professionals in collaborative organisations?

To answer the question of what the role of facility professionals in collaborative organisations is, we have to look at two levels. Firstly, the organisations in the network, and secondly, network that the organisations together form. Organisations of some size have organised facility management support, be it in-house, outsourced or a mix of these. Smaller companies mostly appoint one or more employees who, in addition to their primary activities, are responsible for office management. In addition to the size of the organisation, the accommodation also plays a role. In multi-tenant buildings, the facilities are (partly) arranged by the property manager. If so desired, tenants can make use of full-service accommodation packages.

Stakeholders

So far, nothing new. Where we really expect development is at the level of the network. Many stakeholders are involved in the development of locations such as shopping malls, business parks, campuses and community centres. Stakeholders are the municipality, planners, property developers, property owners, the companies and residents, to name a few. Once the area is developed, it has to be managed. Other, but certainly not less, stakeholders will be involved. It is evident that cooperation can offer benefits for the return on investment of the facilities. Think restaurant, reproduction, security, installation maintenance and joint purchasing; a challenge for facility professionals to take on!
A logical name for the manager of such an area where organisations form some sort of network, is park manager, as a derivative of park management. However, not every area is a (business)park. As we indicated above, it can also involve community functions, a multi-business building or the inner city. That is why we opt for the more neutral term of area manager. The scope of the facility professional is, therefore, an area, rather than a building.

Area manager

There are plenty of reasons to combine facility services on business premises. The most apparent reasons are for branding and to achieve a higher level of quality than that for which the municipality is liable. This could involve security measures, such as security cameras or even a complete fence with barriers and access control. Cross-company risks can only be effectively managed if risk management is also applied at an area level. A second advantage may lie in joint procurement of facility services. Think of maintenance of buildings and installations, catering, security and reception. A business park is a home to various companies with distinctive customs, culture and DNA. This means that the diversity of customers may be considerable. A service level agreement covering all companies will have to have a multitude of options to cater to all needs. Economies of scale and increased efficiency can lead to savings or an increase in quality at the same cost. One step further is to share a single facilities department for the entire business park. This could be set up and managed by a business association; commonly referred to as park management.

Park management

For community centres, as described above, there are also possibilities in this respect. Not so much to economise, but to raise their profile and to be more relevant to society. This is where the so-called sharing economy could manifest itself. The community centre could act as a cooperative, as is customary in agriculture. Members of the cooperative jointly purchase tools and equipment, such as loading bollards, solar panels or mowers. After all, why would every resident buy expensive tools that they don't use much when they can borrow them as well? The person responsible for this might be called district manager. Expanding the functionality of community centres has been suggested for a long time. Still, as far as we know, this has not led to massive application. However, such an expansion of functions will be stimulated by the ageing of the population.

District manager

Many stakeholders and multiple functions in an area mean a multitude of interests, some of which can be conflicting. An important role for an area manager, therefore, lies in the field of relationship management and interdisciplinary cooperation with experts. Contributing to the desired branding and liveability are also essential matters of attention. To this end, the facility professional can look at what works on a smaller scale: in coworking spaces, for example.

6.4.3 What should one be capable of in the role of area manager or district manager?

In this section, we will discuss the competencies of an area manager (in the sense of someone who manages a business park) or a district manager (working for the community). The area manager makes contact with a wide variety of professionals working for different organisations; these may be the members of a Business Improvement District (BID). Besides, the area manager of a business park must be in close contact with the relevant municipal authorities. After all, the municipality plays a decisive role in building permits, environmental legislation, safety, accessibility and layout of the site. The district manager is in contact with the residents of the district as well as with social institutions, care providers and (other) tenants and users of community buildings and other types of social real estate. Naturally, the municipality is also an important partner and client to the district manager.

Although they differ in terms of the target group, both the area manager and the district manager will have to use their entrepreneurial skills to increase their impact and relevance. Moreover, bringing together a multitude of perspectives and needs requires good social skills, empathy, communication skills and negotiation skills. All parties involved have their objectives and interests, their own (organisational) culture and their communication style. Area managers and district managers can have a good conversation with all of these parties by adapting their communication style to the situation and partner(s). And, equally important, they must be capable of distinguishing common interests from individual interests and keep these in mind as the main objective in every decision they make.

Entrepreneurial skills

Social skills

Furthermore, an area manager must be able to process integral costs and revenues (financial and non-financial) in a business case. By clearly communicating costs and revenues and risks involved, he or she supports well-informed decision-making by the partnering institutions.

Business case

A day in the life of Merel de Ruijter, Property Manager at Schiphol Real Estate

Merel de Ruiter is a property manager at Schiphol Real Estate; the organisational unit responsible for real estate, land and grounds - and there are quite a few of them. When we think of airports, we mainly have in mind the terminals - the arrival and departure halls. But real estate also includes the hangars, the parking facilities and all the offices, for example. The majority of these offices are rented out to third parties; not only to KLM but also to other well-known multinationals such as Microsoft and Samsung. Merel is responsible for commercial activities. A colleague takes care of technical management. Merel's work has four cornerstones: Flex, Sustainable and Health, Connected and Smart. Flex and Connected are typical and distinctive for an area manager.
Merel: 'We have noticed that due to the rapid developments in technology, for example, it is difficult for tenants to enter into long-term leases. Tenants need to easily keep up with the company's demands by, for example, temporarily renting extra space (e.g. for a project) or (temporarily) making it available to other organisations.'
For this purpose, the platform Schiphol Spacemaker has been set up. This platform enables tenants to exchange excess capacity on a temporary basis. In this way, utilisation rates go up, and flexible space is widely available.
Schiphol Real Estate believes that companies can best establish themselves in places that meet the needs of their employees to retain and attract talent. By maintaining excellent communication with the tenants, network building can be supported. Merel: 'Schiphol's buildings and facilities stimulate self-development and help you get in touch with colleagues or specialists from other companies at home and abroad. By constantly looking for new work concepts, additional services, meeting places and events, the interaction between the tenants - the community members - is stimulated.'

6.4.4 What are crucial future issues for area management?

As explained in section 2.3, area branding is becoming increasingly important. It is, therefore, obvious that area management will further professionalise in the years to come.

Focus

On the one hand, because areas are expected to focus on a certain expertise or economic activity (which increases the value of the network and the value of the real estate). On the other hand, areas must pay sufficient attention to the combination of living, working and recreational functions to increase the attractiveness and reduce vulnerability to environmental dynamics. Residential neighbourhoods, especially in areas in decline, will need to pay attention to the quality of life. Among other things, by providing the right quality level of facilities and strengthening social cohesion in the neighbourhood. By using marketing techniques, the appeal can be increased. This can contribute to vacancy reduction and the value of the property.

Other trends that will determine the dynamics in an area are climate change, the transition to a circular economy, mobility and accessibility issues, and the evolution of smart buildings into smart cities. In this section, we will briefly discuss these trends in the context of area management.

Heat build-up

Due to climate changes, we are experiencing increasingly hot days during the summer period. In inner cities, with little plants and trees and many buildings with glass facades, this results in intensive heat build-up In turn, this leads to an increasing need for cooling in buildings. Increased cooling triggers higher energy consumption and CO_2 emissions, and results in additional heat emission to the environment. On the other hand, the number of heavy showers is increasing. In urbanised areas high building density and hard paving, this rainwater will have to be drained through the sewers. During heavy rains, the system becomes overloaded, with consequent problems. Plants and trees reduce heat build-up and increase the area's absorption capacity for the excess rainwater. It also contributes to an attractive environment, the preservation of biodiversity and the healthy behaviour of people. After all, the presence of nature in the outdoor area stimulates exercise in the open air. Singapore is an outstanding example of bringing natural elements to a modern city. The city's slogan 'Singapore, the City in a Garden', is absolutely well-chosen.

6

Natural elements need to be well-arranged; otherwise, it impairs people's sense of security and has the opposite effect. The area manager can bring all these perspectives into the planning process and represent a variety of users. Just like the green space in and around buildings, green space in an area also needs to be maintained. In residential areas, the maintenance of public green spaces is a task for municipalities. However, residents are increasingly being called upon to help. In business parks, the area manager can be responsible for this. This offers opportunities to contribute to a more inclusive society by hiring people with a distance to the labour market (see section 2.7.2).

As explained in section 2.4.4, mobility and accessibility are issues of growing concern. Where companies are, people work. And those people have to commute to work and back home. If areas have a combined function of working, living and recreation, the traffic volumes increase, and they are extended to 24 hours a day. An area manager, therefore, plays an essential role in the accessibility of an area, especially if the regular public transport network does not serve it. Crowd management is an important area to ensure that traffic flows can be processed as safely and efficiently

Crowd management

as possible. Big data can help to keep track of the number and type of movements (e.g. via the number of GSM signals from a certain location, possibly in combination with data generated by the use of navigation apps). This insight makes it possible to predict traffic jams and divert them in good time. By promoting the joint use of location-specific facilities such as parking spaces, peaks in demand from one location can be accommodated by nearby excess capacity. This dynamic, effective use of facilities is consistent with the sharing economy (section 2.4.2). It is expected to be boosted by broad real-time availability of utilisation data.
Mobility issues also arise at a district level. Just think of the lack of parking facilities combined with increasing car ownership and a large number of parcel deliverers driving through residential areas. As a result of the sharing economy, shared cars could grow in popularity, provided a person has access to such vehicles at a short distance from home or work. Here, too, it is crucial to think about solutions at the level of the area.

Energy transition

Another important issue is the energy transition from fossil fuels to sustainable alternatives. Especially at the area level opportunities exist. By generating and managing sustainable energy at area level, peaks in supply and demand can be balanced, and economies of scale can be achieved in procurement and maintenance. Joint reflection on residual flows can lead to new solutions for local reuse. What is a residual product for one company may be a valuable raw material for another — even heat. In so-called heat networks, an area uses residual heat from, for example, water treatment plants or nearby companies to heat buildings and water.

For many of these developments, the creation of added value at an area level is primarily made possible by the availability of (big) data: the aforementioned smart cities.

6.5 Open communities (comparable to fitness centres)

Even though independent professionals - the name says it all - are self-employed, they often work in a network of freelancers. A nice craft example is a carpenter with a mobile workshop in a van. In principle, he needs nothing and nobody else. There are exceptions, of course: every carpenter regularly needs a 'third hand' - a person who helps with lifting, for example. The carpenter then makes use of his informal network. He could call a friend (also self-employed) who happens to be working in the neighbourhood. And so, the independent professionals help each other back and forth. In this example, a shared place does not play a role in the network. In many situations, however, a sense of place is a crucial factor for social cohesion, encounters and knowledge-sharing; in so-called coworking spaces and incubators, for example.

Open communities show similarities to a fitness centre. Access is easy; anyone can become a member. There are no requirements for fitness, nor are you assigned to a team with other athletes of your level. There are no fixed training times and no competition schedule; you can come and go as you please. Some fitness centres even have subscriptions that are not tied to individuals, but addresses. A membership card may then be shared within a group of users, and they are allowed to bring guests.

6.5.1 What do we currently see in practice?

Coworking spaces can consist of a single building or part of a building that is at the disposal of several independent professionals. The shared or self-contained workplaces are rented out flexibly. In addition to working at a desk, guests can have meetings and take lunch here. These concepts stimulate meetings, enhance the experience and facilitate cooperation and interaction. The building blocks of coworking spaces can be summarised as:

6

1 *Lounge.* This is a place to flex, plan appointments or enjoy a cup of coffee. Other manifestations are an espresso bar, brasserie or grand café.
2 *Workplaces.* A place to work, ranging from flexible workplaces to large shared spaces to individual workplaces that can be booked for a half-day, day or number of days per month.
3 *Meeting space.* Thematic meeting rooms, coaching rooms or conference rooms, all sorts of formats can be found. An absolute must is catering (see also point 4).
4 *Amenities.* Services are offered on location, for example, catering, printing, internet and parking, but sometimes also coaching and training. Lunch is an excellent source of conviviality and an easy way to interact with others. In some coworking areas, lunch is collectively ordered and shared. Offering lunches through the lounge or grand café is also common.
5 *Community.* Succesful coworking spaces have a vibrant community of people who share knowledge, give each other assignments but also collaborate on projects. The network can be supported by apps that show which users are present and the knowledge they have. For example, Seats2Meet sets virtual check-in as a condition for the use of its facilities; users 'pay' with social capital (their knowledge). Some providers offer an online platform where users are connected, even if they are not present at the location. Events ensure that the place is bursting with energy and that the community members feel connected to each other.

In essence, these are 'hotels without beds', because the only thing they cannot help the user and what a hotel does inherently offer are sleeping places. Just as Booking.com enables booking a hotel room via the internet, there is a platform for temporary workspaces: deskbookers.com. Many of these buildings have long been in use as such by companies such as Seats2Meet and Regus.

Revenue model

The revenue model consists of a mix of paid and free services. For the paid workstations, a flexible subscription (e.g. the right to use the workstations four days a month), voucher or day pass is usually used. Meeting rooms are usually charged by the hour. Some providers offer free, flexible workstations in an open space (e.g. the lounge), and more comfortable workstations for a fee. Other concepts require a fee for each workstation but offer free coffee and tea. Some use a scheme whereby users can earn paid services (e.g. lunch) by helping out with hands-on assistance.

Coworking spaces have a look and feel that suits their target group. The choice of location must satisfy the target group. Urban areas that are still under development, for example, appeal to the more creative and innovative industries.

Although the number of flex concepts has grown strongly in recent years, the market for flex concepts is still in its infancy. According to Emergent Research, the number of coworking members worldwide will increase to 3.8 million by 2020 and the number of coworking spaces to approximately 26,000. This is a twofold increase compared to 2017. The following web message illustrates this market development.
In the Netherlands, serviced offices and coworking spaces currently have a combined market share of 2 to 3%. As in the rest of the world, this share is increasing sharply, especially in the major cities.

● www.smartwp.nl

CBRE launches its own coworking company

Real estate giant CBRE starts offering coworking spaces in New York with the new service Hana Spaces. Hana enters into property management agreements with landlords to provide flexible office space to large corporate clients. CBRE's model allows the company to co-invest in the costs of building and managing coworking spaces. However, the collection of the rent is handled by Hana for the building owner not to have to worry about it. "We already see a lot of interest from building owners looking for a trusted partner to offer flexible offers for coworking spaces," said CBRE President and CEO Bob Sulentic in a statement.
In addition to the service for large companies, called Hana Team, the new company also responds to the traditional coworking models. To this end, it has set up the services Hana Meet and Hana Share. Hana Meet offers conference and event rooms on an hourly basis. Hana Share is more of a traditional collaboration model for smaller customers, offering shared facilities and technology in a shared space. Hana Spaces will start in the United States for the time being. If successful, the concept will soon be introduced in Europe.

6.5.2 What is the role of facility professionals in open communities?

The open community consists of individuals who, depending on their (temporary) projects, collaborate with others. One could say that independent professionals work in a nomad-like style. What can a facility professional contribute here, one might wonder? Well, somehow, such an open community needs a place, either as a network or as a (group of) individual(s). The locations where these (mostly) independent professionals meet should also provide coffee, lunch, cleaning, and so on. Thus, there is also facility management - only with temporary customers. In this case, the facility professional runs a kind of multi-tenant building but with strongly fluctuating tenants consisting of individuals. But there is more. In this concept, the physical environment has a twofold function. It supports the formation of a meaningful community by attracting users who fit the desired profile (branding), while the design of the physical space contributes to people meeting and sharing knowledge. As we described, the community manager is like the linking pin that connects the participants. This is where the added value is created. The more attractive the network or community is for the participants, the more successful the coworking space is.
This is comparable to a sports club. Success is not so much determined by the equipment - although the amenities must meet the expectations and needs of the target group. It's also about the atmosphere and being motivated to maximise personal performance and accomplish one's goals. In a fitness centre, the owner is an entrepreneur who needs to make sure his offer is highly attractive to new and existing users. One of the ways to do this is to create a welcoming, stimulating atmosphere that gives athletes a sense of belonging and the feeling of being supported. A location manager of an open community has a similar role. Sometimes it means coaching for regular members of the community. Within coworking spaces, the success of the concept is determined by the added value the community offers to users. The manager of the location is the hub and can connect and inspire the participants. Events also contribute to this: a mix of community branding, knowledge transfer, motivation enhancement and, of course, networking. This is a role that facility professionals can perfectly play! We call this type the *community builder*.

6.5.3 What should one be capable of in the role of community builder?

As said, the community builder is responsible mainly for the atmosphere at the location. A good sense of hospitality, combined with marketing, event management and entrepreneurial skills are essential for success in this role. Hospitality and accessibility of the location are essential to attract new users and to retain existing users. The hospitality concept must fit in with the (intended) target group and underline the values and the code of conduct of the community. As we described in Chapter 3, behaviour plays a vital role in the experience of hospitality. It is therefore crucial that the community manager coaches his team well about attitude and behaviour, without depriving team members of the freedom to be authentic. Open communities are communities of people. Sincere interest in others and innate curiosity are personal qualities that community managers need to make people feel that they belong; in short: to engage and retain customers.

As a purebred entrepreneur, the community builder keeps a close eye on the market, identifies opportunities and taps into them. A well-designed revenue model ensures a balance between income (e.g. from the rental of meeting rooms, sale of lunches and subscriptions for workplaces) and use of facilities in exchange for social capital or commitment to the community. In the latter case, customers sometimes take on the role of volunteer for a defined chore. In such cases, the community builder will need to be able to flexibly switch between its role as a service provider and the role of the manager in contact with the same customer.

By paying attention to branding, the community builder ensures that he/she appeals to the desired target group and that the location is visible and findable, both offline and online. Marketing is essential: without marketing a community cannot be established in the first place and will quickly cease to exist. By understanding psychology and how to influence people's behaviour through space and services, the community builder can effectively brand and support the community.

A day in the life of Nynke Visser, Site Manager at S2M030

In the heart of Utrecht city, anyone looking for a workspace can visit Seats2Meet Utrecht Central Station (S2M030). The philosophy of Seats2Meet (S2M) can be summarised as: share what you have in abundance. To be successful in a networked society, people need more than one system. The core values of S2M are, therefore: connect, collaborate and grow. 'By connecting people, unexpected, meaningful encounters can take place,' says Nynke Visser, site manager at S2M030. 'And especially meetings with people you don't yet know that can bring new insights and opportunities, and thus contribute to one's success. We call this serendipity'. S2M's business model revolves around balancing income from meeting rooms and services (monetary capital) and facilitating serendipity to enrich systems (social capital). S2M, therefore, provides work and meeting places in exchange for their social capital (the willingness to share knowledge with other attendees). The location of which Visser is manager, has 21 meeting spaces to generate monetary income and 100 social capital workspaces. 'This way, we can create a perfect win-win situation and a culture of cooperation and sustainable growth,' she says.

S2M030 can be characterised as an open community. They call it an open ecosystem. Everyone is welcome here because everyone can be of value to someone else. New users are essential to boost the chance of meaningful encounters. A static group of users reduces the chance of meaningful encounters. The only condition is that you check in on the (digital) S2M Passport. This gives you access to the entire network, and people can find you based on your expertise'.
Nynke is ultimately responsible for the location: 'In particular, I am responsible for maintaining and entering into cooperative relationships with stakeholders and other providers of flexible office space in Utrecht. I also contribute to the optimisation of the S2M booking system. Also, I help the team to discover how they can develop their skills. I do all this with a mindset of sharing knowledge so that everyone becomes stronger as a result'.
When asked which competencies you need to fulfil this role well, Nynke answers: 'Communicative skills, curiosity and a sense of hospitality. Each location is allowed its interpretation of the S2M core values, so above all, you have to be entrepreneurial and be able to deal with freedom in a chaotic-orderly system'.

6.5.4 What are crucial future issues for open communities?

As we described in section 2.5.1, the Netherlands is a frontrunner when it comes to flexible working (mobile working). Also, the number of self-employed professionals (full-time and part-time) has been increasing for years (approximately 13% of the total working population in the Netherlands in 2019). These people may work from home, at clients and from a coworking space. This increases the need for flexible workplaces for self-employed professionals (and mobile workers). It is therefore expected that the market share of flex concepts in the Netherlands will grow by approximately 20 to 30% per year. This means that in five years, serviced offices and coworking spaces will account for about 5 to 10% of the office market. This is a worldwide development, with now even listed international providers of these types of concepts.

As in the 'regular' office market, the concentration will take place within the large and medium-sized office cities. Locations at public transport hubs in city centres (supplemented by the centre directly around them) are expected to have the most robust growth. Also, the distinction between serviced offices and coworking spaces is becoming increasingly blurred: many providers offer hybrid forms (see section 2.5.3 on blurring). A new phenomenon is 'pairing': companies that combine their accommodation with flexible renting of other workstations from coworking providers. This is, in fact, a combination of a closed organisation with an open network of individuals. For private organisations, this can be very attractive as it gives them access to talent. In a market where talent is scarce, and companies increasingly outsource to external parties, this can be of high added value.

The challenge is to maintain the added value of the community through its members without compromising the openness of the community. Dynamics are also important. A community consisting of a steady group of employees and users will behave differently; it will become more closed and less innovative. As a result, the likelihood of chance encounters that are of value decreases. From corporate social responsibility, however, working with only temporary contracts for employees is not appropriate. One way out of this dilemma is temporary employment arrangements in learning and working paths. This way, the limited length of the contract is relevant to the learning process of the employee, and rotation enhances the dynamics sought after. Such a search for win-win solutions fits in well with the community spirit.

If the congestion problem spreads from rush hours to the entire day, organisations will have to start thinking differently about workplaces. Not only do traffic jams lead to a loss of productivity, but they also reduce the recruitment area for new staff in an already tight labour market. Prolonged commuting times lead to a weakened work-life balance. For vitality and well-being, flexible working arrangements are preferable. However, working from home is not always the right solution. Not everyone has a suitable, distraction-free workplace where they can work in an ergonomically responsible manner. One can imagine that it can, therefore, be advantageous for organisations to make agreements with providers of coworking spaces for their employees, for example, in the form of a subscription. Just as mobility cards facilitate and bundle business travel, a workspace card could support working at coworking spaces — an interesting new target group for community builders.

Until now, most communities have mainly been offline communities, in many cases supported with some form of digital environment or app. However, technology will continue to change the way people live, learn and work (together) in the coming decades. This will also impact coworking spaces. For example, what will happen when virtual reality and augmented reality become intertwined with everyday life? Will people still need a physical place to connect with others? Or will that place be replaced by a virtual environment that feels just as real? Is a community builder still needed? Or will algorithms take over that function? The network is a source of capital that the community builder must carefully manage within the limits of the European General Data Protection Regulation.

6.6 Private communities (similar to a golf club)

A private (or exclusive) community is a relatively stable network of independent professionals or micro-businesses that are subject to considerable dynamics. From rapid growth due to a large order to the immediate cessation of activities related to a salaried job; it all happens. This requires flexibility and agility of the location and services offered. Contracts usually have a short term with few restrictions. Yet, the network is one in which an outsider (non-tenant) does typically not intervene.

A private community shows similarities with a golf club. Golf clubs have a ballot committee that determines whether someone is allowed to become a member. Members must meet specific requirements, including at least a demonstrable skill in the game (handicap). Members of the club pay a membership fee and can, in return, make use of the facilities offered by the club as required. It is possible to make external use of the facilities, for example at the invitation of a member, but not without prior registration (and permission).

6.6.1 What do we currently see in practice?

In the care industry, this type of closed network often occurs, for example, as a partnership of specialists or general practitioners and health centres. In addition to the advantages of sharing housing costs, a health centre offers more support options and facilities, such as reception, management and administration. Moreover, assistants, support staff and other care disciplines can be deployed more efficiently. Because care providers work close to each other, problems of replacement during illness, holidays and refresher courses (provided they practise the same profession) are a thing of the past. This also opens up the possibility of part-time work. For the patient, a health centre means more continuity of care and a broader range of services. For care providers and other practitioners, there are mainly more opportunities for peer support and professional development. Such more or less permanent partnerships have also been established in business services (lawyers, architects, consultancy, accountancy, recruiters). In these kinds of private communities, the participants present themselves as one organisation to the outside world. Still, these are independent professionals or small companies that work closely together. Competition between them occurs, albeit not on price but the quality (specialisation, knowledge, experience and affinity).

In the examples as mentioned earlier, the private community already exists and subsequently needs a place to operate. It can also be the other way around, such as in multi-tenant buildings or serviced offices. In that case, there is a place, for which tenants are sought. For example, serviced offices. Serviced offices are turn-key offices that are provided with support staff. Various types of workstations (office units and meeting room facilities) can be used for a flat fee for a flexible period. In such a building several tenants find a place, usually a mix of small companies and independent professionals (whether or not in partnerships). Smaller tenants often look for a location where like-minded people are active. Building owners may present themselves with specific profiles in the rental market. The Mediacentrale in Groningen, the Netherlands, is an example of this. It is a former power station located at the Europe Park office area, that also accommodates the professional soccer club

FC Groningen. This former power station has been transformed into an office building of no less than 18,300 square metres. It houses some 65 (mostly small) companies active in media, marketing and IT. In addition to offering accommodation for companies, the building also offers a conference centre with space for conferences and (external) events. Because tenants are all operating in the same sector, they are more likely to have added value for each other. The nice thing about this example is that it is a network on two levels. Within the building, it is the private network of tenants; on area-level, the Mediacentrale is part of a business park that provides amenities for shared use.

A specific form of or multi-tenant buildings, are the so-called breeding grounds. Incubators (also called start-up labs) resemble coworking spaces, with the difference that the community has a different dynamic. Start-ups are active in the field of technological innovation. Often, they are initiated by students or researchers to develop an innovative idea into a product that can be launched on the market. The incubator is the physical place where start-ups can work on building their business. In addition to a physical workplace, the offer also includes a coaching programme aimed at refining the idea up to and including attracting investment capital and first customers. A start-up track usually takes between six months and a year. Start-ups that succeed in making the first step face the next challenge: scale-up. For this phase too, there are private communities that offer a temporary place and coaching for the company (which now consists of several people), to broaden the customer base. Start-ups and scale-ups are often located on university campuses because of access to knowledge and tools. These locations have a completely different atmosphere than coworking spaces. They are often more basic and have a much less homely atmosphere. The main goal is to strengthen the participants' network with the outside world (because that contributes to success) - not building a network with fellow participants.

Compared to the confined organisation, the dynamics within the network are higher, but by no means as great as in an open community, where the number and type of users per part of a day can vary greatly.

6.6.2 What is the role of facility professionals in private communities?

The facility professional in private networks has to deal with independent, enterprising customers, each with their requirements and wishes. All these customers are at the same level in the hierarchy. This type of facility professional is employed by (or hired by) the property owner.

In traditional multi-tenant buildings, the facility professional is the link between the property owner and tenants. This professional has both financial responsibility (asset management) and substantive responsibility for the building and facilities. For the property owner, the vacancy rate must be as low as possible from a yield point of view. Tenant satisfaction is a determinant for the occupancy rate and financial performance. Tenant satisfaction depends on a suitable range of full-service accommodation, joint facilities tailored to the target groups, contractual flexibility, additional services and the added value that the network offers with other tenants. Thus, a facility professional who can bring together people with complementary activities and products will be successful. By using the right rental mix, the facility professional will be able to strengthen this proactively, but within agreed frameworks. If the financial performance of commercial real estate is part of the responsibilities of the facility professional, we speak of integration of tactical asset management and facility management.

A community manager employed by a collective of tenants (such as in partnerships of lawyers, consultants, etc.), often runs a relatively small location. S/he is often the only (or one of the few) professionals to understand housing management, procurement and service management. This means that the community manager must be able to translate the objectives of the members into an appropriate concept. As a sparring partner, s/he will proactively advise on opportunities for innovation and optimisation, whereby the added value to the primary process must be demonstrated clearly. At the same time, coordination and execution are also part of the job.

To stay in the metaphor of the golf club, you can describe the facility professional in these kinds of environments as the golf course manager. A golf course manager is not only responsible for all facilities, but also for maintaining and expanding the membership base. He has to be proactive but acts on behalf of the board of the club. If we translate this metaphor to the examples mentioned before, then the term *community manager* fits. You can also compare incubators with golf clubs. Golfers decide for themselves when and with whom they play and use the club's facilities for their purposes (relaxation, networking, enjoy the game). For new golfers, however, it is also essential to work on their technique to improve their handicap. Every golf club employs a golf pro, which teaches according to the needs of individual members, just as incubators employ coaches.

6.6.3 What should one be capable of in the role of community manager?

The community manager who is employed by a collective of companies (such as a partnership) is often a loner. Generalists who like to work independently, who find business alignment issues interesting, but who are also not afraid to roll up their sleeves themselves, are well placed here. Members of partnerships are often specialists in their field, used to advising others on what to do. As a rule, these professionals do not just take on anything, not even in the area of facilities. A community manager must be able to adapt to the often divergent needs and wishes of the independent professionals for whom he works. In doing so, he must be able to empathise with the commonality of those entrepreneurs. He needs to see and take up opportunities in terms of synergy.
Moreover, all costs are borne by the members of the partnership. Expenses, therefore, have a direct impact on the income of the members of the collective. Organisational sensitivity, a keen awareness of the added value of housing and services for the primary process and persuasiveness are essential skills for community managers.

The community manager of multi-tenant buildings is hired from a real estate services provider (e.g. CBRE, JLL) or by the real estate owner. Just like the golf course manager, he has to identify opportunities and act proactively, but within the (financial) frameworks set by the board or the owner.

● creative-council.nl

Creative incubating? More complex than imagined!

Creative incubators are occupying an increasingly important position within the creative industry. Various studies show that creative, independent professionals and SMEs are most productive in 'co-located' networks. In these networks, risks, knowledge and skills can be shared, and new business contacts can be established. But what about daily practice? In what way do creative incubators stimulate creativity, innovation and entrepreneurship?
Eight useful tips for increasing the potential of creative incubators are:

1 Facilitate meeting places, preferably utilising a catering establishment.
2 Recognise the importance of a community manager who, as an entrepreneur amongst entrepreneurs, is part of the 'system of creativity'.
3 Act as a 'doorman' and 'matchmaker' for both external assignments and internal cooperation in the incubator.
4 Contribute to boosting a knowledge infrastructure and a learning community.
5 Use a form of selection to create a balance between the diversity of entrepreneurs.
6 Organise internal and external presentation opportunities, for example, through exhibitions and pitches.
7 Contribute to a positive (brand) identity that is recognised and acknowledged both internally and externally.
8 In collaboration with strategic partners, develop innovative business models for the creative incubator that take into account both exploitation and creative objectives.

6

The difference between a community builder (see section 6.6.3) and community manager, is that the former is much more an entrepreneur. If a community builder is not successful, he or she often runs a financial risk. In the case of the community manager, this risk lies with the owner(s).

Especially in the case of incubators, the community manager must have a vision of the opportunities that exist in interdisciplinary cooperation. This vision and skills also come in handy when selecting participants. Moreover, he must have or be able to build up a network with external parties that are of value to the starting entrepreneurs. The community manager uses his own entrepreneurial and networking skills to inspire and coach the 'residents' of the incubator - the community. An incubator's community manager is a pure-bred networker who, in addition to an entrepreneurial mind, also has coaching skills. This type of community manager is a born entrepreneur full of initiative and creativity, who can engage people in a discipline that they can further develop through synergy with other users. At the same time, he or she must be able to survive in an environment that is anything but entrepreneurial. As mentioned above, an incubator is usually associated with a university or college of higher education. These institutes classify as bureaucracies that, for the sake of efficiency, strive for predictability and standardisation. It requires organisational sensitivity and excellent negotiating skills to be allowed to do things differently - in an entrepreneurial way - within such an institute.

6

A day in the life of Thom Wernke, CEO at StartDock

As soon as you enter one of StartDock's locations, you feel the atmosphere of a community. It's the total of a monumental building, an inspiring working environment and a collection of young entrepreneurs who feel connected. The concept is in line with the revolution taking place in office buildings. We are in the 'flex-working era'. The permanent office is disappearing. New workplace concepts that meet the challenges of the new economy are gaining in popularity.
One of StartDock's leading principles is the coworking principle. This involves a shared working environment with multiple users who jointly use on-site facilities. Thom Wernke, CEO and one of the founders of StartDock, has a bachelor's degree in Facility Management, but in his current job, he mainly feels like a community manager. 'My role is to bring together entrepreneurs. At StartDock, we deliberately don't talk about 'customers', but about 'members' (also known as Dockers). The members can be startups, scale-ups, freelancers, self-employed or entrepreneurs who like to work from a lively community. As a facilitator, we focus on creating and strengthening the community in which we try to connect and engage the individual parties. By matching supply and demand within the community, we create added value for all parties', says Thom.
StartDock can be classified as a private community. The closed character expresses that we are a warm community of entrepreneurs that act in a safe, non-competitive environment. This strengthens the mutual connection and stimulates the sharing of knowledge. There are also private models known in which, for example, only women are admitted or organisations that are specifically active in the field of sustainability'.
StartDock cannot be compared to open communities such as Seats2Meet because of its confinement. The community spirit is much less prevalent there because the composition of attendees changes every day. We also rent out meeting rooms, but this is not yet our focus. StartDock's primary business is renting out offices and workplaces. We apply different rates, from a simple workplace to a luxury room, including all kinds of additional services.'
After the launch in Amsterdam, StartDock extended its working area to Rotterdam. When asked what competencies a community manager needs to fulfil this role well, Thom answers: 'You have to understand how service works, be service-minded, understand the wishes of starting entrepreneurs, have above-average social skills and an interest in entrepreneurship'.

6.6.4 What are crucial future issues for private communities?

As we described in the introduction to Chapter 2, we live in exponential times which creates major environmental dynamics. Resilient organisations are organisations that can adapt flexibly to ever-changing circumstances. In combination with the rise of the sharing economy - in which use is more important than possession (see section 2.5) - this explains the increasing popularity of serviced offices. Short-term leases offer the possibility of absorbing the decline and growth of the organisation. This shifts the financial risk of a vacancy from user to owner. As we have described in section 3.2.6, requirements will be set for the energy performance of office property. In the case of serviced offices, the obligation to invest in

sustainability will be shifted to the owner. An important issue for owners is, therefore, the future value of the property. The future value is determined by three factors: adaptability, sustainability and yield.
The risk of vacancy can be reduced by making buildings suitable for multifunctional use and by choosing a mix of functions. For example, catering, retail or social facilities in a building's plinth, with a combination of residential and work on top. Building for redevelopment (e.g. the possibility of transforming offices into apartments) increases the future value considerably. A non-sustainable property has no future value and only limited residual value. The yield is partly determined by the degree of letting and the term of the leases of current tenants. This is directly influenced by the community manager of multi-tenant buildings. By keeping in touch with the wishes and needs of tenants, and knowing what developments are taking place in the market, the community manager can always ensure an optimal service concept. Satisfied tenants are less inclined to relocate, which benefits the yield.

As we described in section 6.2, hybrid forms of multi-tenant buildings and coworking concepts are emerging. In these hybrid-forms, part of the building is used as a traditional multi-tenant building (lease of square metres), and part is used as coworking space (e.g. in a building's plinth). It is important to properly incorporate the impact on the appearance and attractiveness of the building for tenants in the choices that are made. Coworking can increase the attractiveness of a location for tenants (because it offers a form of flexibility and control). Still, the difference in dynamics can also cause an experienced nuisance to permanent tenants. Facility professionals are thus confronted with increasingly diverse customer groups and customer needs. Smart-building technology enables tenants to pay for what they use instead of a uniform service charge that is the same for all tenants. Thus, it becomes possible to provide custom services to individual tenants without the need for extensive administrative processing and the consent of all tenants.

In knowledge-based economies (see section 2.4.1), innovation is essential for economic prosperity. That is why governments value the strengthening of entrepreneurship among students and researchers. However, new ideas or insights from research cannot be sold directly; product development must first take place. The role of incubators is to support such development. You could say that this makes incubators vital for economic growth. Yet their existence is not guaranteed. Incubators depend on external parties for funding. The start-ups earn little or no money yet and can pay no (or very limited) rent. Currently, incubators are mostly funded with government subsidies. Subsidies are always temporary, and their availability is cyclical (see section 2.4). This makes the continued existence of incubators vulnerable. Attracting sponsorship funds from large companies or successful former participants is only in its infancy. Community managers of incubators should consider the branding of the incubator as well as ways of attracting sponsorship funds from the market.

6.7 The confined organisation (comparable to a football club)

The confined organisation is the most traditional of the four segments in the matrix. The most prominent feature of this type of organisation is the limited access to premises for outsiders. Someone needs a registered appointment to pass the reception and security gates. This type of organisation is still the most common - think of large corporates such as insurance companies, banks, ministries and accountancy firms.
The confined organisation has characteristics of a football club. With its strong culture and identity, members are recognisable by their outfit (as employees of organisations conform to a dress code or uniform). Also, there are clear rules and agreements about who is expected where and when (the training and competition schedule). Members of the football club are divided into teams. They have a more or less permanent position within their team, such as employees of closed organisations work for a specific department with a clear task and job description. The coach determines the strategy to win from the opponent - just like the Board does to beat the competition. Team members put that strategy into practice by working together, just like employees execute the strategy within an organisation.

6.7.1 What do we currently see in practice?

New forms of collaboration (including agile working and scrumming) change the demand for the functionality of accommodations. The need for space is also evolving due to the introduction of mobile working and the automation of processes. Thus, large organisations housed in large office buildings are thus increasingly confronted with an excess of square metres. To reduce costs, some organisations choose to let vacant floors or building sections. What ensues is a shift from a confined organisation to a multi-tenant building with one primary user who also has ownership.

The County Hall in Utrecht is an example of this. The Province of Utrecht owns an office building of which several floors have been leased to external entities. All users share the parking facilities, the reception, the security, the catering facilities and the meeting centre. Such shared-use reduces the Province's housing costs, but it does lead to a blurring of roles. In addition to being an administrative body, the Province has also become a landlord. For government bodies, (the impression of) a conflict of interest must be avoided at all times; hence a careful selection of tenants.

Another option for organisations with excessive space is to let vacant space to providers of coworking spaces, such as Spaces, Knotel and WeWork. A confined organisation would then become a hybrid form with an open community, whose users are not always known. Besides the economic benefits, such a combination offers the flexibility to absorb peaks in occupancy. The shared-used of the building influences the risk profile, and thus the security measures.

6.7.2 What is the role of facility professionals in confined organisations?

As mentioned before, the confined organisation is traditional in nature, which also applies to the role of facility management. The focus of FM is creating a fit-for-use environment for the organisation by connecting space, technology, services, processes and people. Operational facility management is often outsourced. The stage of development of the facility function depends on the maturity and positioning in the parent organisation. The more professional the facility management organisation, the more strategic its positioning is and the more the focus is on delivering strategic added value for the customer (see section 1.5.2).

Most professional football clubs have a facilities department that takes care of the management and maintenance of accommodation and sports grounds. Besides, they take care of catering, security and crowd management at major events. This department contributes demonstrably to the success of the club, even if it is less visible than a scoring striker. The quality of the pitch partly determines how well players are performing, the hospitality services generate income for the club, and without adequate risk management, the safety of visitors is compromised. Efficient use of resources (e.g. by using LED lighting) frees up budget for the purchase of new players. The facility professional who is ultimately responsible for all this is commonly referred to as the facility manager, just like his colleague in other confined organisations. However, because technological developments broaden-up the concept of the work environment and the collaboration with other disciplines is becoming increasingly important, we choose to name this role: *the workspace manager.*

● www.facto.nl, based on Gerard Dessing, 2019

City Hall Alkmaar after thorough rebuilding: meeting and working together

In 2016, the Council of the Municipality of Alkmaar expressed its ambition to become the most innovative 100,000+ municipality in the Netherlands by 2021. This involved a change in the organisational structure and the way of

working: from departments with permanent teams to project-based collaboration in dynamic teams assembled based on people's knowledge and skills. To support this change, the work environment had to change as well, says Remko Huitema, Facilities Consultant. Together with his colleagues in Real Estate and IT, he was - under the direction of a steering committee - integrally responsible for the successful execution of the renovation.
A major constructional intervention was necessary to strengthen interdisciplinary networks of colleagues: open connections between the floors were created by removing segments of floors. On the upper floors, the fixed workplaces made way to a flex concept with a tight workplace factor of 0.3. Many informal seating areas, booths for private conversations, lounge sofas, armchairs and the like have been installed throughout the entire building to strengthen interdisciplinary networks of colleagues. Huitema emphasises that the primary aim was to provide a work environment that would support the cultural change and new working methods, not a reduction in square metres.
Huitema is satisfied with the result achieved, especially because of the innovative technology applied in the building. 'In total, there are some 300 sensors installed that control various technical installations in the building. These sensors also make it possible to find colleagues via an app. It is an opt-in feature; after permission, one can be found by colleagues via a search function. The app also makes it possible to book meeting rooms and to check real-time whether, when and where specific meeting rooms are available. Even more relevant to many users: through the app, personal coffee preferences can be sent to the coffee machine.
Furthermore, smartphones play an important role. Certain parts of the building can be accessed by employees using the smartphone. Registered visitors receive a QR code that allows them access to the network cafe via their smartphone. FAFS officers are paged via their phone and no longer need to carry a beeper. Together, this contributes to high convenience for the building users.

The next steps we are thinking of are paying for sandwiches exclusively via the smartphone, printing, copying and scanning with the smartphone and access security for the entire building via smartphone control. There are plenty of new possibilities for us to explore.'

6.7.3 What should one be capable of in the role of workspace manager?

The workspace manager directs facility services and facilities. S/he is indispensable for the smooth running of the primary process. The workspace manager possesses organisational sensitivity and situational leadership skills that enable him to attune his behaviour, attitude and communication style to what is appropriate in diverse situations. Like no other, the workspace manager translates changes in demand into appropriate facility solutions (the supply), taking into account the indirect consequences in addition to the added value of facility management. Facility management is people work, and therefore in addition to a good relationship manager, the workspace manager must also be a people manager at heart. As we consider this to be the standard facility managers role, we will not elaborate on what a facility manager should be able to do. After all, that's what this whole book is about.

A day in the life of Michel Tobé, Facility Manager at the Municipality of Capelle aan den IJssel

Michel Tobé is Head of Facility Services, Real Estate and Information Management for the municipality Capelle aan den IJssel. As with many cities, the responsibility for schools, theatres, sports halls, swimming pools, community centres and, of course, that of the Town Hall lies with the facility manager.
As far as Michel is concerned, his focus lies on community service. Michel manages three services: facility services (the soft services), information management (both process-related and as a client for ICT) and property management (the hard services). The aim of public real estate is to

strengthen social cohesion in the municipality, which is why it is an advantage for him also to be responsible for real estate.
Michel: 'In addition to the social aspects, I attach great importance to hospitality. That's why I took the initiative to develop a hospitality concept together with other departments of the municipality, such as public affairs and social affairs. We have joined hospitality training sessions. I think that interdepartmental collaboration is essential, for otherwise there will be no support for the concept'.
Michel's work is very multi-disciplinary. For example, he is also responsible for the flow of information and archiving at the municipality. In doing so, he must comply with archive and privacy legislation. No two days are the same, especially for the dynamics municipal politics bring. For example, the unexpected news report about the harmful effects of rubber granulates on the sports fields seriously disrupted Michel's agenda. He is also responsible for the management and operation of the sports fields, and the Council required immediate action.
Besides this, Michel earns the most appreciation for his efforts to boost people's talents, making them perform above their expectations. This is one of the reasons why the traditional image of the civil servant is changing, Michel observes. He truly believes that inspired people can create a better world. And facility management, in particular, can contribute to that. In an inspiring working environment, people will experience more enthusiasm for their organisation and the work they do. And if civil servants feel passionate about their work, this positively impacts society.

6.7.4 What are crucial future issues for of confined organisations?

The culture of private organisations has long remained stable but is now subject to changes under the influence of technology, increasing diversity and environmental dynamics. This creates new challenges, many of which are described in Chapter 2. Not all social developments will be equally relevant, and industry-specific events may have higher impacts than social change. Much depends on the context of the organisation. However, several developments affect all confined organisations, as a result of which there will be a shift towards partially collaborative organisations and private communities. This shift increases the importance of risk management and security.

Several developments that can be grouped under the term 'networked economy' stimulate the opening up of organisations (see section 2.4). Team collaboration and knowledge sharing are gaining importance due to the increased complexity and knowledge work. These -mostly temporary- teams consist of a mix of employees from a strategic core (often from different departments) and temporary hires. The choice for a combination of a stable core and temporary staff may have several reasons; most often a combination of the need for agility in unpredictable markets, a tight labour market and the availability of independent professionals.
A logical consequence of this is that traditional business structures are shifting towards matrix and network organisations. Collaboration is

becoming more multidisciplinary, and also takes place at external locations and with external parties — this dynamic use results in a fluctuating demand for space and services. Big data analysis can give insight into patterns that - to some extent - enable predictions of use.
Interdisciplinary collaboration is also reflected in the facilities function. More intensive cooperation with HRM, IT and real estate management contribute to achieving strategic goals in the field of productivity, attractive employment and corporate social responsibility. Some predict that in the future, these departments will merge into one shared service centre (SSC). However, this has been voiced for more than twenty years.
Location policies of organisations also adapt to this new reality. More and more people are working (also) in external hubs (third workstations or coworking spaces); these hubs are (usually) managed by specialist parties.

Finding, retaining and capturing talent (both for the strategic core of long-term employees as well as for temporary hires and partnerships) is becoming increasingly crucial for the success of organisations. An important area of attention with a view to branding is the development and maintenance of a single overarching corporate culture that is consistent with the organisation's core values. Facility management plays an essential role in this task, which is not very easy due to the dynamics of working with temporary employees. On the other hand, the choice for a small and stable strategic core in combination with a dynamic, flexible shell is also visible in the sourcing of facility activities. This explains the growing popularity of the directing model and the rise of the demand management organisation (as the most far-reaching form of outsourcing, see section 1.5.2).

Another development that forces closed organisations towards openness is the transition to a circular economy (see section 2.4.5). Closing bio-based and technological cycles are not possible without entering into partnerships with suppliers and more intensive cooperation (and data sharing) in the supply chain.

A final significant development that moves closed organisations towards open networks is technology-driven. As described in section 6.4, smart buildings can be linked together to create smart cities. Sharing information on a regional level can help to maintain or improve the sustainability, quality, safety and liveability of areas.

6.8 General future expectations that will affect the FM-profession

The future of the profession is not unambiguous, but rather multiform. Therefore, in this section, we will describe some general developments that are expected to be relevant for all facility professionals. We will use the middle segment of the cone elaborated in Chapter 1: Facility Management (see Figure 6.2) as a structure. This circular segment contains the following internal areas of attention: identity, policy, marketing, quality management, procurement, environmental management, risk management and information management.

FIGURE 6.2 The focus of this chapter

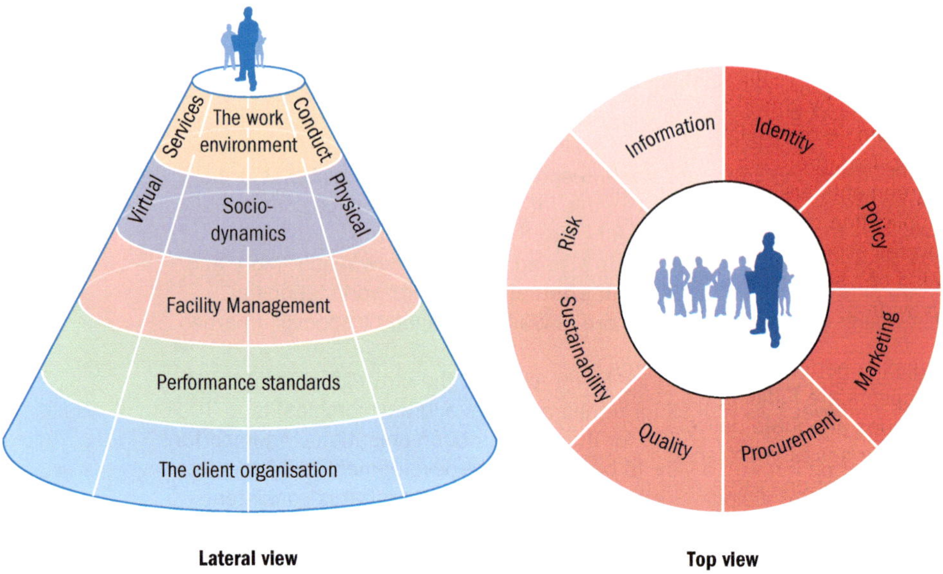

6

Based on these focus areas, we will describe possible future changes in the facility management domain. In doing so, we build on Chapter 4, inwhich all the focus areas from Figure 6.2 are explained earlier.

6.8.1 Developments in the field of policy making

As we already explained in Chapter 2: every facility professional has to - and wants to - anticipate the rapidly succeeding changes. Multi-annual planning is giving way to a multi-track policy for the short term, guaranteeing manoeuvrability. After all, only organisations that can adapt smoothly to the ever-changing environment will be successful. Flexibility and innovation are essential. The format of policy documents is also changing: from multi-page text policy plans to an overview of action points in table form, an animation or infographic, i.e. forms aimed at communicating the key points effectively both attractively and fast.

Expressing and giving meaning to core values is also becoming increasingly important. After all, the so-called purpose economy is emerging, in which people will increasingly pursue their passion and ideals (purpose) that give real meaning to society, including through work and enterprise. In a networked economy, organisations and independent professionals are increasingly reliant on each other. By being clear about their core values, organisations and individuals can find like-minded people to work together to achieve shared goals.

6.8.2 Developments in the field of branding and FM-marketing

Customer expectations will continue to rise under the influence of what (technological) comfort they are accustomed to at home. Means are not infinite, while these days, customers are accustomed to quality being free of charge. Managing customer expectations and maintaining customer satisfaction, therefore, requires more innovation and entrepreneurship in facility management.

Today's economy has many faces: sharing-economy, circular economy, purpose economy, experience economy, to name a few. The latter makes it clear that people are inclined to make non-rational purchases. Industry responds to this by focusing on storytelling rather than on price itself. Storytelling is a powerful way to evoke an experience with the customer. Here are opportunities to boost the image of the facility management profession with positive emotions and to communicate the added value of facility management to stakeholders in an appealing way. Thus, storytelling can help facility professionals to build their professional image: as individuals, as ambassadors of an organisation, and as a professional group.

In the past, marketing & communication departments had full control over communications with the outside world. Nowadays, this control is limited to the official statements of the organisation through formal channels. Business becomes more and more personal. How employees present themselves on social media (both commercial and private) and in contact with others, can strengthen them, but also weaken the image of organisations. People are mostly aware of the potential impact of business communications, but much less of the impact private actions may have in their professional lives. Here, too, a clear mission and core values and clear rules of conduct about what is and what is not desirable and what is essential. These guidelines help people judge whether an expression they make as a private person (but which does reflect on the company) is appropriate and desirable.

6.8.3 Developments in procurement management

Compared to the construction world, facility purchasing is in its infancy. For many years, the construction world has had a high degree of standardisation: of specifications, cost budgets and contracts, for example. A building specification always has a fixed chapter and paragraph classification used by every architect, structural engineer and contractor. With the publication of the ISO 41012 Facility Management: Guidance on strategic sourcing and the development of agreements in 2018, a classification for contracts, schedules of requirements and service level agreements is available for facility professionals. Standardisation will be an important theme for the years to come.

Also, the focus of procurement is expected to change. NEVI (2017) provides a comparison between current and future business drivers. Business drivers are aspects that define the operational and financial results of companies most. Table 6.1 shows that NEVI expects sustainability/CSR to become most important in the future. Not surprisingly so, as a circular economy entails that by definition, sustainability will be part of every programme of requirement. Moreover, the relative weight of sustainability indicators in the evaluation of bids will increase.

TABLE 6.1 Change of priority of business drivers (source: NEVI, 2017)

Current Business Drivers (2017)	Future Business Drivers
1 Cost reduction	1 Sustainability/CSR
2 Sustainability/CSR	2 Innovation
3 Innovation	3 Cost reduction
4 Quality	4 Ethics and integrity
5 Ethics and integrity	5 Quality

NEVI notes that over the past ten years, the procurement agenda has changed from a focus on cost reduction only to a balance between cost reduction, risk management and value creation. However, this change is not yet a fact of life for all parties involved. Awareness of this has, however, increased in organisations (and not only in procurement department). NEVI's other interesting findings relate to developments in the procurement profession in the near and distant future. Table 6.2 shows the effect of the shift in business drivers and the influence of social trends and developments on the purchasing profession (such as technological development, increasing dynamics of the environment, and labour market tightness).

TABLE 6.2 Short- and medium-term developments in the procurement profession

Top 10 Developments in Purchasing	
The next 2 to 3 years	*In 5 years and beyond*
1 Cost reduction	1 Socially responsible procurement
2 Automation (P2P, SAP, IoT)	2 Digitisation
3 Supplier management & relationship management	3 Innovation
4 Socially responsible procurement	4 Scarcity
5 Innovation	5 Agility
6 Business driven procurement (see Table 6.1)	6 Business intelligence & big data
7 Further professionalisation of the procurement process	7 Supplier management & relationship management
8 Value creation	8 Supply chain management
9 Contract management	9 Value creation
10 Best value for money	10 Talent development

Blockchain technology will have a major impact on contract management. This technology enables contracts to be signed, stored and made available to users without the risk of fraud. All actions in a blockchain can be traced back to users and changes can only be made with the consent of all authorised parties.

6.8.4 Developments in quality management

When you think of quality, you think of working conditions, the environment, and whether a product or service meets expectations and agreed-upon results. Quality management is all about expectation management. Under the influence of the purpose economy, people start focussing on bringing meaning to life. This changing mindset also changes the interpretation of quality.

In addition to people changing, technology is also changing the way we think about quality. Technology allows us to signal the smallest deviations from agreed results and expectations in real-time. Thus, continuous quality control will take place everywhere: on all equipment, in all rooms, on indoor climate, on response and waiting times, to name but a few. Fully integrated quality measurement in all processes will become the norm — and

continual monitoring comes with a vast amount of data. The real-time availability of big data makes it possible to map the effect of interventions on behaviour, use and consumption of raw materials. An example: an organisation wants to stimulate its employees to be more active, for example, by taking the stairs. The question is which incentive is most effective. Do more people take the stairs when the speed of the lift is adjusted downwards? And what does this do to customer satisfaction? Having this kind of data allows you to experiment and make evidence-based decisions based on the results; thus, research skills increase in importance.
In technological development, there is the danger of what is called the technology push, which means that advanced, efficient technical applications are always perceived as something good. However, this is not true in all situations. For example, automating reception with robotic solutions can be efficient, but for some customer groups, highly ineffective as they value personal attention and interpersonal contact. A key question for the future is how organisations can make technology work for them, with the hospitality performance optimally attuned to the intended goal, the target group and the circumstances at that time.
The last development we will discuss is how quality is reported. In a quality culture, all employees are focused on doing their job a little better every day. Minor improvements are easy to achieve, and many small adjustments together make a big difference. To do so, they must know the possibilities for optimisation. Visual management is a useful lean-methodology tool. It offers attractive and straightforward visual insight into the state of affairs, thus stimulating a quality culture while preventing information overload.

6.8.5 Developments in the field of environmental management

Of course, the transition to a circular economy will have a significant impact on environmental management. To make such a transition, governments use legislation and regulations as a steering mechanism. This means that organisations (and citizens) will be forced to invest in sustainability if they do not do so of their own accord. Although this is seen by many as far-reaching government interference, it does ensure a level playing field. Moreover, the intended transition of the economic system is only possible if everyone participates.
At present, the operating costs of companies that invest in sustainability are higher compared to competitors who do not; which creates a barrier and slows down the achievement of climate targets. If everyone invests in sustainability, these barriers will disappear. Moreover, market demand drives innovation. We know from the past that this increases the quality of solutions, while the price of measures decreases. Just look at the developments in the field of PV cells (solar energy).

Because the circular economy has already been extensively discussed in section 4.9, here, we will focus on the sustainable development goals that will apply until 2030. In 2016, the United Nations compiled so-called SDGs, or global sustainable development goals (see section 4.8.2). All UN members have pledged to work towards achieving these goals for a better, healthier and fairer world. How countries will work with the goals is not defined. Each state sets priorities based on the situation in its own country and chooses a politcally-supported approach. It is not a sanctioned

Global sustainable development goals

agreement, but a joint intention. Of the seventeen sustainable development goals, a number belong to the (core) areas of facility management:

- good health and well-being;
- gender equality;
- clean water and sanitation;
- affordable and clean energy;
- decent work and economic growth;
- industry, innovation and infrastructure;
- reducing inequality;
- sustainable cities and communities;
- responsible production and consumption;
- climate action;
- partnership for the goals.

As we indicated earlier, facility managers play a crucial role in sustainability. Wouldn't it be nice if the professional community came up with a concrete elaboration of SDG-proof facility management?

6.8.6 Developments in risk management

Not every risk can be avoided. But the facility manager will have to be well informed about which preventive measures should be taken from a logical and reasonable perspective. It is essential to realise that the risks to which an organisation is exposed are subject to change. New developments, applications or collaborations always bring along new risks. At the same time, we live in a society in which risk-taking is less and less accepted. Profound (new) risks are related to cybercrime, terrorism and climate change. The impact of these can be enormous. So even if the likelihood of such an event or disaster is small, a facility manager must do everything possible to minimise its consequences. Risk management could, therefore, become a vital area of attention for every facility professional.

A facility manager working at a multinational accountancy firm gave us insight into opportunities and risks that technology brings. S/he claims he knows a lot more about the behaviour, habits and needs of accountants, tax specialists and other employees than their directors. Much information comes to the facility manager, such as orders for lease cars, requests for literature (subscriptions and one-off expenses), study trips (conferences and training courses), times of working from home, room reservations and office attendance. If he would be allowed to enter this data into a learning database, he will be able to predict and thus better anticipate users' needs. But the question is to what extent he is allowed to store and use this information in the light of privacy legislation. And that is a good thing.
The fact that the new information technology is penetrating the field shows from the collaboration between one of the largest facility service providers in the world and IBM's Watson IoT Platform. This collaboration promises to be able to integrate and analyse data from millions of devices and sensors. They expect to gain more insight into people's use of the work environment, through sensors placed in buildings, including doors, windows, chairs, and meeting rooms. But what will happen if the building management system is hacked?

6.8.7 Developments in the field of information management

Figure 5.3 shows dashboard meters, which we use as a starting point in this section. Information technology developments that will influence the underlying information structure of these meters include:

- the use of sensors on equipment and in rooms in combination with the internet of things;
- the analysis of big data with artificial intelligence, enabling self-learning systems and automatic diagnosis and prediction-based control;
- the use of blockchain technology, which will provide scalable and reliable integrated information, including finance, agreements and contracts.

We follow the dashboard meters in Figure 5.3 from left to right and discuss service level; costs; contract review; and customer satisfaction.

Service level

Availability, response time, assortment and finishing are examples of service features that determine the service level. Service levels are not unambiguous classifications; a service level can be defined and interpreted in many ways. A standard for service levels will benefit the profession. An additional advantage is that due to its predictability, service levels can be translated into corresponding budgets.

A service level is generally defined in specifications to which the service contract refers. Nevertheless, often these are perceived differently in the various documents that mention service specifications, such as the PSC and SLA. This common shortcoming can be avoided by using a learning database, in which similar descriptions in various documents are recognised and automatically matched to each other.

Costs

The blockchain technology can play a significant role in cost control. A blockchain is a collective system that is accessible to all those involved, making it impossible to falsify the recorded data by corrupting one central point. Fraud is, therefore, no longer possible. A blockchain can contain property acts, agreements, personal messages or other data, which can be linked to financial transactions. It is possible to track the underlying contracts of each transaction. But also, interim agreements and adjustments are directly related to the whole and can be observed by all parties involved. This will make it easier to control expenditure and deviations from the budget.

Contract review

In smart buildings, sensors provide extensive information about, for example, temperature progression, humidity, movement (presence), use, consumption, deliveries, drainage, light and energy. These sensors provide all the input needed to assess contracts. And due to a much higher degree of standardisation, the arrangements will be minimised by referring to standard service descriptions.

Customer satisfaction

Nowadays, customer satisfaction is generally measured by surveys. However, more and more (small) touch screens are appearing throughout buildings, allowing the user to give a score for various test criteria on the

spot. It will become increasingly simple to perform statistics. By this, we mean that by using big data and artificial intelligence (a learning database) it will become possible to identify those factors that have an impact on satisfaction. Because of the advancing information technology possibilities, facility managers will know what the dissatisfiers, satisfiers and delighters are (see section 4.5.2). In this way, they will be able to better adapt the service to the wishes of the user, at the lowest possible environmental impact and cost. As a result of the knowledge thus created, facility managers will ultimately be able to optimise the level of service.
A widespread trend is the use of visual management in the transfer of information to employees. Visual management is a lean management tool that presents information in such a way that it is clear at a glance what the message is and what needs to be done. Colours and symbols show the status and deviations from the planning or standard. Because it is well applicable to support dynamic processes, it is used in agile working methods (such as scrum). Examples are a planning board, a scrum board and an A3 (named after the size). Visual management reduces the risk that important information (e.g. about the progress of the goals or the quality of the work) will be lost or not noticed in time.

6

Visual management practices

6.9 At a glance

In the previous sections, we have provided a wealth of information. That is why we will conclude this chapter with a summary. Figure 6.3 is a visual representation of the quadrants.

FIGURE 6.3 Summary of characteristics

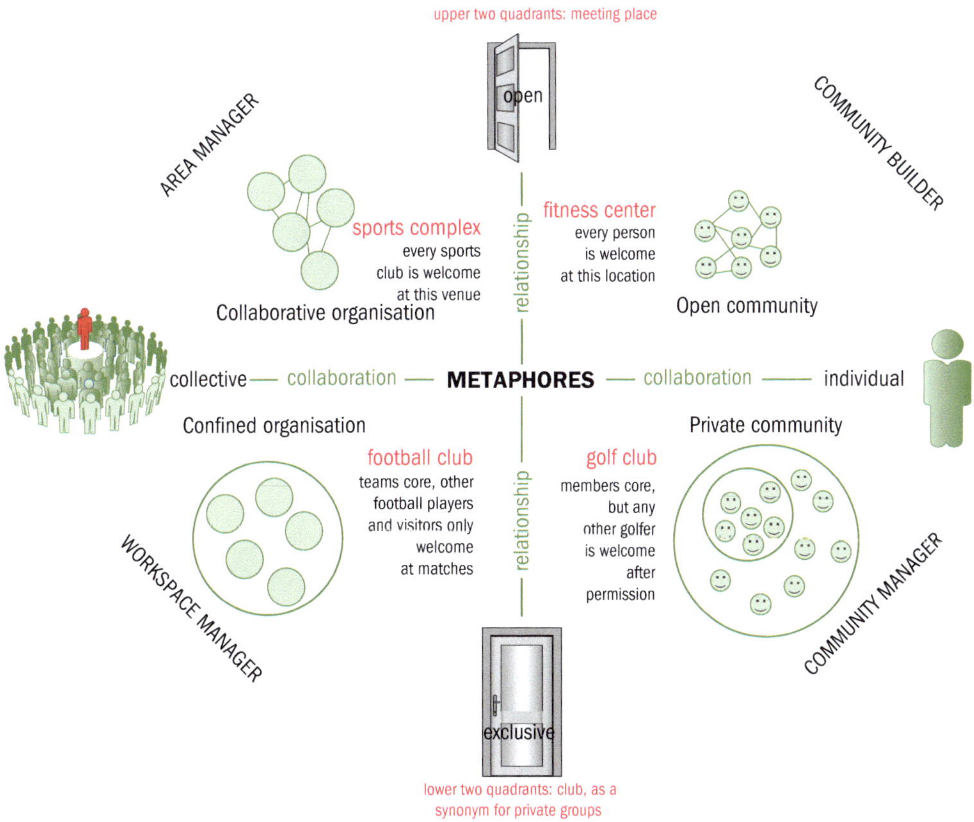

Table 6.3 lists the most prominent characteristics of each manifestation of the FM-profession described in this chapter. For some aspects, we use a five-point scale to indicate the extent to which something is essential or occurs. It is an ascending scale, where 5 equals the highest intensity or the highest importance.

6

TABLE 6.3 Comparison of the Roles of Area Manager, Community Builder, Community Manager and Workspace Manager

	Area Manager	**Community Builder**	**Community Manager**	**Workspace Manager**
Physical Environment	Defined area (business park, campus, shopping area, (city) district)	Distinctive location in an easily accessible area, such as public transport hub, in the city centre or on campus	Location, nature and size depending on the intended use and target group	Location of the organisation
Clients	Association of owners, business association, BID, municipality, real estate investors	Not in case of a (self-employed) entrepreneur, or the entrepreneur who owns it	Property owner (multi-tenant buildings), a joint venture (partnership or board of directors)	For the facility manager: board of directors/management[en]For providers: the facility manager
Target Group(s)	Multiple organisations: all stakeholders in an area.	Individuals: self-employed professionals, mobile workers, customers using meeting facilities.	Individuals or smaller companies: tenants or owners (in case of partnership), members of the community, their visitors and clients.	Organisation and everyone who works for the organisation, visitors and customers.
Revenue Model	Budget out of contribution; not for profit (but the area manager has to generate a payback).	Rental income; (also) profit objective.	Budget of client, budget via tenants, based on service costs, or (commercial) letting of facilities, grant; depending on form and purpose, profit/no profit motive.	Budget: not for profit or based on an offer; for profit: (providers).
Examples of Added Value	Image (branding), sustainability (including on-site joint energy production), risk management, innovation, the value of real estate, attractiveness (safety, quality, accessibility), cost control (exploiting economies of scale).	Innovation, flexibility, productivity, satisfaction, social capital of the community.	Highly dependent on target group, but in any case: satisfaction, image (branding), flexibility, risk management, sustainability, productivity/innovation support, cost control, and for commercial real estate also the value of real estate.	Supporting culture, image (branding), productivity (supporting innovation, and vitality), flexibility, sustainability, cost control.
Role	Connector, consultant, service provider, purchaser	Entrepreneur, connector, event manager	Service provider, connector, consultant, purchaser, depending on the target group and goal: entrepreneur	Consultant, service provider, purchaser and/or entrepreneur (supplier)
Number of Stakeholders	Large and known	Large and partly unknown	Average and known	Small and known
Dynamics	Average	High	Average	Predominantly low, but highly dependent on the type of organisation
Importance of networking (1-5)	4	5	3	2
Importance of branding (1–5)	3	5	4	3
Degree of Entrepreneurship (1-5)	4	5	3	2 (providers: 5)

Scenario-planning helps to interpret future developments yet gives a limited representation of all the possibilities that might be available. If one chooses other driving forces (e.g. artificial intelligence and transition to circular economics or not), then other opportunities and challenges for the facilities industry become visible. It is explicitly not a question of either/or, but of and/and. Also, within the chosen driving forces, it is possible to make nuances. Figure 6.4 shows that all kinds of hybrid forms occur, in which user networks are a mix of organisations and individuals that differ in terms of openness. It demonstrates that the matrix, the quadrants and the axes, as shown in Figure 6.3, should not be interpreted too strictly.

FIGURE 6.4 Matrix with additional hybrid forms

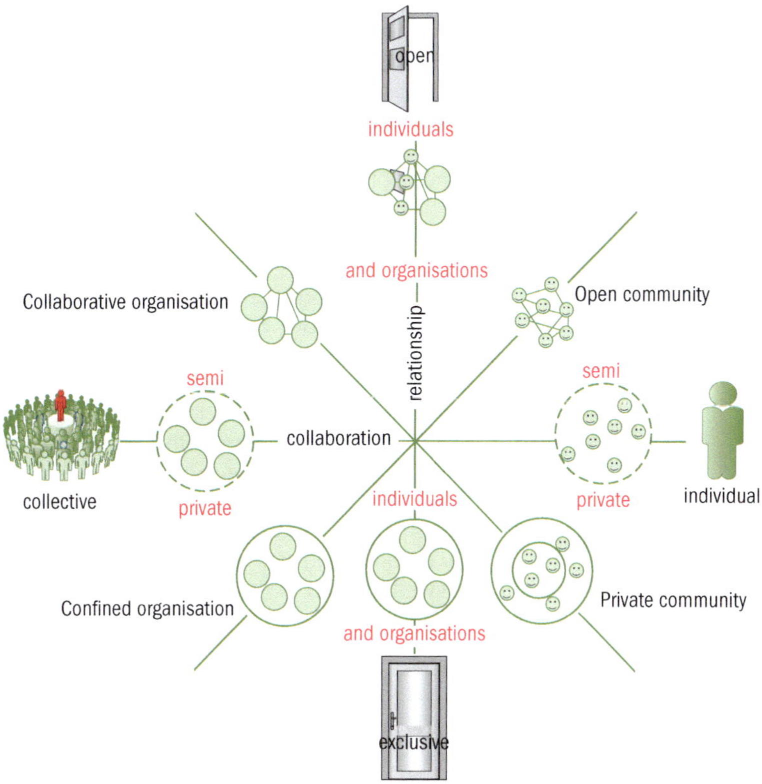

The following web message on 'gig-spaces' shows that the distinction between an organisation and self-employed professionals can be diffuse in reality and that there can be more than one dispute. The employees in the fixed core form a closed organisation. However, that core depends on employees who work for the organisation on a flexible basis (often not exclusively). Although there are reasons not to fill these positions with permanent employees, there is a need to ensure that they become part of the community - without this being at the expense of the overall security level.

6

● www.smartwp.nl 23-01-2019

From workplace to 'Gig-Space'

The future of our work is connected to the breakthrough of the platform economy, according to real estate specialist Colliers International in its most recent white paper.
After the boom of companies such as Uber and Deliveroo, the platform economy is now going to shake the labour market of office workers. Work will become flexible in time and place and on-demand. Three characteristics that do not fit in well with the way offices are set up. For companies, it is essential to use gig spaces for gig workers to create a gig workplace.
Deliverers and drivers do not use an office, but this is the case for app developers, data specialists, marketers, designers, call centre employees, copywriters and administrative staff. Some of these functions become part of the gig structure and are therefore no longer a permanent part of the core of the company.
Real estate managers now have the task of housing these gig employees in the right place within the organisation. The solution is to set up three different zones in the office building: public space, gig space and core space.
The gig space is the zone between the public space and the workplace for the permanent employees and is specially designed for gig employees. Without a lengthy screening procedure, they will have access to facilities and the systems needed to carry out their work. This increases the speed and agility of the organisation.
Housing gig employees in coworking areas is not a real option for excessive rent is then paid for flexible workplaces that are, in fact, permanent. Moreover, placing these activities outside the office is detrimental to the collective identity of the company. The gig workplace is, therefore, the structural housing solution. Gig employees are 'in house', making them part of the community of the organisation. This increases their involvement, ensures a closer collaboration with other employees and contributes to the pursuit of collective success.

Young Professionals in Facility Management

Charlotte de Block, Facilities Organisations Project Coordinator for Germany, Austria and Switzerland at Tesla, graduated Bachelor of International Real Estate and Facility Management at BUAS, The Netherlands, in 2018

Always accept new challenges, even if you are not sure how you will manage

When I was young, I often worked in my parents' bakery when I came out of school. I liked to help them with organising the business, to be hospitable and to support all the processes. When I started the study program of 'International Real Estate and Facility Management' at the NTHV/BUAS, I realised that the course suited me well. The combination of managing different teams, working on the customer experience and the business' administrative part, in some way are a reflection of working in the family business.

When I graduated, I immediately received a job offer at the company I fulfilled my thesis internship. The step from being an intern to a hired employee proved to be much bigger than I had expected. As an employee, you are more independent and bear responsibilities without the protection of a manager. However, it worked out fine. After being responsible for three locations in the Netherlands for five months only, I received the opportunity to switch focus to sites in Germany, Austria and Switzerland. Although based in Amsterdam, every day I am in direct contact with different countries and cities. The different cultures I deal with daily make my job a fascinating one. A challenge is building good working relationships with people from a distance, especially with people from another culture.

Since my graduation, I learned that it is essential to remain open to new challenges and tasks. It is often better to first say yes and see how things work out in practice than to decline because of any fears you may have. It is always possible to address the workload once you have experienced that it has become too heavy. Keep in mind: you will grow with every new challenge you accept!

Summary

- Developments in society manifest themselves in how companies operate and how people work - directly influencing location-choice, the demand for housing, the use of the workspace and the need for facility support.
- The development of businesses focusing on a stable core with a flexible shell goes hand in hand with the steep rise of the number of independent professionals.
- Every working person is connected to one or more organisations - whether that organisation is small or large. The nature of that relationship determines the future role of the facility professional.
- A network is what we call the total of relationships. A network can have an open, partly open and closed character.
- The networked economy offers opportunities for facility professionals to engage in facility activities in new contexts: in the role of area manager, community manager or community builder.
- These additional roles bear similarities and differences. Analogies are the need for entrepreneurial skills, the focus on networking, and the importance of branding - albeit differences in the emphasis. Differences lie in the physical environment, the type of client, the target group, the added value at stake, the role, the financial model and the number of stakeholders.
- The role of in-house facility managers will also change, resulting, for example, from the changing labour market and increasing automation and digitalisation. Strategic goals in the field of productivity, attractive employment and corporate social responsibility are drivers for a more intensive collaboration with HRM, IT and Real Estate Management. The workplace is a business resource for these goals but requires an integral approach. That is why we think the job title 'workspace manager' better suits the scope of the facility management function.
- Policies are increasingly contained in concise plans with a short planning horizon, using visual management to get the message across. Clear core values can help to find like-minded people (employees, suppliers or network partners).
- Customer expectations will continue to increase under the influence of what they are used to in their private lives. Means are not infinite, while customers are accustomed to free-of-charge quality.

- Social developments have an impact on the procurement agenda.

- In quality management, big data contributes to real-time measurements of effects and insight into influencing factors for customer demand and behaviour.

- Of the 17 global sustainable development goals, a few belong to the (core) areas of facility management. Facility managers play a crucial role in sustainability.

- Three significant challenges in terms of risk management are cybercrime, terrorism and climate change.

- Information management is changing as a result of the internet of things, the analysis of big data by artificial intelligence and by blockchain technology.

- Scenario planning helps to interpret future developments but gives a limited representation of all the possibilities that exist. In practice, all kinds of hybrid forms of the scenarios will occur.

Questions and Assignments

6.1 Shirley is a facility manager at a medium-sized company with flexible working practices. The management of the company has decided that the organisation will focus on its core activities because it is increasingly difficult to predict the market. This unpredictability increases the importance of agility of the organisation. It is expected that in time 40% of the current workforce will consist of independent professionals who will be hired for individual projects or assignments. These independent professionals will then work temporarily at the corporate location, for example, as a member of a project team. Shirley says that the workplace concept will have to be revised. The management doesn't see the problem; there is a flexible work concept - and soon, many desks will be empty. Do you know why Shirley suggests that changes are necessary?

6.2 Social developments influence the procurement agenda. In section 2.7, we described that sustainable business operations are about finding a balance between the three Ps: people, planet and prosperity (or profit). Table 6.4 shows that purchasers expect this balance to shift in the future. Can you explain why buyers predict that sustainability will become more important than costs?

TABLE 6.4 Change of priority of business drivers (source: NEVI, 2017)

Current business drivers (2017)	Future business drivers
1 Cost reduction	**1 Sustainability/CSR**
2 Sustainability/CSR	2 Innovation
3 Innovation	**3 Cost reduction**
4 Quality	**4 Ethics and integrity**
5 Ethics and integrity	5 Quality

6.3 Table 6.5 is a section of Table 6.1. Here you can see that the extent to which entrepreneurship is essential for a workspace manager strongly depends on his or her employment. How can you explain this?

TABLE 6.5 Importance of entrepreneurship per role

	Area Manager	Community Builder	Community Manager	Workspace Manager
Degree of Entrepreneurship (1-5)	4	5	3	2 (providers: 5)

6.4 Community builders and area managers are the roles in which entrepreneurship skills are fundamental. Yet tasks and responsibilities are quite different. Which aspects of the role of area manager determine the importance of entrepreneurial skills?

6.5 All facility professionals have to deal with complex customer relationships. The stratification of the FM-customer has already been discussed at length in this book. Table 6.6 shows the clients, the revenue model and the number of stakeholders of the area manager, the community builder and the community manager. For each of the roles, indicate what drives the complexity of customer relationship(s).

TABLE 6.6 Clients, stakeholders and revenue model

	Area Manager	Community Builder	Community Manager
Clients	Association of owners, business association, BID, municipality, real estate investors.	Not in case of a (self-employed) entrepreneur, or the entrepreneur who owns it.	Property owner (multi-tenant buildings), a joint venture (partnership or board of directors).
Number of Stakeholders	Large and known	Large and partly unknown	Average and known
Revenue Model	Budget out of contribution; not for profit (but the area manager has to generate a payback).	Income from rental of facilities; (also) profit objective.	Budget of client, budget via tenants based on service costs, or (commercial) letting of facilities, grant; depending on form and purpose, profit/no profit motive.

6

7

Research in Facility Management

In this chapter, we will address the following questions:

- What is research?
- What is the difference between research conducted by research universities and research conducted by universities of applied sciences?
- How is research organised within universities of applied sciences and how can students participate?
- What contribution does research make to the further development of the discipline of facility management?
- What are the topics that are the subject of research?

◄ Through the eyes of the facility manager:
Museum Naturalis employs more than a hundred researchers. It is a research institute and a museum combined, where you can experience the wonderful world of nature. And all this in one building: hospitality for museum visitors versus a restricted laboratory environment.'

Expert Column

Contributing to a better World

What thrills me as a researcher is the opportunity to gain deep insights into the working life of FM professionals, and to help them see innovations that increase the added value of what they do. In particular, I am proud of the work I do to speed up sustainable transitions in the field of FM. I am motivated to make an impact by participating in research and development projects. And by sharing the lessons learned at conferences, in publications or by other means of communication, I hope to inspire others to make similar efforts and achieve even more significant impacts for a better world.

Every student project has the same opportunities based on a real-life problem: to investigate, to discuss appropriate strategies and to develop pragmatic solutions that are ready for implementation. In my experience, there is a genuine and positive attitude towards researchers and students. Working together with research and educational institutions, our collaboration partners reflect upon strategies and learn new approaches.

The world of facility management is multidisciplinary, and therefore, we need to combine know-how and theories from various disciplines to produce new knowledge that can change the world. When we do this, we should do it with full transparency of the methodologies we use. Anyone wishing to contribute to FM innovation and knowledge development should, therefore, adopt the basic skills of research.

I hope you will enjoy reading Chapter 7!

Susanne Balslev Nielsen
FM Expertise Director in NIRAS A/S and former researcher/educator at the Technical University of Denmark and Oslo Metropolitan University, Member of ISO Technical Committee 267 Facilities Management

7.1 Introduction

This chapter explains what research is and why research is vital for the further professionalisation of facility management. This first section defines several general concepts that you need to know to understand what research is. Sections 7.2 and following deal with the organisation of research in higher professional education and the main themes of workspace research. To illustrate this, FM-Research groups of the Dutch universities of applied sciences present and explain their focus.

Research meets information need. Entrepreneurs, managers, specialists, organisations and even society at large, seek answers to questions, for example, to be able to make a decision. Sometimes the information that is required is not available. In such cases, it may be necessary to conduct research. Research always starts with an information need and is carried out to find facts on which to base the answers. Here are a few examples to illustrate the importance of data for decision-making:

- A facility manager has set the goal of improving the quality of service. To achieve this, (s)he will first have to establish the baseline quality, to be able to demonstrate any improvement.
- A facility provider wonders how a new service concept can best be put on the market. First of all, it is necessary to understand the concept thoroughly and to describe it in such a way that it will be able to convince potential customers. Thus, it is essential to investigate the actual benefits as perceived by the customer, for example, by asking consumer panels.
- A facilities company wants to know what the users think of a particular service to be able to show that their service meets the service levels agreed to in the contract. The supplier will have to use a measurement method to gather reliable (real-time or periodic) information about customer satisfaction.
- The facility manager of a university of applied sciences wants to know the average utilisation rate per classroom and lesson to assess whether extra capacity is needed. To this end, data from the scheduling programme and the room reservation system will be analysed for planned occupancy. On the other hand, registering attendance on site will provide insight into actual utilisation. Based on these sources, it is possible to assess the extent to which the capacity is sufficient.

But questions can also be more general:

- On what aspects do guests base their first impression?
- What determines the cleanliness experience of users of sanitary facilities?
- Which service aspects are most decisive for the satisfaction of the internal customer?
- How can the customer journey of hospital visitors be described, and how can hospitals provide a more welcoming experience?
- What is the effect of workplace concepts on the productivity of employees?
- What contribution does the introduction of time- and location-independent working make to the sustainability of business operations?

7

Asking a question

In other words, research is always about asking a question and searching for reliable answers to that question in a systematic and verifiable manner. Research is a process that proceeds according to preconceived steps.

There are two common misconceptions about research:

1. *Every thorough investigation always leads to a satisfactory result.* Expectations of research are sometimes too high. And often the study is only an intermediate step towards the final result. The result of the study is insight, in the form of an answer to the research question. On this basis, solutions can be devised, and advice can be given. In a thesis, you will also recognise this division. Most theses start with a management question. To give advice, make a design or come up with a plan, you first need information. This is the goal of the field research. The field research provides part of the building blocks for the management advice, design or plan requested.
2. *Anyone with common sense can do research as a side task.* The skills and time required to carry out research properly are regularly underestimated. Research is a labour- and knowledge-intensive activity. To do proper research, you need a down-to-earth, critical attitude that leaves room for new ideas. This requires thorough training in research methods and techniques. However, you also need sufficient knowledge of the field you are investigating. Moreover, proper research involves training. This is also the reason why (almost) every course pays attention to research methods and techniques as early as the first year.

Every student in higher vocational education and at universities faces research: in school (learning how to do research), internships and projects (practising how to set up and carry out small research) and during the final year of graduation (thesis). As a result, research universities and universities of applied sciences produce many publications, such as articles, research reports and graduation projects - not only by students but also by professors, researchers and lecturers. Institutional repositories have been set up to support open access to research results. These are an essential gateway to studies carried out by universities and universities of applied sciences and are a valuable source of information for students on a wide range of subjects. The European network of institutional repositories is called DRIVER (Digital Repository Infrastructure Vision for European Research). DRIVER provides access to over 2,500,000 scientific Open Access publications from around 250 repositories in 33 countries (see: www.driver-repository.eu). Via NARCIS (see: www.narcis.nl) all Dutch repositories of universities of applied sciences and universities can be searched, including the HBO Kennisbank.

7.2 Why do institutes of higher professional education conduct research?

All countries must invest in knowledge development and innovation if they want to maintain economic growth. For Europe, investing in the knowledge economy is necessary because Europe cannot compete on labour costs. The production of labour-intensive goods and services is rapidly of-shored to countries where wages are lower, such as China and India. This is not limited to manufacturing, such as cars or computers, but also, for example, about call centres. In terms of knowledge and innovation, Europe can compete with its relatively highly educated population and research tradition.

7.2.1 Research as second core task for universities of applied sciences

Since 2001, in addition to providing education, applied research has been a core task of universities of applied sciences. Because of the strong relationship with practice on the one hand and education, on the other hand, practice-based research is characterised as 'connective research'. Universities of applied sciences can help to further the professional field by providing answers to specific questions that prevail there, thereby complementing the existing range of knowledge available. In this way, universities of applied sciences act as knowledge partners that contribute directly to innovation. The results of this research are directly applicable in practice. This is in contrast to so-called fundamental research, which aims to develop new knowledge and insights without (immediate) application in practice. This is the type of research carried out at research universities (which also carry out applied research).

7.2.2 Research groups

Professors head research groups. One or more research groups can form a knowledge centre. Some research groups provide minors (electives) for which students can apply in the main phase of their studies. Other research groups offer thesis projects for students. In section 7.4, four professors responsible for research in the facilities domain explain the research they carry out and how students are involved in it.

Research is expensive because it is labour-intensive. Moreover, technical research often requires costly equipment. Many research programmes are grant-funded for a limited period of, for example, two years. Another source of income is contract research for companies or institutions that have a specific

7

demand. These are often short-term projects; the challenge for research groups is to attract partners from the market who are interested in long-term collaboration. Some research groups collaborate with research universities to bring the results of fundamental research into practice, and vice versa, to underpin results with more fundamental insights from meta-analysis of multiple studies.

Knowledge and insights are not only passed on to practice but also to students and lecturers through education. Students who conduct research learn to ask valid questions, collect and analyse data and then to draw conclusions from it. This is important, because we live in an information society and the quality of the decisions people make depends on the accuracy and topicality of the information they have at their disposal. Facility managers can not only rely on their observation and gut feeling but will also base their decisions on the results of research. It is important that they have enough knowledge of research methodologies to be able to assess whether the results are reliable and useful in the context of their own question.

7.3 Research in facility management

In 2017, Wim Pullen, director of the Center for People and Buildings (CfPB), and Professor Rianda Appel-Meulenbroek of Eindhoven Technical University have defined the four most essential research themes relating to the work environment. They worked together with experts in the field. The aim of these top 4 is to give an impulse to a combined research agenda for research by both real estate management and facility management. The starting point was a shortlist of six research themes based on an earlier study by the CfPB in 2016. In order of importance, they established the following research themes for the coming years:

1 *User behaviour.* The central question is: how can the use of the work environment be designed, in such a way that users can work productively and be satisfied with the facilities offered? It is crucial to gain more insight into how people make choices and how they (can) give feedback about their experience. This issue concerns both individual users and teams. On an individual level, it is about connecting the workplace concept to a diversity of users as well as work style preferences. It is also about the proper use of the workplace concept in practice. At the team level, supporting social cohesion, cooperation and team spirit are important. At the same time, the organisation strives for agility to be able to respond quickly to ever-changing circumstances. Moreover, the concept of a team is also fluid. Previously this only referred to the department where someone worked; nowadays people increasingly work in temporary, cross-departmental project teams, whether or not in combination with external parties.

2 *Business cases.* The central question is: how can we make balanced business cases that take into account the interests of all stakeholders in both the short and long term? An integrated business case needs a broad approach, but it must be manageable in practice. Supporting impact (other than financial) is also a challenge in many cases. After all, what tangible contribution does the work environment make to employee vitality? How do you measure it, so that you can calculate with it? How do you determine the share of the work environment in the overall employer-attractiveness? And

how do you determine positive or negative effects of the work environment on employee productivity? To make the situation even more complicated, not every stakeholder has the same interests and priorities. How do you deal with these differences in building a business case? This theme underlines the importance of further research into the added value of facility management and real estate management.

Themes 3 and 4 are an elaboration of two of these added values.

3 *Productivity*. The main question is: how can the work environment support and increase the productivity of knowledge workers? This question is related to the second research theme. As almost all organisations try to control productivity, it has been included as a separate theme. Through experiments, the facility and real estate domains try to find out what works, what does not work, and why it does or does not work within which context. As many factors influence the productivity of knowledge workers, it is necessary to gain more insight into the determining factors, their relative importance and the connections between them. For thesis projects, these studies are too large. Still, you may be researching possibilities to increase the job satisfaction of employees. Because we know from current research that workplace satisfaction is one of the factors that influence productivity, the underlying goal of such an assignment may be to increase the productivity of employees. From earlier studies, we know that the degree to which knowledge workers are satisfied with the support of their personal productivity depends on their satisfaction with personal control and choice of the work environment, possibilities for (informal) communication, and possibilities to work in isolation.

4 *Vitality.* The central question is: how can the work environment positively support the health and well-being of employees? This concerns the effect of the physical work environment on the physical and emotional well-being of employees. Over the past decades, research has mainly focused on illnesses related to the work environment. An example of this is the sick building syndrome associated with the indoor climate. Nowadays, the positive effects of the work environment on people's well-being and stress levels are given more attention. Examples of positive outcomes are the effect of artificial light that mimics the natural course of daylight on blood pressure, and the effects of nature on concentration, relaxation and comfort. This research theme is of interest. Not only because the workforce is ageing, but also because the labour market is tight and less absenteeism due to illness has a positive effect on the productivity of the organisation as a whole. Another reason is that employers who care about their employees are perceived as attractive by potential employees. Attention to vitality, therefore, serves many purposes.

Research results are used to make well-considered and informed decisions, further professionalising the field. This is the reason why many professional associations research the most important trends and developments in the facility management market. As an example, the main findings of Dutch and Belgium facility and workplace trends research 2019 are summarised in the box below. Although this is not a research agenda, it is possible to deduce from these trends which issues the facilities industry in these countries will be dealing with in the years to come.

As you can read in the box, sustainability and technology are critical issues. However, the four main research themes of the CfPB and Eindhoven University of Technology do not explicitly mention them. This may seem strange, but can be explained by the fact that sustainability was not an essential trend in FM-market research in 2017 either. These rapid swings once again show the dynamics around the world of work. Given that the environment is continually changing, new questions will always arise.

FACILITY & WORKPLACE MANAGEMENT IN 2019 - RESEARCH INTO THE DUTCH AND BELGIUM MARKET SIZE, TRENDS AND DEVELOPMENTS

The Facility & Workplace Management Market Research Report 2019 is part of a biennial trend survey carried out by consultancy firm Twynstra Gudde on behalf of FMN, the Dutch professional association for the facilities domain. It is the sixth time that this survey has been carried out. In the study, over 500 facilities professionals indicated which considerations play the most significant role in their policymaking. The respondents mentioned sustainability as the most critical trend for the first time (55%), followed by customer focus (44%) and technology (39%). Similar research in Belgium is conducted by IFMA Belgium Chapter and consultancy firm PROCUS Group. In 2019, they published their biennial study on the trends and developments in de Belgium FM Market for the third time. The three most important topics influencing the work of Belgian facility managers at that time were: (1) change in the strategy of the client organisation (24%); (2) sustainability (19%); (3) the integration of new technologies in facility management (15%).

Changing strategy of the parent organisation. This is the result of increased dynamics of the business environment (as discussed in Chapter 2 of this book). The lesser the flexibility of facilities organisations themselves and the working environment, the more effort it takes to adjust to changes in the client organisation.

Sustainability. In the previous Dutch survey, only 15% of the respondents found sustainability to be the most important trend. According to the researchers, the Paris Climate Agreement, Urgenda and UN climate targets have broken the pattern. However, the step from good intentions to practical plans has yet to be made. For example, 73% of the respondents indicate that their organisation has sustainability ambitions. Still, only 60% of them also have specific sustainability goals for the facilities organisation. Only 51% have drawn up a plan of approach specifying those objectives. In Belgium, the emphasis in the previous market survey was on energy efficiency, and this continues to be necessary. After all, in Europe, 40% of the CO2 emissions from buildings come from energy consumption. It will take much more to achieve a climate-neutral society in 2050. How facility managers can contribute to this varies greatly. Anyone who not only uses buildings but also owns them will have to improve their energy efficiency and make them more sustainable systematically. At the moment, no less than 97% of all buildings do not comply with the requirements for energy-neutral buildings. With a renovation rate of 1% per year, it is clear that without taking firm action, the desired result will not be achieved. Sustainability will remain an essential point of attention for the time being.

Customer focus. The trend customer focus is also a new entrant in the Dutch top 3 and even this trend scored low (18%) in 2017. According to the researchers, the increase in this trend seems to be related to the decreased attention for cost management. Cost management was still in the top 3 in 2017 at 33%, now it has dropped to place 7 (19%) and is expected to drop to position 9 in the next three to five years. According to the researchers, the increased popularity of customer focus can be partly explained by the recognition of the added value of FM for the success of HRM policies. Customer focus offers an opportunity to be distinctive in a tight labour market where the war on talent is in full swing.

Technology. Technology seems to be becoming one of the most critical trends in the Dutch FM-market. According to the respondents, it has become clear in recent years that it's not about the technology itself, but rather about how you use it. According to the researchers, big data and internet of things (IoT) are essential technologies for the facilities market. With 82 and 73% respectively, they leave the other techniques - 3D printing, augmented reality, biometrics, cloud computing, artificial intelligence, robotics and 3D speech technology - far behind. Respondents see the most value in big data for attractive workplaces and qualitatively better workplaces and see the most value in IoT for sustainability and efficient use of space. In Belgium, the integration of new technologies in the FM-domain is the focal point. The three most important topics in the field of technology are big data in buildings/smart building technology, Integrated Workplace Management Systems (IWMS) and the use of mobile apps. The integration of building information model/management (BIM) and FMIS are to expected to become topical in 2022.

7

Research results lead to new insights, new methods, new forms of contracts and contract management and new ways of managing facilities organisations. And to further questions. For example, based on the section on sustainability, you can conclude that many organisations are still looking for ways to concretise their sustainability ambitions. In the growing attention for customer focus, you can see the link to the added value of facility management. Many researchers, market parties and students around the world are researching this issue. These insights further the professionalisation of the discipline; illustrated by the declining position of cost management to the increasing attention for customer experience. Thinking in terms of added value is also reflected in the discussion of technology in facility management as a means of reducing the ecological footprint or supporting attractive employment practices.

7.4 Facility research by universities of applied sciences

Facility management is a business discipline that includes elements of economics, engineering, construction, behavioural sciences, information and communication technology and legal affairs. You could say that there is no such thing as FM-research, but rather that research takes place within a facility management context. And it is the context which determines whether the results are useful to the facility manager. Various universities of applied sciences have set up a professorship that focuses on research and education within the facilities domain. At the time of publication of this book, there are four Dutch 'facility professors', according to the FMN professional association. They play an essential role in knowledge development and knowledge exchange. In this section, they explain what the most critical research themes of their professorship are and what role students can play in their research, as an illustration of connective research in FM.

Professor dr. Joop van Duren – research group Facility Management, Zuyd UAS

Facility Management is responsible for creating, maintaining and managing 'fit for use' environments and associated services by aligning people, processes, space and technology.
It is a broad field that is in a state of flux. Sustainability, reallocation, sensor-controlled installations and the internet of things are just a few examples that drive this development.
Facility Management training courses aim to deliver sought-after professionals with relevant and up-to-date knowledge and skills. To be able to do that, education needs to tap into new insights that arise from addressing issues in practice. And that is precisely the purpose of the professorship:

1 to carry out applied research that takes practice a step further (which, incidentally, makes an active contribution to community building);
2 to embed the research results in curricula (to help keep the education up to date);
3 to engage students and lecturers in research so that both groups become more proficient.

In determining the research agenda for the second term, the research group 'Facility Management' has been strongly guided by what is vital in the Limburg region. This region suffers from a high level of vacancy; vacancy in so-called public real estate, in particular, poses a threat to the quality of life. Hence the first research theme: vacancy and reallocation.
The second theme is also relevant to this region: quality of life in districts, neighbourhoods and villages. This quality of life is under pressure in many places due to demographic decline, and technological and economic developments. How can we maintain or improve this quality of life? And what contribution can facilitators make in stimulating community initiatives, connecting interests, coordinating supply and demand and, above all, empowering citizens to take the lead?
The third and last theme concerns circular procurement. Purchasing in such a way that 'lifecycle wise' smart solutions are chosen and that after the period of use materials are not lost but remain in the cycle by giving them a new destination.
Research on these themes requires a broad approach. That is why the research group works together with research groups in Healthcare (District-based Care) and the Built Environment (Smart Urban Redesign) on the theme of Smart Living. Smart Living is about challenges in the built environment and the living environment. Given the complex issues and challenges facing us, a multidisciplinary approach is essential in these areas.

We hope you will further broaden your view and that you will learn to look beyond the boundaries of your field of expertise. After all, it is at the intersections of traditional knowledge domains that most innovations are born.

For more information, please check: www.zuyd.nl/onderzoek/lectoraten

Professor dr. Brenda Groen – research group Hospitality, Saxion UAS

Man is a social being and needs interpersonal contact, a sense of belonging. Everyone is, to a greater or lesser extent, part of social and business networks. Networks can vary from strictly personal to strictly

company and from narrow (small number, intense and close, such as family, good friends) to weak (many but vague acquaintances).

Within networks, people share knowledge and acquaintances - resulting in increased social capital. Social capital is the sum of the actual and possible sources in and from your network of contacts. Social capital comes from the people in your network, and the knowledge, skills and things they possess and that you may use. In a social sense, networking is vital for wellbeing, diversity and inclusiveness, but also for stimulating business activity and innovation. This applies to corporates as well as to independent professionals and small and medium-sized enterprises.

The Saxion professorship 'Hospitality' is interested in factors that stimulate or hinder networking within and between organisations or companies. The professorship also looks at the added value of these - hospitable or otherwise – networks in three research programmes: Hospitality, Safety & Security; Realising Hospitality; Hospitality, Buildings & Collaboration.

A variety of factors determines the ease with which you get to know and maintain contact. Examples of these factors are your personality, the availability of means of communication, the way you work, the culture and the physical environment. The physical environment can encourage or hinder meeting, sharing, collaboration and community building.
Facility managers play a definite role in this. For example, networking in an office building is stimulated when there is sufficient opportunity to meet each other both planned and spontaneously. Encounters with your close ties, but above all with new people or people with whom you have a weak connection. It is precisely these people that one depends upon for fresh and new ideas. The building, the hospitality-concept, and the way of working of individual employees and organisations play a role in this.
This applies not only to corporates who own accommodation but also to independent professionals, micro-businesses, start-ups and scale-ups based in concepts such as coworking spaces, incubators, serviced offices and multi-tenant buildings. They too have a particular interest in network and community building.

Facility managers are responsible for work environments, for 'the bricks and mortar' and for services. They can facilitate meetings, create a hospitable building and stimulate encounters and networking. This is how they contribute to the ultimate goal: happy employees who feel part of a company, or community of independent professionals that share knowledge. And above all, people that are productive, for their employer or - when self-employed - for themselves.

'The working environment is increasingly becoming a network of people and places.'

For more information, please check: www.saxion.nl/onderzoek/areas-and-living/hospitality

Professor dr. Frans Melissen – Professor in Sustainable Experience Design & Management Education for Sustainability, Breda UAS

Many people think of sustainability as green, environmental, more expensive and less luxurious; often something that is necessary, but not really attractive. However, sustainability is about much more than saving energy and being wise with water. The underlying question is how to ensure that future generations can have a (pleasant) life. It's about people, about a fair distribution between here and there, between now and later. That's why my personal definition of sustainability is a lot more positive: taking good care of people and things around you! Who can object to that? Who wouldn't want that? And if you use this guideline for behaviour and decisions, it will automatically lead to the necessary savings and redistribution of resources.

My professorship 'Sustainable Experience Design' focuses on the question of how companies and other organisations, in all kinds of industries, can contribute to the desired sustainable development. Within this context, two important developments come naturally together: (1) the increasingly urgent need to take major steps in the field of sustainability and (2) the increasingly important role that experiences play in the lives of many citizens, certainly in their roles as consumers and employees. The second development is often referred to, at the intercession of Pine and Gilmore, as the rise of the experience economy. Whatever name you give it, what is clear is that we as consumers and employees are not satisfied with just any product, service or interpretation of our work. We live in a society where identity is more and more a personal choice (and not a derivative of upbringing or origin) and personal branding is a must. Products, services and work/career are increasingly a vehicle for something we crave: experiences that help us shape our identity, that enable us to communicate our identity to others and that give us access to the groups we want to belong to.

Against this backdrop, creating experiences that can make sustainability part of these desires and demands of today's consumers and employees plays a crucial role in promoting sustainable development. How do you, as a company or organisation, create experiences that appeal to the consumer and your employees and at the same time give an impulse to sustainability? By making the experience itself less wasteful or by making the experience a stimulus for sustainable behaviour by that consumer and employee, also in the longer term. That is the challenge that my lectureship explores and tries to find and share solutions for.

7

It is clear that facility management can play a crucial role in this within organisations. This research group therefore likes to work together with facility managers to give substance to this in practical situations.

For more information, please contact: melissen.f@buas.nl.

Professor dr. Mark Mobach – research group Facility Management, Hanze UAS Groningen, also senior lecturer in Facility Management at Wageningen Research University

Facility management often distinguishes two sub-areas: space and infrastructure (including buildings, systems, layout, workplace and interior) and people and organisation (including human resources management, marketing, procurement and logistics). The applied research of the research group 'Facility Management' at Hanze UAS focuses on an integrated approach to these sub-areas: a link between business issues (people and organisation) and spatial and technological issues (space and infrastructure). How, for example, can you design a space in such a way that the organisation performs better as a result? And can we prove that this is because of space?

The research of this research group, therefore, focuses on facility management in the broadest sense: a meeting of space and organisation. Our practical research is also aimed at improving the actions of the facility manager, given specific practical problems and issues. The research results always lead to better knowledge about the possibilities of using facility management to influence business performance positively. Think, for example, of simple changes in care or in the workplace that can generate demonstrable business benefits, such as stress reduction in patients and healthier employees in the office.

Currently, the research group consists of twenty members, eleven of whom are PhD candidates. We always collaborate with universities in the Netherlands and abroad. Currently, we have four innovation labs: 'campus design', 'healthy cities', 'health space design' and 'healthy workplace'. In an innovation lab, we link PhD research to FM subjects, and we combine teacher-research with student research. In graduation projects, we always work in a practice-oriented and preferably multidisciplinary way, to find better solutions. The learning experiences and results find their way to education through publications, presentations and discussions.

For more information, please check: www.hanze.nl/nld/onderzoek/kenniscentra/kenniscentrum-noorderruimte/lectoraten/lectoraten/facility-management

7.5 Research under the influence of trends and developments

In the field of facility management, there is still a lot to investigate. And the dynamics of the domain change the direction of such research. Sustainability, the transition to the circular economy and the potential and impact of technology (including big data) are subjects that require research, as the market research described above shows. Developments such as 'as-a-service' concepts also raise new research questions. Finally, risk management, welfare and vitality, and productivity are research areas that require continued attention. Here, too, there is a relationship with trends and developments. As we pointed out in Chapter 2, as a result of demographic developments, the labour market is tight, and people retire at a higher age. This explains the interest of companies and researchers into optimal support of vitality. The increasing occurrence of extreme weather, the advancing digitalisation and the increased level of threats (terror) lead to new risks - and thus to questions in the field of risk management and security. Besides research in universities, this information-need is also evident in the programming of professional conferences and whitepapers by leading providers.

Because the changes are happening so fast, another interesting research question pops up: how can the facility manager anticipate the future? That is why we have devoted a separate chapter to it (Chapter 6). A question like this is quite abstract and requires input from many different perspectives from other sciences, such as philosophy and sociology, but also technology. This is why the relationship practice <–> practice-oriented research <–> fundamental research is so important.
Intensive collaboration between professionals, students, lecturers and professors for the professional field, universities of applied sciences and research universities contributes to more precious insights and higher levels of innovation.

Young Professionals in Facility Management

Thomas Klein Bramel, procurement consultant at Tender Services Group in The Hague, the Netherlands - graduated bachelor (hons) Facility Management at Hanze UAS in Groningen, The Netherlands in 2018.

From Graduation Project to European Tendering Consultant

During your studies, you will face several choices. For example: which specialisation within Facility Management would suit me, which minor helps to build my profile, where would I like to graduate? The combined result of all of such choices forms the foundation upon which you will develop your career. At the time, I decided to take a minor in procurement management and do my graduation internship at a procurement consultancy firm in The Hague, the Tender Services Group. As a Northerner, this was quite a change of scenery for me. Upon the successful completion of my graduation period, I joined the Tender Services Group in the position of procurement consultant.

As a consultant, I am currently responsible for the tendering process of bidding parties (our clients) and for writing winning quality documents for these tenders. The (European) bids I am involved in vary from cleaning services and ICT to legal advice. Thus, my work is highly diverse. Moreover, the stakes are high, as it revolves around millions of euros and involves significant interests for the bidding parties. For example, when the incumbent supplier does not win the tender, he loses his business. These factors also motivate me to do everything in my power to submit a winning bid in close collaboration with my clients. Nothing beats celebrating a successful tender with the client.

In addition to our office in The Hague, we now started a branch in Groningen. This expansion offers many opportunities for both our organisation and for me personally - as a colleague and I share the responsibility for its success.

Summary

- Research meets information need by searching in a systematic and verifiable way for the most reliable possible answers to a research question.

- Universities focus on fundamental research in which new theories are often the starting point. Universities of applied sciences focus on practice-oriented research in and for the market. The starting point is a question or problem that is urgent in the market.

- Applied research is the second core task of universities of applied sciences. This core task has been assigned to professorships. A professor heads a professorship or research group.

- Research is essential for the further professionalisation of the field of facility management. It contributes to new insights, new methods, new forms of contract and new ways of managing facility organisations. In today's information society, the quality of people's decisions is dependent on the accuracy and topicality of their information.

- For the coming years, the most important research themes for facility management are diverse: enhancing user experience, strategic alignment, holistic business cases, productivity support, vitality support, healthy work environments, sustainability, circular economy and circular procurement, the impact of new technology on facility management, liveability in neighbourhoods, vacancy and redevelopment, and supporting community building through the physical environment.

- FM-research does not exist as a separate discipline. It is the context in which the research is carried out that determines whether the results are useful for facility professionals.

Questions and Assignments

7.1 Suppose, as a near graduate bachelor in Facility Management, you can opt for a thesis project commissioned by a company or for a thesis project commissioned by a professorship. Would that make a difference? Research is research, or isn't it?

7.2 Research contributes to the further professionalisation of the discipline, but research is not always necessary. In which situations is it not useful to conduct research?

7.3 Read the text below.

In future, the office will be everywhere
M. de Hoog

The 'experience offices' are a hit with the employees. But does it also deliver something? At Samas, an office furnisher, they say a definite yes. Absenteeism has fallen sharply, and in the first year after implementation, people felt better. Van den Heuvel: 'And if you like your work, you work better too. Productivity skyrocketed. At Microsoft, they also see greater employee satisfaction and lower absenteeism. However, the consequences of the new way of working cannot be captured in other figures. There is not yet much scientific evidence that the New World of Working (NWoW) increases output. Peter van Baalen of Erasmus University Rotterdam (EUR) is investigating the New Way of Working at Microsoft, among others, by interviewing employees. They appear to be very satisfied with their new work environment. But productivity has only increased slightly.
In any case, Microsoft and Samas are delighted with their new way of working. Productivity is difficult to capture in figures', says Hans Bos of Microsoft. However, we notice that it has increased.

Samas indicates that productivity boomed after the introduction of NWoW and Microsoft and also noted that productivity has increased. Researcher Van Baalen of the EUR has, however, scientifically proven that the productivity of employees increases only slightly when applying the Innovative Workplace. Can you explain this difference in findings?

Glossary

Accommodation	Literally: providing shelter. Almost everything that has to do with buildings is included.
Accommodation management	Accommodation management revolves around the optimal coordination of all accommodation-related activities. These can be internal activities, but also activities concerning business strategy, business philosophy, a (changing) housing need or the real estate market in a general sense.
Account management	The intermediary function between the client and the facilities organisation. An account manager is responsible for the smooth running of the process of matching supply and demand. An account manager is often expected to be result-oriented and communicative.
Activity	Task or tasks that contribute to the completion of deliverables.
Activity-based working	Workplace concept in which employees always choose a workplace that suits the task they are performing at a particular time. In general, you will find concentration areas, informal meeting spaces, areas to work together, and areas where confidential (telephone) conversations can be held.
Actual costs	The actual costs are formed by the actual expenses.
Added value	Refers to the impact of facility services on the functioning of the organisation, for example by creating a working environment that supports the productivity of employees or a hospitality concept that strengthens the organisation's image.
Aggregate (to)	Building up information, from detail level to more abstract level.
Agile	Agile is an organisation's ability to timely change direction or focuses in response to an emerging opportunity or threat.
Agile working	Agile is a collective term for ways of working that offer maximum scope for interim adjustments based on progressive insight. The starting point of agile working is that dynamics and change are part of the project. Scrum is a specific form of agile project management.

Applied research	Research conducted to answer specific questions from the market/work field, aimed at direct application. Applied research is a core task of universities of applied sciences institutions. Also referred to as practice-oriented research.
Area manager	Management function responsible for the facilities in a specific area, e.g. a business park.
Asset	Item, thing or entity that has potential or actual value for an organisation.
Asset management	Coordinated activity of an organisation to increase the value of assets.
Audit	Systematic, independent and documented process for obtaining audit evidence and evaluating it objectively to determine the extent to which the audit criteria are fulfilled. The purpose of internal FM-audits is to ensure that the FM system conforms to the management system standard; that it is achieving its objectives, conforms to its planned agreements, that service levels have been adequately implemented and maintained, and opportunities for improvement are identified.
Authorisation	Authorisation is the right to make decisions necessary for the execution of a task, or the right to access information.
Back office	See Front and back office.
Balanced Scorecard (BSC)	The Balanced Scorecard is a management system aimed at translating an organisation's strategic goals into a set of performance objectives that, in turn, are measured, monitored and changed if necessary to ensure that the organisation's strategic goals are met. The Balanced Scorecard combines four perspectives: customer and stakeholder satisfaction; financial performance; internal business processes; knowledge and innovation.
Benchmarking	Process of internally or externally comparing processes, performances and quality against practices of the same nature, under the same circumstances and with similar measures to improve efficiency and increase competitiveness and optimising processes.
Block plan	Drawing up walls on a floor plan after finalising the bubble plan.
Blurring	New combinations of products and services across industries, such as a catering establishment in a department store or co-working spaces in a theatre lobby.

Bottleneck products	Products purchased from a limited number of suppliers and that are important for the continuity of the organisation. The supplier has much power in negotiations about these types of products. The purchasing strategy is, therefore aimed at securing the supply of such resources.
Branch	Part of a larger structure, for example: a branch of a bank or multinational. Can also indicate a specialisation within a larger domain, such as social media marketing is a branch of marketing.
Brand	Identity of a company, distinguishing it from its competitors.
Branding	Giving meaning to the brand to increase its attractiveness, e.g. employer branding to become more attractive as an employer.
BREEAM	Assessment method to determine the sustainability performance of buildings. The abbreviation is for Building Research Establishment Environmental Assessment Method.
Brief	A document containing the specifications for a product or service to be delivered to avoid different interpretations during the execution. Also called Programme of Requirements (PoR) and Specifications (specs).
Bring Your Own Device (BYOD)	This means that everyone brings/chooses hardware according to personal preferences; instead of the tools being imposed by the IT department.
Bubble diagram	A bubble diagram is a schedule in the form of a floor plan, in which all functions are arranged based on the desired mutual relationships and required minimum areas. This plan consists of one or more floor plan drawings in which the use of space per department or function is indicated by 'zoning' outlined by a 'bubble' on the plan. The level of detail is not yet very high.
Budgeted costs	Costs that the facilities company forecasts and that are calculated each year in advance and, after approval, laid down in the annual plan.
Building Information Modelling (BIM)	A process for managing the information produced during a construction project, in a standard format, from the earliest feasibility stage through design, construction, operation and finally demolition, to make the best and most efficient use of that information, involving the generation and management of digital representations of physical and functional characteristics of places. It is also known as 'digital twin'.
Building maintenance	Building maintenance comprises all building-related activities to maintain and optimise the functional value of the building.

Building Management System (BMS)	Computer-aided control systems (including hardware and software) to collect and monitor parameters and performance data of a plant, equipment, systems and elements, either at a source or remotely and that enable corrective action.
Building operations management	All building-related services aimed at maintaining the functional value of the building.
Built environment	Collection of buildings, landscaped areas, infrastructure and other construction, works within a space.
Business continuity	Capability of the organisation to continue the delivery of products or services at acceptable predefined levels following a disruptive incident (ISO 22300).
Business Continuity Management (BCM)	Holistic management process that identifies potential threats to an organisation and impacts to its business operations. It provides a framework for building organisational resilience by being able to continue critical business processes in the event of a calamity at a previously agreed level and possibly at another location. The ISO 22301 offers guidelines.
Business Improvement District (BID)	A defined area within which businesses are required to pay an additional tax (or levy) to fund projects within the district's boundaries. The services provided by BIDs are supplemental to those already provided by the municipality. Examples are improving traffic facilities, signposting, green areas, waste collection, lighting, cleaning, maintenance, damage repair, fire safety, graffiti removal and camera surveillance.
Business model	The elements of the business that create and deliver value. These elements normally include the value proposition, the customer segments and their associated relationships, the channels used to take products, services and solutions to market, the revenue & cost streams, partners, and the critical resources and processes of the organisation.
Business Model Canvas (BMC)	Tool for defining the value proposition and mapping the business model of an organisation or organisational unit and, consisting of nine building blocks in four main focus areas: customers, supply, costs and revenues and infrastructure.
Business processes	Sequential clustering of activities to the production of goods or services (see process).
Business services	Companies that offer services to other companies or individuals for profit (profit sector). For example, banking institutions, insurance companies, accountancy and law firms.
Carbon footprint	See Footprint.

CEN	European Committee for Standardisation: a public standards organisation whose mission is to foster the economy of the European Union (EU) in global trading, the welfare of European citizens and the environment by providing an efficient infrastructure to interested parties for the development, maintenance and distribution of coherent sets of standards and specifications.
Circular economy	An economic system designed to maximise the reusability of products and raw materials and minimise value destruction. The circular economy has two cycles of materials. A biological cycle in which residues flow safely back into nature after use, and a technical cycle in which products or product components are designed and marketed in such a way that they can be reused at a high(er) quality level (up-cycling).
Circulation areas	Space used for in-door traffic, for example, the corridor, the stairwell and the central hall.
Cleaning services	All services and resources aimed at cleaning and maintaining the property and inventory of an organisation in a clean condition.
Client	Internal client at a strategic level in the organisation; that has the power to delegate authority and allocate resources within the organisation. Within the ISO standard also referred to as 'demand organisation'.
Code of conduct	Guideline that indicates how employees or companies deal with situations. Can also apply to a specific profession. Dutch examples of this are the Buyer's Code of Conduct (drawn up by NEVI), the Code of Responsible Market Conduct for the cleaning, catering, security and removal industries.
Code for responsible market conduct	Dutch code for countering excessive market forces that threaten the continuity of suppliers' business operations because they compete too strongly on price. The signatories of the code undertake to take responsibility for realistic quotations and the responsible awarding of work.
Collaborative organisations	Companies and institutions that (intensively) collaborate in a network whether or not within a particular geographical area. This type of organisation is also known as a network organisation.
Commercial catering	Service model in which a commercial catering provider is invited by an organisation to come up with its own proposals for the method of catering and the range on the basis of which it believes it can make a profit.

Commodity	Product or service with a high routine content. The execution of the service is usually described in manuals. Low savings potential because the costs are low. The purchasing strategy focuses on simplifying the purchasing process and reducing the number of suppliers. Procurement activities are focused on standardisation, framework contracts and fixed order, order and invoice handling processes.
Company catering	All services related to the consumption of food, refreshments and beverages, also known as foodservices. It mainly concerns: lunch provision (in the company restaurant), workplace and meeting service, drinks and snacks provision (usually with vending machines) and catering for events (such as receptions). This also includes the management of the company restaurant, including the kitchen.
Computer Aided Facility Management (CAFM)	Systems Applications and tools that automate functions needed to support the core business in its efficient and effective use of facilities. See also: Facility Management Information System (FMIS) and Integrated Workplace Management Systems (IWMS).
Consumer	Recipient of facility-related services. An internal customer at the operational level of an organisation also called end-user or benefitting customer.
Contract	Agreement under which two parties undertake to exchange a product or service for payment.
Contract management	Managing, implementing and enforcing contracts. Managing contracts involves archiving contracts and managing relevant supplier data. Implementation is aimed at announcing the agreements and explaining the agreed working method.
Contract register	System in which contract data is stored. This system makes it possible to order online from a digital catalogue. It follows the administrative procedures that are necessary for handling the invoices under the agreements in the contract.
Control information	See Management information.
Control principle	How the management wants to manage the organisation and how responsibilities and powers are delegated to lower levels in an organisation.
Controllable fixed costs	Fixed costs that can be altered in the short term, for example by changing contracts.
Coordination	Mechanism that ensures that existing work is distributed among workers and that work is coordinated to ensure that the work is carried out efficiently and effectively.

Core activity	Strategic tasks that improve customer value and drive profits and to which an organisation owes its existence and success. In principle, core tasks are not outsourced.
Core business	Entity from which needs are derived.
Core processes	See Primary processes.
Core values	Principle(s) that an organisation views as being of central importance and in which it wants to excel. These guiding principles dictate behaviour and can help employees understand what is expected of them.
Corporate Social Responsibility (CSR)	A way of doing business in which a company, besides its economic performance, takes account of the impact of its activities on the environment (planet) and human aspects inside and outside the company (people). The aim is to find a balance between the three Ps.
Cost centre	A way of organising facility management in which the main job of the facility manager is to provide services at the lowest possible cost. The budget is set at management level, and internal customers are obliged to 'purchase' services from its own facilities company.
Cost of capital	Cost of external funding; mainly interest costs for bank loans.
Coworking	Coworking involves a diverse group of people who don't necessarily work for the same company or on the same project, working alongside each other, sharing the working space and resources.
Coworking spaces	Workplaces in shared areas that everyone can use flexibly (for a fee) as third workplaces (see Coworking).
Cradle-to-Cradle	The core of the cradle-to-cradle policy lies in the concept: 'waste is food'. All the materials used can be put to good use for another product at the end of their useful life in one product. No quality loss is allowed, and all residual products must be reused or environmentally neutral. See also: the circular economy.
Critical Success Factor (CSF)	Business variable that is critical to achieving the business strategy or important activity; that in which the organisation must excel to be successful. A CSF describes the condition and helps to determine whether the organisation is doing the right things (effectiveness). A CSF is qualitatively described, such as 'speed of delivery'.

Crowd management	Activity that focuses on the safe management of crowds and flows of people, for example, at events. Part of crowd management is crowd control. Crowd management can also be seen as controlling a group and crowd control as directing a crowd.
Culture	See Organisational culture.
Customer	Internal customer at the tactical level in the organisation; the paying customer.
Customer excellence	A combination of cost leadership (operational excellence) and customer intimacy (customer intimacy) value strategies. This value strategy focuses on maximising customer satisfaction and efficient process management. It well suits organisations that apply lean management.
Customer journey	Model for mapping the guest's experience from the very first contact up to and including departure.
Customer focus	Thinking and acting from the perspective of the customer, anticipating customer needs and a high priority on service and customer satisfaction.
Customer-driven organisation	Organisation that approaches the customer's question with personal care and attention. The employee is the problem owner. See also Customer-oriented organisation.
Customer-focused organisation	Organisation that approaches the customer's question by means of rules and regulations. Management remains the 'owner' of the problem. See also Customer-driven organisation.
Customer intimacy	A value strategy aimed at long-term relationships with customers. See also Value proposition.
Customer Satisfaction Survey (CSS)	Instrument to collect information on the extent to which the service meets the needs and expectations of the customer.
Defining customer, paying customer and benefitting customer	Customers of the facility division stratified based on the hierarchical level in the organisation: strategical level (defining customer), tactical level (paying customer) and operational level (benefiting customer).
Delighters	(Aspects of) products or services that the customer does not expect and with which he is pleasantly surprised. As he does not expect them, he is not dissatisfied when they are not there. For example, a bowl of fresh fruit at the counter, a bottle of drinks for visitors when they leave or a beautiful bouquet of flowers. See also satisfiers, dissatisfiers and KANO-model.

Demand	Stated requirement for services or products to be delivered.
Demand organisation	Entity which has a need and the authority to incur costs. Also called client.
Demand management organisation	Organisations with a far-reaching form of FM-outsourcing with a high management-operational split. Characteristics are minimal FM-organisation that has the primary task to define the demand of the organisation and to allocate it to suppliers acting as partners. Also known as a demand management organisation.
Depreciation costs	Estimated or expected decline in value of an asset. In accounting: to spread the costs of an asset over several years.
Directing organisation	Principal organisation form of facility organisations, in which the operational services are carried out and managed by external parties. The facility manager is responsible for translating the customer's demand into a suitable range of services, their purchase and contract management.
Dissatisfier	Elementary provision that is necessary but does not contribute to satisfaction. In general, dissatisfiers do not stand out to a customer or guest until they are not in conformity with the standard, such as coffee that is not tasty or a beamer that does not work. See also Satisfiers, delighters and KANO-model.
District manager	Responsible for the quality and quality of life in an area, for maintaining relations and conducting consultations with the administration, higher authorities, subsidy providers, citizens, market parties and the external representation of the municipality concerning area development.
Diversity and inclusiveness	The extent to which the people within the organisation recognise, appreciate and utilise, the characteristics that make individuals unique. It can relate to age, beliefs, class, ethnicity, physical abilities, race, religion and sexual orientation.
Document management	Control over the generation, distribution, storage and archiving of information. Initially, it only involved copying, printing and archiving, but nowadays it's all about the entire process.
Driving forces	Core uncertainties, a trend and countertrend that determine the future of an organisation. For example, higher or lower economic growth.

Ecosystem	A fundamental principle of an ecosystem is interdependence, i.e. something that happens in one part of the system may affect other parts within the system. In the context of an organisation, there are many factors external to it that affect how it operates, but over which it has no control. These can include government policy, the economic and social make-up within its region and neighbourhoods, the general religious and cultural expectations of its communities, demands for sustainability and available financing.
Effectiveness	Extent to which planned activities are realised and the planned results achieved.
Effort-based contract	Traditional form of contract in the cleaning industry in which the cleaning frequency is specified on aspect level. Also referred to as input-specified contract.
Employer branding	See Branding.
EN 15221	Standard developed by the European standardisation body CEN to standardise facility management at the European level. The EN 15221 consists of five parts after part 1 and part 2 were withdrawn (replaced by ISO): EN 15221-3 Processes in FM; EN 15221-4 Taxonomy of FM; EN 15221-5 Facility processes and organisational improvement; EN 15221-6 Space measurement and surface determination and NENEN 15221-7 Benchmarking.
End-user	See Consumer and Defining customer.
Environmental management system	Structured and coordinated approach to dealing with the environment in the most optimal manner possible within a company, according to specific standards. Which legal rules apply depends on the organisation's activities.
Environment	Society consisting of all parties and stakeholders.
Ethical behaviour	Ethics are well-founded standards in a culture by which a person's actions are classified as right or wrong. They influence behaviour and help an individual to make the right choices and act responsibly from its cultural or social perspective.
European tendering	According to the European Procurement Act, there are five ways in which procurement can take place for public procurement. (1) The open procedure; (2) the restricted procedure; (3) the negotiated procedure with prior notice; (4) the negotiated procedure without prior notice and (5) the competitive dialogue. The contracting party is free to choose between the open procedure and the restricted procedure with prior selection. The other methods are subject to conditions.

European Result Oriented Quality Measurement System (ER-QMS)	Quality measurement system to measure the quality of cleaning work per room. This level ranges from 1 to 6. The higher the agreed quality level, the fewer errors (in jargon 'disruptions') you may find after cleaning. This system meets the European standard EN 13549 and is increasingly used in output-specified cleaning contracts.
European tender	Supplier selection according to European legislation and guidelines. The starting point is that every supplier throughout the European Union has an equal chance of acquiring an order from the government. See European tendering.
Executive level	See Operational level.
Exponential times	Increasingly substantial social changes that succeed each other ever faster.
External customer	Customer, guest or visitor of the organisation: visitors and patients, but also hotel guests, holidaymakers, visitors to events. For the facility manager, this is the client of the internal client. For commercial facility providers, the external client is the client.
External developments	Factors that the organisation itself can influence to a lesser extent.
Facilities	Tangible assets that support an organisation.
Facilities main process	Facilities process described in outline from input via throughput to output.
Facility Audit	Determining the extent to which the provision of an existing facility matches the needs within an organisation. The audit usually is performed by an (external) inspector (auditor) or mystery guest.
Facility Excellence Model (FE)	A variation on Total Quality Management (TQM), designed for facilities organisations. The FE-model is based on four development phases: (1) skilled FM-organisation; (2) introverted FM-organisation; (3) extraverted FM-organisation and (4) enterprising FM-organisation.
Facility management	Organisational function which integrates people, place and process within the built environment to improve the quality of life of people and productivity of the core business. Management of all services related to the work, care and professional living environment and that are needed to support people in their pursuit of added value for the organisation.
Facility management division	Business unit that is responsible within the organisation for an optimal working environment, both physically and virtually.

Facility Management Information System (FMIS)	Integrated package of linked software applications, intended to support facilities organisations in managing and executing processes.
Facility metrics	Ratios are metrics that enable the assessment and comparison of facility costs and performance on a tactical level. See performance indicators.
Facility service desk	Point of contact (virtual or physical) for requesting information and action in response to facility-related needs; for requesting services and reporting faults and complaints.
Fit-for-use	Literally: suitable for the intended use, i.e. tailored to the wishes and needs of the organisation and the users.
Fixed costs	Costs that recur periodically and do not change in principle, such as for subscriptions, coordination of services by external parties and all services insofar as the costs thereof are contracted at a fixed price.
Fixed price	Set price for which the contractor enters into an obligation to achieve a result. If more people or material are required to achieve the agreed result than anticipated, the additional costs will be borne by the provider or consultant (hence the risk premium).
Flexible working arrangements	Work programs wherein employees are given greater scheduling freedom in how they fulfil the obligations of their positions. The most commonplace of these programs is flexible time arrangements, which gives workers far greater leeway in terms of the time when they begin and end work, provided they put in the total number of hours required by the employer. Other standard flexible working arrangements involve telecommuting, job-sharing, and compressed workweeks.
Flex-working	Workplace-concept in which employees do not have a fixed workplace, which increases the occupancy rate of workplaces. Working from home is also part of flexible working (location-independent work).
Floor Area (FA)	The floor area of buildings is the sum of the area of each floor of the building measured to the outer surface of the exterior walls including the area of lobbies, cellars, elevator shafts and in multi-dwelling buildings all the common spaces. Surface space of balconies is excluded.
Flow chart	Graphical representation of successive activities and information flows within an organisation, in chronological order. A flowchart gives an overview of the interrelationships between processes in organisations.

FM-procurement	Purchase of all services and resources related to the work environment and not directly intended for the primary production processes. Form of indirect procurement (see Indirect procurement).
Footprint	A measure to compare the environmental impact of different types of consumption behaviour (lifestyle). It is an indicator of the sustainability of certain consumption patterns. Also referred to as carbon footprint.
Freelancer	Also referred to as an independent contractor or independent professional. Terms commonly used for a person who is self-employed and is not necessarily committed to a particular employer long-term. Freelance workers are sometimes represented by a company or a temporary agency that resells freelance labour to clients; others work independently or use professional associations or websites to get work.
Front and back office	Organising principle for processes that separates customer contact and execution. A front office is primarily responsible for the interaction with customers and users. It is the central contact point for all questions and comments from customers. The back office is responsible for the production and delivery of products and services and resolves complaints and malfunctions.
Front office	See Front and back office.
Functional quality	Quality in which the how-question prevails: how did the service process proceed? What was the response time? Is the environment where the service is enjoyed pleasant?
Fundamental research	Scientific research for obtaining new knowledge and insights, theoretical in nature. Fundamental research is a core task of research universities.
Furnishing plan	A plan that provides information about the placement of furniture in the various rooms. The furnishing plan is the next step after the adoption of the block plan.
Gamification	The application of game-design elements and game principles in non-game contexts, for example, to stimulate the desired behaviour of the user.
General Data Protection Regulation (GDPR)	A regulation in EU law on data protection and privacy for all individual citizens of the European Union (EU) and the European Economic Area (EEA), that also addresses the transfer of personal data outside the EU and EEA areas.

Gig space	Zone between the public space and the workplace for employees who are not employed by an organisation, but who do provide services to the organisation and need a workplace to do so. Without a lengthy screening procedure, these workers gain access to facilities and the systems required to carry out their work. This increases the speed and agility of the organisation.
Global Sustainable Development Goals (Global SDGs)	The United Nations' 2030 Agenda for Sustainable Development provides a shared blueprint for peace and prosperity for people and the planet, now and into the future. At its heart are the 17 Sustainable Development Goals. They recognise that ending poverty and other deprivations must go hand-in-hand with strategies that improve health and education, reduce inequality, and spur economic growth – all while tackling climate change and working to preserve our oceans and forests.
Governance	A framework of authority and control within an organisation used to help it fulfil its legal, financial, ethical and general stewardship obligations.
Gross Floor Area (GFA)	Area of all floors together, measured from the outside of the facade. The external walls, corridors, stairwells and elevator shafts are therefore also included in the GFA.
Gross Internal Area (GIA)	Area of a room or a group of rooms, measured at floor level, between the bounding upright partition constructions of the individual rooms.
Guest cycle	Customer journey in four steps: pre-arrival, arrival, stay, departure.
Hard services	All services relating to buildings, fixtures and fittings and installations (space and infrastructure). See also Soft services.
Health care industry	Economic sector made up of companies that specialise in products and services related to health and medical care, such as hospitals and nursing and care institutions.
Hospitality	The art of offering sincere attention, safety and comfort to the customer, resulting in a customer's experience of hospitality.
Hostmanship	The art of making people feel welcome.
HVAC	The abbreviation for Heating Ventilation & Air Conditioning
IT	The abbreviation IT stands for information technology. This includes everything related to software and hardware.

Identity	'DNA' of an organisation, which is expressed in branding, communication, behaviour and performance.
Indirect procurement	The procurement of all goods and services for a business to enable it to maintain and develop its operations. Products and services classified as indirect procurement are commonly bought for consumption by internal stakeholders (business units or functions) rather than the external customer or client. The procurement of FM services belongs to indirect procurement (see Facility procurement).
Interdepartmental charging	Allocating costs to internal customers (budget holders) on the basis of actual costs, a fixed rate or a combination of these.
Image	Appraisal of the organisation by customers and other stakeholders, in other words: the reputation.
Impact	Indirect result that follows from the output of a process. Also called outcome.
Impact analysis	Analysis in which, for each scenario, the probability and impact are determined.
Inclusive society	A society in which everyone, regardless of limitations, can equally participate. This is the result of the UN Disability Convention of July 2016. This treaty establishes the rights of people with disabilities.
Independent contractor	See Freelancer.
Independent Professional (IP)	See Freelancer.
Industry	A particular sector of a country's economy is the part connected with that specified type of industry.
Information	Composite and structured data, which are related to each other in such a way that they can be controlled.
Information management	A process that ensures that the information needs of organisations are translated into information provision, concerning the accessibility and quality of information for work processes.
In-house emergency services	Team responsible for safety at work, providing first aid in case of accidents; firefighting starting fires; alerting and evacuating all persons present in the building in case of an emergency; supporting other first-aid organisations, such as the fire department and ambulance service.
In-house service provision	See In-house sourcing.

In-house sourcing	Manner for organising facilities services in which the services have been merged into one company. The operational services are carried out and controlled by the facilities organisation. The traditional management organisation usually works with a budget set by the management, which may or may not be recharged to the users.
Innovation	The development of new products, services or practices.
Input	The input for facilities processes consists of real estate, technology, staff and knowledge.
Input-specified contract	Contract based on input specifications, in which the service provider essentially delivers some predetermined resources within a specific quality and time frame at an agreed price.
Integration of functions	Integrating FM functions involves seeking and identifying potential synergies for reallocating task responsibilities, multi-skilling personnel and scheduling activities so as to improve productivity of FM personnel; developing consistent information and data standards; developing consistent workflow processes for all service lines; implementing FM technologies to support the consistent delivery and measurement of FM services.
Integrated Facility Management (IFM)	In integrated facility management, (most) services are outsourced to one external supplier. Processes are outsourced at a (partly) strategic level (directing), tactical (structuring), and operational (performing) level. IFM entails one contract with one supplier. IFM is also called total facility management (TFM) or total facility services (IFS).
Integrated Facility Services (IFS)	See integrated facility management.
Integrated Workplace Management System (IWMS)	A software platform that helps organisations optimise the use of workplace resources, including the management of a company's real estate portfolio, infrastructure and facilities assets.
Integrity	See Ethical behaviour.
Internal service provision	See In-house sourcing.
Internal Customer	The facilities customer is diverse in position and power and consists of employees (end-users, consumers), managers/ budget holders of business units (customers) and the company leadership (decision-makers, clients).
ISO	ISO is an abbreviation of International Organisation for Standardization. ISO is an organisation that sets global standards, which mainly focus on procedures. It concerns the working method in the organisation and the standardisation of work processes (management system standards), and the unambiguous definition of terms (technical standards).

ISO 14001	Environmental Management Systems – requirements with guidance for use. This management system standard specifies the requirements for an environmental management system that an organisation can use to enhance its environmental performance. The standard is applicable to any organisation, regardless of size, type and nature, and applies to the environmental aspects of its activities, products and services that the organisation determines it can either control or influence considering a life cycle perspective.
ISO 22001	Global management system standard in the field of food safety through Hazard Analysis Critical Control Point (HACCP).
ISO 22301	Business Continuity Management Systems Requirements - Global management system standard in the field of Business Continuity Management (BCM).
ISO 26000	Guidance on Social Responsibility - Global management system standard in the field of Social Responsibility. This guideline provides support for the implementation of CSR. ISO 26000 applies to all types of organizations, large and small, public and private. ISO 26000 offers internationally agreed definitions, principles and core themes in the field of CSR, as well as guidelines for the implementation of social responsibility.
ISO 31000	Risk Management Guidelines - global management standard in the field of Risk Management. Guideline for managing the risks in organisations. The application of these guidelines can be customised for each organisation and its context.
ISO 41001	Facility Management - Management systems - requirements with guidance for use. Management system norm that enables FM organisations to develop tools to address the needs of the demand organisation and to improve the delivery of FM services in a proactive, cost-effective and sustainable manner. The requirements specified in this document are non-sector specific and intended to be applicable to all organisations, or parts thereof, whether public or private sector and regardless of the type, size and nature of the organisation or geographical location. The ISO 4100 management standard is supported by technical norms ISO 41011, ISO 41012, and ISO 14013.
ISO 41011	Facility Management Vocabulary - a technical standard which, together with ISO 41013, replaces NEN-EN 15221-1.
ISO 41012	Facility Management - Guidance on strategic sourcing and the development of agreements - technical standard replacing NEN-EN 15221-2.
ISO 41013	Facility Management: Scope, Key concepts and benefits - a technical standard that replaces NEN-EN 15221 together with ISO 41011.

ISO 9001	Quality management systems Requirements - a quality management system for all organisations to demonstrate their ability to consistently deliver products and services that meet the requirements of the customer and the applicable laws and regulations.
Kano-model	Matrix that combines two axes: the extent to which a particular customer need is met, and the degree of satisfaction that this brings to the customer. When mapping needs, you can then distinguish between three different factors: basic factors (dissatisfiers), performance factors (satisfiers) and the so-called wow factors (delighters). Meeting some customer needs can lead to higher or neutral satisfaction and failing to meet other customer needs can lead to lower or neutral satisfaction.
Key Performance Indicator (KPI)	A quantifiable measurement of strategic activities, that demonstrates how effectively a company is achieving key business objectives. It is supported by metrics (see performance indicators). Together, (key) performance indicators form the dashboard of the (facility) organisation.
Knowledge workers	Workers whose main capital is their knowledge; their job involves thinking for a living. Knowledge worker's productivity cannot be measured in the number of pieces they have produced in period of time. See also: productivity.
Lean six sigma	Improvement methodology aimed at setting up processes in such a way that an organisation only does those things that the customer really wants (effectiveness) at the lowest possible process costs (efficiency).
LEED	Abbreviation that stands for Leadership in Energy and Environmental Design. LEED is a scoring system to measure the sustainability performance of buildings.
Leesman index	Measuring instrument that measures how well the working environment supports the employee in his work. The Leesman Index consists of a digital survey with ninety questions about four main elements: the activities that someone performs at work, the physical facilities in the work environment, the facilities services that are offered and how often someone changes places.
Leisure	Recreational Industry.
Lettable Floor Space (LFS)	Floor Area (FA) increased by the horizontal traffic area (corridors/entrance), the area of the non-load-bearing internal walls and the free-standing (construction) parts smaller than 0.5 square meters. All parts of the building not comprising common parts which from time to time are either occupied by the Landlord for the purposes of its business or by a tenant or tenants for office purposes or are so constructed or adapted as to be capable of being so occupied.

Leverage products	Products with a low supply-risk and a major impact on the financial result. The buyer then has a strong position in negotiations with suppliers. The strategy of competitive positioning is appropriate to make maximum use of the competition in the market.
Long-Term Maintenance Plan (LTMP)	Plan in which the necessary maintenance and the required budget are distributed over a more extended period (five to ten years).
Logistics	Organisation, planning, control and execution of the flow of materials and goods.
Make, buy, cooperate	See Sourcing.
Management information	Information needed to evaluate and adjust or improve the service, also known as management information.
Management-operational split	Distinction between management activities and outsourced execution in an organisation. A high management-operational split refers to a small in-house facilities management team that focuses on strategy and demand management; tactical management is outsourced to providers. A low management-operational split refers to a situation in which the in-house team is responsible for all tactical and strategical management.
Management system	A set of coherent or mutually influencing elements of an organisation to establish a policy and objectives, plus the processes to achieve those objectives. The components of a management system include the organisational structure, roles and responsibilities, planning and execution.
Management system standards	Methodology for organising and optimising an organisation's management system.
Management processes	Processes that regulate the planning, evaluation and adjustment of the organisation.
Manager	Function responsible for initiating and directing the actions of people in an organisation. A manager is often responsible for the (financial) result of a business unit or department and leads a group of employees.
Managing	Directing the processes that take place in an organisation.
Marketing mix	Also known as the 5 P's: Product, Price, Place, Promotion and People. In the facilities field, a **sixth P,** that of Process, is of great importance for delivering services at the right quality/price level.

Master Service Agreement (MSA)	Contract or document in which the strategic policy of an organisation is translated into frameworks of what is delivered in the facility field; at the strategic level in an organisation.
Megatrends	Trends that are expected to be global influences in the coming years. These are the trends that an organisation should heed if it is to remain relevant; regardless of its size, reach or private/public sector status.
Metered billing	See Pay-Per-Use.
Metrics	See Performance indicators.
Mission	Description of the product-market combinations and the way in which they can achieve a structural competitive advantage.
Monitoring	Determining the status of a system, a process or an activity.
Mystery guest	Someone who checks the contract compliance (called an auditor) by observing 'incognito'; a kind of undercover agent.
Nanotechnology	Technique that enables us to work with particles in the order of one nanometre (one billionth of a metre). Nanotechnology can modify the properties of products, from food to materials.
Need	Expectation, specific or abstract, from the demand organisation, which is essential to enable the achievement of the core purpose and key objectives.
Net Internal Area (NIA)	Sum of the total floor area of all buildings on a site. It does not include walls, the area of stairs, loading bays, accessways or car parking areas or any area occupied by machinery required for air conditioning, heating power supply or lifts. Also called Usable Floor Area (UFA).
Net Promoter Score (NPS)	Way to measure the reputation of organisations (commercial and internal). The NPS uses only one question: 'How likely is it that you would recommend organisation X to a friend or colleague?'. Detractors (opponents or critics): these respondents give a score between 0 and 6 and are most likely dissatisfied with the organisation. Passives (passively satisfied): these respondents give a score of 7 or 8 and are satisfied with the organisation. Promoters: these respondents reward a score of 9 or 10 and are very enthusiastic about the organisation.
Net zero (building)	Energy-neutral building with an energy performance coefficient (EPC) of precisely zero. Only the annual building-specific energy consumption is taken into account.
New Way of Working (NWoW)	Time and place-independent working. Employees decide for themselves where and when they can work and have the freedom to organise their work as they see fit.

Non-critical products	Products that are not critical for business continuity, for example, office supplies. The purchasing strategy focuses on simplifying the purchasing process and reducing the number of suppliers. Purchasing activities focus on standardisation, framework contracts and fixed order, order and invoice processing processes.
Norm	See Standard.
Objective	Result to be achieved.
Occupancy rate (workplace)	Average presence of employees at the workplace.
Operating result	Difference between sales price and production cost.
Operational excellence	Value strategy aimed at achieving the lowest possible cost price. See also Value proposition.
Operational information	Information needed to keep the processes going. This information is also called process information.
Operational level	Level in an organisation at which activities are performed routinely. This involves carrying out activities such as staffing the reception desk, maintaining the sites, supervising the swimming pool and offering catering facilities. Operational management relates to the question of how the services are provided, by whom and when — also referred to as executive level.
Operational procurement	The last three phases of the purchasing process: ordering, monitoring and aftercare. See also Tactical procurement.
Organisation	Person or group of people that has its own (shared) functions with responsibilities, authorities, and relationships to achieve its objectives.
Organisational culture	The specific collection of values and norms that are shared by people and groups within an organisation that influence, over time, the way they behave with each other and with stakeholders outside the organisation.
Organisational structure	Way in which tasks, powers and responsibilities in an organisation are divided, and mutual relationships are arranged. Which structure suits an organisation depends on its strategy and environment.
Outcome	See Impact.
Output	Result of a process. Facility output consists of basic facility products and services and customised solutions.

Output-based contract	Form of contract that specifies the customer needs and expected outcome in terms of service levels and quality, and in which providers can decide as to how they organise the services to provide the result desired. See Input-based contract.
Outsourcing	Transfer of certain activities, functions or processes to external organisations. Whether outsourcing is the right choice depends on the organizational strategy and culture. The executive activities of support departments, in particular, are outsourced.
Outsourcing rate	Percentage of services that are outsourced.
Out-task	Provision of a single service from an external service provider.
Partner	An external party with whom the organisation strategically chooses to work with to achieve common objectives and provide long term mutual benefit.
Partnership	A durable working relationship between the organisation and one or more partners that creates and shares greater value for both parties. Partnerships can be formed with, for example, suppliers, distributors, customers, educational establishments, consultancies or research organisations. Partnership by outsourcing: vested contract.
Pay-Per-Use	Periodic invoicing based on actual consumption, also known as metered billing.
PDCA-cycle	See Planning and control-cycle.
Performance	Measurable result that can relate to either quantitative or qualitative findings.
Performance-based contract	See Output-based contract.
Performance indicator	Quantifiable metrics that periodically assess the performances of processes or employees at a tactical level. Performance indicators (metrics) support strategic Key Performance Indicators (KPI's). Performance indicators must be defined in a way that is understandable, meaningful, and measurable. An indicator can only measure what happened in the past tense. See Key Performance Indicator.
Performance level	See Service level.
Performance management	Directing the organisation by systematically defining a mission, a strategy and objectives of the organisation, and translating these into all levels of the organisation in a structured manner and making them measurable with the help of critical success factors and (key) performance indicators.

Persona	Detailed customer profiles to 'bring to life' the knowledge about target groups.
Personnel costs	Salaries and additional costs of all facility staff; of everyone who is employed by the facility company or the facility provider.
Planning and control-cycle	Control process by which the management controls the organisation. It is a cyclical process with fixed moments for planning policy, and for reporting financial and non-financial results. Also referred to as PDCA-cycle (Plan, Do, Check, Act).
Policy	Intensions and direction of an organisation as formally expressed by its top management, that guide goals and strategies for the business units (e.g. facilities policy) or business functions (e.g. procurement policy).
Predictive maintenance	Installations send out a signal when parts are approaching the point at which malfunctions may occur due to frequent use. Instead of waiting for these to take place, maintenance is (automatically) scheduled before then.
Primary activities	Activities that constitute the distinctive and indispensable competencies of an organisation in its value chain.
Primary process	Process to realise the product or service of the core activity for the external customer. Also called core process.
Process	Set of interrelated or interacting activities which transforms inputs to outputs; sequence of activities that create value for the customer. Typically, there are three types of processes: main or 'core' processes, management processes and support processes.
Process description	Schematic representation of how input is processed into output. The basis for standardised and measurable services. Flow charts and service blueprint are the methods discussed in this book to represent processes schematically.
Process information	See Operational information.
Procurement	Activity of acquiring goods or services from suppliers. A distinction is made between direct and indirect procurement. FM-procurement belongs to the latter (see indirect procurement and FM-procurement)
Procurement process	The process that leads to the acquisition of products and services. Essentially it is a cyclical process that consists of four phases: procurement analysis, procurement process, implementation of the contract, management phase.
Product leadership	Value strategy aimed at distinguishing by product innovation (see also Value strategy).

Product range	Products purchased by an organisation.
Product and Service Catalogue (PSC)	Means of communication at the operational level for communication with customers and users. A PSC describes (based on SLAs) the various products and services provided by the facilities organisation and contains at least the following information: a description of the facilities product, the quality levels available, the ordering method, the delivery conditions, the price, information on complaints and after-sales service.
Productivity	Ratio between the amount of output (goods, products or services) and the input required for it (labour, materials, machinery). There are three ways to increase productivity: (1) to produce more output with the same input; (2) to produce the same output with less input; (3) to increase the output stronger than the input increases. The productivity of knowledge workers is measured as the perceived productivity (how productive a person is according to his or her assessment).
Professional roles	The roles of facility manager, specialist, consultant and entrepreneur, as distinguished by the The Dutch FM-Education Board (LOOFD). A role is not the same as a function, and more roles are possible than these four.
Program of Requirements (PoR)	See Brief.
Public-Private-Partnership (PPP, 3P or P3)	A cooperative arrangement between two or more public and private sectors, typically of a long-term nature. Public-private partnerships are primarily used for infrastructure provision, such as the building and equipping of schools, hospitals, transport systems, and water and sewerage systems.
Public procurement	Public procurement rules govern the way public authorities and their agencies purchase goods, works and services. For fair competition in an open and transparent market, laws and regulations require the procuring authority to issue public tenders if the value of a bid exceeds a threshold value. Government procurement is also the subject of the Agreement on Government Procurement (GPA), a plurilateral international treaty under the auspices of the World Trade Organisation (WTO).
Quality	Degree to which characteristics of a product or service are corresponding to the requirements and wishes of consumers. The quality experienced by the customer arises during and after the service. Quality consists of three factors: technical quality (what), functional quality (how) and relational quality (who).

Quality assurance	Set of planned and systematic actions necessary to demonstrate that the product meets the requirements.
Quality management	Continuous process of setting goals (plan), providing the service (do), measuring whether the performance meets the objectives (check) and if not, adjusting the service or the work process (act).
Real estate	Property consisting of land and the buildings on it; also used when referring to buildings or housing in general. Also: the business of real estate; the profession of buying, selling and letting land, buildings and housing.
Real estate management	Developing, managing and disposing of buildings and land, usually to achieve financial returns in the longer term.
Recalculation of costs	Invoicing on the bases of actual consumption.
Reception	First point of contact for visitors and customers of the organisation, both physically (at the counter) and by telephone (via the general telephone number of the organisation).
Relational quality	Quality that revolves around the relationship with the customer. If there is intensive contact between the customer and the service provider, the employee must be friendly, have empathy with the customer, be competent and capable. This contributes to the quality experience of the customer.
Relationship diagram	Part of the brief which indicates which functions should and should not be clustered.
Remote working	Working outside the traditional office.
Requirement	Need or expectation that is stated, generally implied or obligatory (see Demand).
Research	Research is always about asking a question and in a systematic and verifiable way searching for the most reliable answers to that question.
Revenues	Earnings from e.g. services provided to external clients.
Responsible procurement	A process by which environmental, social and ethical considerations are taken into account when making a purchasing decision. It is also known as sustainable/green procurement, environmentally preferable purchasing [EPP] or sustainable/responsible purchasing.
Risk	Chance of damage or loss due to a specified uncertain event. Risks that play a role in a working environment, for example, due to fire, burglary and power failure.

Risk analysis	Analysis carried out periodically, for example, every three years to identify risks. To keep the data up to date a limited annual update can be made.
Risk Assessment & Evaluation (RA&E)	A list of all health and safety risks involved in business operations and a plan for their solution.
Risk Management	Risk management tries to protect the organisation against disasters, calamities, vandalism, terror, crime and the like. The purpose of risk management is to guarantee the continuity of business operations (see business continuity management).
Risk matrix	A tool for mapping risks. The horizontal axis shows the probability of occurrence (once in the ...); the vertical axis indicates the severity of consequences (impact). Both the likelihood and consequences are qualitatively characterised.
Robotization	Trend that (mainly routine) tasks are taken over by robots.
Safety	Involves the safety of users in a building; in-house emergency response (FAFS: first aid and fire safety) plays an essential role in this.
Satisfier	Basic product or service for which user satisfaction increases with increasing quality. See also: dissatisfier, delighter and KANO-model.
Scenario	Vision of the future that can be distinguished based on current developments. Scenarios are used by organisations to make policy choices.
Scenario matrix	A scenario matrix is a matrix created by confronting two driving forces in a horizontal and vertical axis. Each of the four quadrants that result is a scenario.
Scenario planning	Steps taken to arrive at scenarios for the future. The process consists of mapping trends and developments, determining the driving forces for changes that affect the organisation, drawing up scenarios matrices and translating the scenarios into the organisation.
Scrum	Scrum refers to an iterative framework for managing intricate work in projects. Scrum is built around the given that customers will change their minds about what they want or need, and the occurrence of unpredictable challenges—for which a predictive or planned approach is not suited. As such, Scrum focuses on how to maximize the team's ability to deliver quickly, to respond to emerging requirements, and to adapt to evolving technologies and changes in market conditions.
Secondary Process	See Supporting process.

Sector	See Industry.
Security	Ensuring the optimal physical and social safety of a site. A combination of organisational, structural and electronic security measures is applied.
Self-service	Digital (online) ordering of services, booking rooms, filing complaints, etcetera by filling in digital forms in the FMIS or via an app.
Service	A service is a time-perishable, intangible activity, process or collection of processes to achieve an agreed facility result.
Service blueprint	A service blueprint is a visual representation of a service process from the user's point of view.
Service design thinking	Method to develop services from the customer perspective, based on five principles: customer experience is the starting point; the customer journey; co-creation with all stakeholders; physical elements to support service experience; holistic (involves as many senses as possible).
Service level	Complete description of requirements of a product, process or system with their characteristics, expressed in terms of intensity of use, range and use of materials.
Service Level Agreement (SLA)	Document that describes the agreed performance, measurement, and conditions of service delivery following policy frameworks, between the demand organisation (client) and a service provider.
Service provider	Organisation that delivers one or more facility services. Can be internal or external to the client (demand) organisation.
SERVQUAL-model	Abbreviation of SERvice QUALity. Diagnostic model to identify causes of customer dissatisfaction. The model identifies causes for gaps or lapses between expectation and experience that result in dissatisfaction.
Shared Service Centre (SSC)	Business unit in which several support services are accommodated, often a combination of FM, ICT, HRM and the financial administration.
Single service supplier	Facility provider that limits itself to one type of service, for example, cleaning, catering or security - but not the combination thereof.
SIVA-model	Model to help determine the marketing mix based on the needs of the customer. The abbreviation stands for solutions (that meet the customer's needs), information, value (for the customer, price-quality), accessibility.

Slash workers	People who work for multiple clients at the same time, whether or not in a combination of employment and self-employment.
Smart building	Building that allows installations to communicate with each other via a building management system (BMS), enabling the building to react (and even anticipate) to use.
Smart city	A smart city is a city, village, district or region, which is managed and controlled by high-level technology. The aim is to improve the quality of life of citizens through efficient operations and by bridging the gap between inhabitants and the administration.
Social costs	Costs to society as a whole; the private costs plus the costs of any adverse external effects.
Soft services	Everything related to user-related services, such as catering, cleaning and reception. See also Hard services.
Sourcing	Practice which identifies, evaluates and engages internal and external service providers to deliver a service or products to meet a specification. It's about the choice between doing it yourself, partially outsourcing or completely outsourcing (make, buy and ally). Make: the facilities organisation carries out all the services itself. Buy: the facilities organisation outsources services unless there are convincing arguments for performing a particular service itself. Ally: the facilities organisation outsources the execution and management in full. If there is a partnership, this is called vested outsourcing.
Space management	Comprehensive term for various activities: inventory and allocation of space (housing), management of communal areas, design of workplace facilities, signage, landscaping and art management. Also referred to as Workspace Management.
Specification	Detailed description of the essential performance and technical requirements for services or products set out by the demand organisation to make clear to the service provider the criteria to be fulfilled. See Brief.
Stakeholder	Person or organisation that can affect or perceive itself to be affected by a decision or activity.
Standard	A standard is a document with recognised agreements, specifications or criteria about a product, a service or a method. Standards can be set within a company or organisation, within a consortium of organisations or by recognised standardisation organisations (also referred to as a norm).
Strategic management	Developing the strategy and taking care of its implementation.

Strategic organisation level	Highest level of the organisation where the Executive Board and the Board of Directors determine policies. This level is also referred to as the C suite. This is where the course and objectives for the organisation are set, in line with developments in the market.
Strategic products	Products with a significant impact on the financial result, for which it is difficult to change suppliers because there are often only a few suitable alternatives. A good match with the supplier is essential, so an obvious strategy is a cooperation.
Strategy	Strategy is a plan indicating how the organisation intends to achieve its objectives.
Stratification of the customer	Refers to a distinction between internal customers of facility management, based on their hierarchical position and role in an organisation. Directing customer (client at strategical level), paying customer (at a tactical level), consumer (end-user, or benefitting customer at an operational level).
Subcontracting	Process of engaging a subcontractor.
Subcontractor	Organisation engaged by the service provider to perform a specific portion of a facility service.
Supplier	Provider of services or goods.
Supplier performance assessment	A retrospective assessment of a supplier's performance to the contractual obligation(s). This assessment includes determining whether the requested products and services have been delivered and whether compliance with the related agreements concerning response times, invoicing, and accessibility has been achieved.
Supply chain	The system of organisations, people, activities, information, and resources involved in delivering a product or service; from suppliers to end-users.
Support process	See Supporting process.
Supporting process	Non-primary activity delivered in support of core business. Also knows as support process or support service.
Support service	See Supporting process.
Sustainable Development Goals (SDGs)	See Global Sustainable Development Goals.

Tactical organisation level	The organisational level responsible for planning and managing the specific mechanisms and resources for operational delivery of products and services. Here decisions are made about the range of services on offer, the product range, sourcing, the organisation of the work processes and the deployment of personnel.
Tactical procurement	First five phases of the procurement process: orientation, specification, selection, contracting, implementation. See also Operational procurement.
Technical quality	Objective quality of the product or service provided.
Technical standards	Standards that aim to arrive at an unambiguous comprehension of all aspects of the field of study.
Tendering	A selection in which supplier respond to a request for proposal or request for quotation by a bid. Based on pre-defined and weighted selection criteria, the best-scoring offer is selected. See also European tendering procedure.
Third workplace	Workplace in a different place than at work or at home, for example in a business centre, in the library or at a co-working location.
Triple P	People, Planet, Profit/Prosperity, representing the three elements that must be balanced for sustainable development. When the combination is not harmonious, the other elements will be affected negatively.
Total Cost of Ownership (TCO)	Costs and revenues during the entire lifecycle of a building or installation.
Total Facility Management (TFM)	See Integrated Facility Management (IFM).
Total Quality Management (TQM)	Business philosophy that implies that everything an organisation does must be aimed at satisfying all stakeholders in the most efficient way possible.
Transfer of Undertaking Act (TUA)	A transfer of undertakings occurs when a business or part of a business is taken over by another employer as a result of a merger or transfer (outsourcing). When a transfer takes place, there is a legal obligation on the new employer to take on the existing staff of the business or the part of the business concerned. According to this European Act, the employees follow the work and in principle retain their right to their employment conditions, even if these are outsourced to a supplier (for the first time). In the case of subsequent outsourcing of labour-intensive work, the law usually no longer applies.

Usable Floor Area (UFA)	See Net Internal Area (NIA)
Value chain	Porter's model of the value chain that provides, among other things, insight into the value that the primary and secondary activities of a company add for the customer.
Value proposition	The differentiating value that the organisation's products, services and solutions offer to customers. Treacy & Wiersema (2002) distinguish three value strategies: cost leadership (operational excellence); product leadership and customer intimacy.
Variable costs	Costs that vary with production volumes or sales. For example, costs that depend on time spent (work by third parties) or on the number of participants (in an event) or the number of goods.
Vendor costs	Costs of hiring staff and purchasing goods and services. All costs of deliveries for which the service company is billed.
Vision	A vision is a description of what the organisation is attempting to achieve in the long-term. It is intended to serve as a clear guide for choosing current and future courses of action and, along with the organisation's Purpose, it is the basis for strategies and policies.
Vitality	Relates to energy level and resilience of employees.
War for talent	This refers to the competition between organisations to attract scarce talent in a tight labour market.
Waste management	Preventive action against waste and for the promotion of recycling and re-use of materials within the organisation. Waste or garbage are substances, materials and products that the user wants to dispose of. An authorised waste collector must collect the waste. This waste collector chooses a processing method depending on the characteristics and origin of the waste.
WELL building standard	WELL is a performance-based system for measuring, certifying, and monitoring features of the built environment that impact human health and well-being. Ten aspects of the built environment are included in its scope: air, water, nourishment, light, fitness, comfort and mind. WELL is a global tool that is utilized in more than 50 countries.
Work instruction	Work plan, in which successive activities and actions that should take place, are described.
Work process	Part of the main process described in detail. Different employees or departments are involved in a work process; otherwise, it is called a work instruction.

Workplace	Physical location where work is performed.
Workplace management	All activities, services and resources aimed at administration, advice and coordination in acquiring, operating and disposing of space.
Workstation	Location containing furniture and supporting equipment, specially designed or suitable for work-related activities, and which is suitable for permanent use.
Workwear and linen services	The care of uniforms, linen and, in residential care, personal clothing. The provision and care of uniforms is an important task of facility management, especially in public (enforcement) functions, industry, retail, hospitality (including wellness) and care.
Zero-base budgeting	Budgeting method that involves determining the facility budget each year from scratch by making agreements with internal customers about the services to be provided in the coming period.

Bibliography

Books

Atkin, B. & Brooks, A. (2015) *Total Facility Management*, 4th ed., New Jersey, USA: John Wiley And Sons Ltd.
Bakker, I. (red.) (2010). *De breinwerker*. Naarden, The Netherlands: FMN.
Beckers, R. & Roelofs, V. (2016). *Zo maak je een facilitair plan*. Groningen/Utrecht, The Netherlands: Noordhoff Uitgevers.
Bekker, I. de, Pluijm, S. van der, Ruys, H. & Visser, H. (2012). *De facility manager onderneemt.* Alphen aan den Rijn, The Netherlands: Kluwer BV.
Boer, Brouwers & De Keijzer (2016). *Bedrijfseconomische thema's*. Groningen/Utrecht, The Netherlands: Noordhoff Uitgevers.
Bolier, D.G.L. (1988). *Hotelmanagement, theorie en praktijk van frontoffice en housekeeping*. Zoetermeer, The Netherlands: SVH Uitegeveri
Centraal Planbureau voor de Leefomgeving (2015). Toekomstverkenning Welvaart en Leefomgeving Nederland in 2030 en 2050. The Hague, The Netherlands: Planbureau voor de Leefomgeving en Centraal Planbureau
Drion, B., Marée, G. & Melissen, F. (2710). *Workin' Wonderland*. Utrecht, The Netherlands: Arko.
Drion, B., Melissen, F., Stassen, D. & Thoma, M. (red.) (2008). *Een bijdrage aan Duurzaamheid vanuit facilitaire optiek*. Amersfoort, The Netherlands: Hospitality Consultants.
Gelderman, K. & Albronda, B.J. (2017). *Professioneel inkopen*. 5e druk. Groningen/Utrecht, The Netherlands: Noordhoff Uitgevers.
Gerrtisma, M. & Grit, R. (2009). *Zo maak je een Risicoanalyse*. Groningen/Utrecht, The Netherlands: Noordhoff Uitgevers.
Grönroos, C. (2015). *Service management and marketing*. Chichester, UK: Wiley & Sons Inc.
Gunnarsson, J. & Blohm, O. (2003). *Hostmanship, the art of making people feel welcome*. Stockholm, Sweden: Dialogos Forlag.
Hillenius, P.A. (2013). *De 13 gaps van facility management. Op weg naar een betere klantexperience*. Deventer, The Netherlands: SeLA-management.
Hoendervanger, J.G., Wijnja, J. & Voordt, T. Van der (2017) *Huisvestingsmanagement – Van strategie tot exploitatie.* 2e druk. Groningen/Utrecht, The Netherlands: Noordhoff Uitgevers.
Jensen, P.A. & Voordt, T. Van der (eds) (2017) *Facilities Management and Corporate Real Estate Management as Value Drivers – How to Manage and Measure Adding Value.* 1e druk. Oxfordshire, United Kingdon: Routledge.
Jong, de B., Van Diepen-Knegjens, Kok, H. e.a. (2013) *De facilitaire regieorganisatie in bedrijf – Veranderen naar een demand- en supplyorganisatie.* 1e druk. Deventer, The Netherlands: Vakmedianet.
Koetzier, W. & Epe, P. (2017). *Management accounting – berekenen, beslissen, beheersen.* 5e druk. Groningen/Utrecht, The Netherlands: Noordhoff Uitgevers.
Maas, G.W.A e.a. (2013) *Facility Management – Over organiseren en maatwerk binnen een veranderende context.* Deventer, The Netherlands: Vakmedianet.

Maas, G.W.A. & Pleunis, J.W. (2006). *Facility management*. Deventer, The Netherlands: Kluwer.
Marcus, J. & Dam, N. van (2019). *Handboek Organisatie en Management.* 9e druk. Groningen/Utrecht: Noordhoff Uitgevers.
Melissen, F., Rest, J.P. van der & Josephi, S. (2018). *Hospitality Experience, An introduction to Hospitality Management*. 2e druk. Groningen/Utrecht, The Netherlands: Noordhoff Uitgevers.
Melker, G.J. (2010). *250 managementbegrippen*. Groningen/Utrecht, The Netherlands: Noordhoff Uitgevers.
Ministerie van Infrastructuur en Waterstaat (oktober 2018) *Kerncijfers Mobiliteit 2018*. Den Haag, The Netherlands: Kennisinstituut voor Mobiliteitsbeleid.
Poiesz, T. (1999) *Gedragsmanagement, waarom mensen zich (niet) gedragen*, Wormer, The Netherlands: Inmerc B.V.
Ruigrok. K. & Bosschers, E. (2007) *Functioneel Beheer.* Uitgeverij Thema.
Schreuder Peters, R.P.I.J. (2012). *Methoden en technieken van onderzoek, principes en praktijk*. 3e druk. The Hague, The Netherlands: Sdu.
Thuis, P. (2017). *Toegepaste organisatiekunde*. Groningen/Utrecht: Noordhoff Uitgevers.
Weele, A.J. van, Berkel-Schoonen, J. van & Walhof, G. (2017) *Inkoop in strategisch perspectief – shoppen voor professionals.* Deventer, The Netherlands: Vakmedianet.
Westerveld, E., Zeeuw, A. de & Bakker, R. (2000). *Handleiding Facility Excellence Model*. Utrecht, The Netherlands: Berenschot Business Management.

White papers, reports and news items

Appel-Meulenbroek, R. & Pullen. W. (2017) *Input to the FM and CREM research agenda*. eFMI October 2017, p. 46-51.
CfPB (2017) *Nieuwe werkomgeving, betere prestaties – 15 jaar CfPB*. via: www.cfpb.nl
Colliers International. *Occupier Cost Index 2018*. via: https://www2.colliers.com/en-NL/Research/OccupierCostIndex
Creative Council (2015) *Creatief broeden? Complexer dan gedacht!* De ontwikkeling van creatieve ondernemingen in broedplaatsen. Via: http://www.creative-council.nl/_docs/150305_UC_brochure_final_design_online_%281%29.pdf
Cushman & Wakefield (2018) *Nederland Compleet– kantoren- en bedrijfsruimtenmarkt medio 2018*.
Cushman & Wakefield (2019) *European Coworking Hotspot Index*
EFMC (2019) *EFQM Model Brochure*, via: https://www.efqm.org/index.php/efqm-model-brochure-whn64kt8/
IFMA & PROCOS Group (2019) *FM-Trendrapport België 2019; cijfers, trends en ontwikkelingen van de Belgische Facility Management Markt*, Belgium: IFMA/PROCOS Group
Global FM (2019) *Global Facilities Management Market Report 2018.* Dockland VIC, Australia: Global FM.
Kleinnijenhuis, J. (04-04-2018) *Robotisering treft helft van de banen*. Trouw.
Lettink, N., Walstijn, M. van (2019). *De veranderende manier van werken*. Utrecht: Yunno.
Hago (z.j.) *De Toekomst van Schoonmaak*. www.hago.nl/media/162726/whitepaper-de-toekomst-van-schoonmaak-hago.pdf
Hendriks, H., december 2006. *Visiedocument Dienst Facilitaire Bedrijven*, Universiteit van Tilburg.
Hospitality Group (2019) *FM Provider Performance Onderzoek 2018*
IFMA/PROCOS (2019) *FM-Trendrapport België 2019 – cijfers, trends en ontwikkelingen van de Belgische Facility Management markt.* Antwerpen, Belgium: IFMA/PROCOS group
Integron (2019) *Facilitybeleving in Nederland 2018-2019*. Via: https://www.facilitybeleving.nl/rapport-facility-beleving-in-nederland-2018-2019/
LOOFD (2017). *Landelijk opleidingsprofiel hbo-bacheloropleiding Facility Management.*

OECD (2019) *Outlook 2019: The Future of Work*. https://www.oecd.org/employment/future-of-work/
Osterwalder, A. & Peigneur, Y. (2010) *Business Model Generation*. New Jersey, USA: Wiley.
Reitsema (2014) De verantwoordelijkheid van de facility manager in de beheersprocessen rond veiligheid van personen in de SW-sector. [master thesis]. Master Facility and Real Estate Management. University of Greenwich, London, UK.
Rijksoverheid (2016) *Rijksbreed Programma Circulaire Economie*. Via: www.rijksoverheid.nl/documenten/rapporten/2016/09/14/bijlage-1-nederland-circulair-in-2050
NL Smart City Strategy – The future of Living. Via: https://www.thehaguesecuritydelta.com/media/com_hsd/report/114/document/NL-Smart-City-Strategie-.pdf
Ruys, H, Visser, H. & Wiedemeijer, M. (2009). *De facility manager regisseert*. Alphen a/d Rijn/Diegem: Hospitality Consultants/Kluwer.
Saxion Facilitaire Service Organisatie (2018). *Visie en Strategie FSO*. Deventer: Saxion.
TomTom (2019) *World Traffic Index 2019*. www.tomtom.com
Twynstra Gudde (2019). *Facility en Workplace Management 2019*. Amersfoort: TG/FMN.
World Green Building Council (2019) *Bringing embodies carbon upfront*. Via: www.worldgbc.org.

Internet

Bialos, J.P., Nolan, F.X., Cambre, B.M. & Park, E.D. (2019) Public Procurement United States Via: https://gettingthedealthrough.com/area/33/jurisdiction/23/public-procurement-united-states/
BREEAM (n.d.) *BREEAM in-use*. Via: https://www.breeam.com/discover/technical-standards/breeam-in-use/
Business Today (15-01-2019) *India likely to be larger economy than US by 2030: Standard Chartered*. https://www.businesstoday.in/current/economy-politics/india-likely-larger-economy-us-2030-standard-chartered/story/310153.html
Chief Acquisition Officers Council (n.d.) https://www.cao.gov/cao-home
Circularity Check (n.d.) https://ecopreneur.eu/circularity-check-landing-page
Climate Action Tracker (n.d.) https://climateactiontracker.org/
Code Verantwoordelijk Marktgedrag. www.codeverantoordelijkmarktgedrag.nl
Cushman & Wakefield (08-04-2019) http://www.cushmanwakefield.nl/nl-nl/research-and-insight/2019/european-coworking-hotspot-index
Damme, A. van (03-06-20180] *Festivals willen afvalvrij zijn, maar krijg de bezoeker maar eens mee*. www.NOS.nl
Dessing, G. (02-10-2018) *Zonder FMIS mis je helderheid en transparantie*. www.facto.nl
Dessing, G. (05-11-2018) *Een facilitaire ronde lopen? Vergeet je AR-bril niet!* www.facto.nl
Dessing, G. (09-04-2019) *Stadskantoor Alkmaar na grondige verbouwing: ontmoeten en samenwerken*. www.facto.nl
Dijk, D. van & Slooten, S. van (05-11-2018) *Schoon maakt (g)een verschil*. www.facto.nl
DNB The Facility Group (29-02-2016) *Drie dingen die facilitair kan leren van de deeleconomie*. Via: www.db.nl/news/drie-dingen-die-facilitair-kan-leren-van-de-deeleconomie/
DNB The Facility Group (08-11-2019) *Data-driven FM: business case van de nabije toekomst?* www.facto.nl
Dolmans (28-02-2019). *Vacature Facility Manager regio Utrecht*: https://werkenbij.dolmans.com/vacatures/vacature-facility-manager-dos00221-regio-utrecht-87414-11.html
EFQM (2019) *EFQM-model*. www.efqm.org
Emergent Research (2019). https://genylabs.typepad.com/emergent_research/
European Commision (n.d.) Smart cities – Smart living. Via: https://ec.europa.eu/digital-single-market/en/smart-cities
European Environment Agency (10-12-2019) *Europe's circular economy still in its infancy*. www.eea.europa.eu
Eurostat (2015) Distribution of % of fatal and non-fatal Accidents across Industries. www.ec.europa.eu/eurostat

Eurostat (2020) *Share of part-time work in Europe*. www.ec.europa.eu/eurostat
European Committee (n.d.) European Tender Directives. https://ec.europa.eu/info/index
Facto (25-10-2018). *Nederland in 2030 koploper flexibel werken*. www.facto.nl
Facto (10-04-2018) *Samen de afstand tot de arbeidsmarkt verkleinen*. www.facto.nl
Facto (29-01-2019). *Gom en Ibis Styles Haarlem City Hotel winnen Best Practice Award 2018*. www.facto.nl
Foster Fuels Mission Critical (n.d.) www.fosterfuelsmissioncritical.com
Gebouwvanhetjaar.nl (z.j.) *Prachtige duurzame renovatie van een oud, donker en lelijk pand* https://www.gebouwvanhetjaar.nl/entry/renovatie-hoofdkantoor-a-s-r/
Global CR Rep Track (2018) *Global CR RepTrak 100*. https://insights.reputationinstitute.com/australia/2018-global-csr-100-reptrak-data
Kok, H, (6-2-2019) *Wat is dat toch met klanttevredenheid?* (blog) www.facto.nl
Jongejan, R. (27-08-2017) *Klantgericht handelen: 5 manieren om je afdeling klantgerichter te maken*. www.facto.nl
LEAN (n.d.) www.lean.org
LEED (n.d.) *Better buildings make better lives*. Via: https://new.usgbc.org/
Leesman Index (n.d.). www.leesmanindex.com
Measuremen (n.d.) *Workplace Occupancy Study – Netflix*. https://www.measuremen.io/cases/netflix/
Mindspace (26-03-2019) *Ruim kwart jongeren wees potentiële werkgever af vanwege uitstraling kantoor of gebrek aan faciliteiten*. Via: https://www.f-facts.nl/topics/facilitaire-dienstverlening-gebouwbeheer/nieuws/ruim-kwart-jongeren-wees-potentiele-werkgever
NEVI (29-12-2017) *De inkoopscenario's voor de toekomst*. Via: www.nevi.nl
NFC Index (2018) *Facilitaire benchmark awards 2018 naar A.S.R en Hogeschool Utrecht*. www.nfcindex.nl
Pool, I. Van der (11-08-2016) *FM dichterbij primair proces door nieuwe ISO 41000*. www.facto.nl
Procurement Forum (n.d.) https://procurement-forum.eu/about
SWP. (05-11-2018) *CBRE lanceert eigen coworking Company*. www.smartwp.nl
SWP. (23-01-2019) *Van werkplek naar 'Gig-space*. www.smartwp.nl
Tenders Electronic Daily (n.d.) https://ted.europa.eu
UNESCO (n.d.) *Global Sustainable Development Goals*. https://en.unesco.org/sustainabledevelopmentgoals
United Nations Industrial Development Organization (UNIDO, 2019) *Corporate Social Responsibility*. www.unido.org
WELL Building Standard (n.d.) www.wellcertified.com
WELL Building (7-10-2019) *Investing in Inclusion Drives Global Returns*
https://resources.wellcertified.com/articles/investing-in-inclusion-drives-global-returns/
Werken2.0 (mei 2018) *Professionals generatie Y en Z toch meer gefocust op geld dan geluk*. www.werken20.nl
World Green Building Council (WGBC) (n.d.) www.worldgbc.org
Worldstats (May 2019) https://www.internetworldstats.com/stats.htm
Worldometer (n.d.) Urbanisation levels, www.worldometers.info
Zakenreis.nl (19-11-2018) *Integratie hotels en coworking spaces groeit snel*. Via: www.zakenreis.nl
Ziggo Zakelijk (16-05-2018) *Hoe overtuigde schoonmaakbedrijf MAS het Rijksmuseum?*
www.ziggo.nl/zakelijk/inspiratie/in-zaken/mas/aanbestedingen/

Appendix 1

Facility Management associations

Global associations:
- Building Owners & Managers Association (BOMA; www.boma.org)
- Global FM Association (www.globalfm.org)
- International Facility Management Association (IFMA; www.ifma.org)
- Royal Institute of Chartered Surveyors (RICS; www.rics.org)
- CoreNet Global (www.corenetglobal.org)
- World Green Building Council (WGBC; www.worldgbc.org)

European associations:
- EuroFM (www.eurofm.org)
- IFMA Europe (www.ifmaeurope.org)
- CoreNet Global Benelux Chapter (https://benelux.corenetglobal.org/home)

In Northern America and Asia, many cities or regions have a local FM Association, such as:
- IFMA New York City, USA (www.ifmanyc.org)
- IFMA Toronto, Canada (www.ifma-toronto.org)
- IFMA Shanghai Chapter

A full list is provided on www.ifma.org

Many countries have a national professional association, such as:
- Chinese Office Building Owners & Managers Association (BOMA China)
- Dansk Facilities Management netværk, Denmark (DFM; www.dfm-net.dk)
- Facility Management Association of Australia (FMA; www.fma.com.au)
- Facility Management Nederland, The Netherlands (FMN; www.fmn.nl)
- German Facility Management Association (GEFMA; www.gefma.de)
- IFMA Austria Chapter (https://www.ifma.at)
- IFMA Belgium (www.ifma.be)
- IFMA Czech Republic Chapter (http://ifma.cz)
- IFMA Hong Kong Chapter (www.ifma.org.hk)
- IFMA Italy Chapter (http://www.ifma.it/)
- IFMA Nigeria Chapter (www.ifmanigeria.org)
- IFMA Singapore Chapter (www.ifma.org.sg)
- IFMA Spain Chapter (www.ifma-spain.org)
- IFMA Sweden Chapter (www.ifma.se)
- IFMA Switzerland Chapter (www.ifma.ch)
- IFMA Poland Chapter (www.ifma.pl)
- IFMA UK Chapter (www.ifmauk.org)
- Institute of Workplace and Facilities Management (IWFM), UK (www.iwfm.org.uk)

- Japan Facility Management Organisation (JFMA: www.jfma.or.jp)
- Korea Building Owners & Managers Association (KBOMA)
- Macao facility Management Association, China (www.macau-fma.org.mo)
- Norwegian Facility Management Association (NBEF; www.nbef.no)
- South African Facilities Management Association (SAFMA; www.safma.co.za)
- Russia National Association of Facility Operators (NAFO)

Besides these, there are many industry or trade associations.

Appendix 2

Flowchart symbols

Process/function symbols

	process or activity	Indicates process or action.
	alternative process	Indicates alternative process or action.
	manual editing	Displays actions that are not automated.
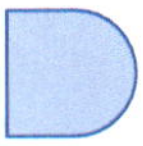	procrastination or delay	Displays waiting time that is part of the process.
	preset process	Indicates process steps described elsewhere. Often used in subprocesses.
	preparation	Shows the preparation that precedes the process.

Offshoot and directional symbols

	connection or inspection	Used as a replacement for arrows, often marked with capital letters A, B, or AA. In process structures symbol for inspection.
	connecting link to other page	Shows the continuation of the scheme on another page.
	dividing line or start and end signal of a process	Shows beginning or end points in a process; used as a starting point, it shows an event that triggers action.

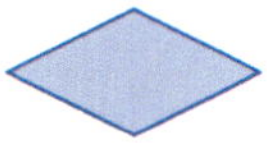	decision point	Usually displays a yes/no question.
	clustering	Shows various subprocesses that converge in one process.
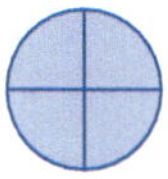	or	Indicates that a process deviates or splits.
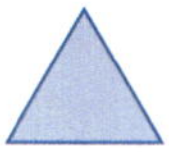	collect or download	Shows that a process is split up into parallel paths. If symbol includes an M this indicates measurement.
	merge	Shows the aggregation of various processes or information.
	sequence or timeline or line	Indicates the direction of the flow chart.

Input and output symbols

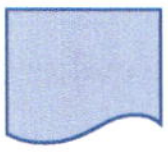	document	Displays a document used in diagram or process.
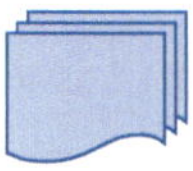	multiple documents	Displays multiple documents used in diagram or process.
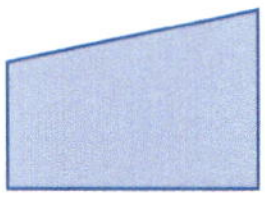	manual entry	Shows a process step where the user has to enter information manually or a process step that is not automated.
	display	Displays a process step where information is shown to a person.

File and data recording symbols

	recorded data	Generally shows a process step where data is stored.
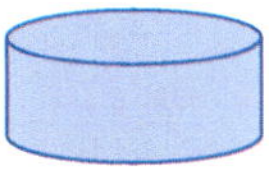	database	Indicates database.
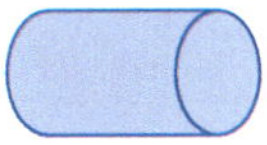	records with direct access	Often used to indicate hard drive.
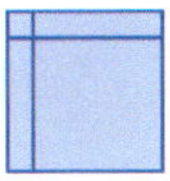	internal storage	Shows that data is stored in memory, not in file.

Data processing symbols

	collect data	Shows a process step that requires organising data, information and materials.
	sort data	Shows the sorting of data, information or materials in a predetermined sequence.

Index

About the authors

Hester van Sprang has been working in the facilities industry for over twenty years. Since 2010 she has been one of the leading lecturers in the Master Facility & Real Estate Management at Saxion University of Applied Sciences/the University of Greenwich, and a senior researcher at the Saxion Research Group on Hospitality. Hester has published in various academic journals and has contributed to many international congresses. Moreover, she the co-author of the successful Dutch Basisboek Facility Management, of which the third edition was published in 2020. Nowadays she deploys her vast knowledge of the interaction between people and buildings as an advisor at MeerWaarde Huisvestingsadvies and as managing director of FaM-CoR B.V.

Bernard Drion worked in the facilities industry for over thirty-five years. He has held various management positions at consultancy firms, including Hospitality Group in Amersfoort, and from 2008 he combined this position with that of professor of Facility Management at Breda University of Applied Sciences (BUAS). With his enormous knowledge and experience, Bernard pioneered numerous initiatives in the field, such as the former NEN 2748 and Netherlands Facility Cost Index and the Facility Performance Standard. Bernard is co-author of the book Workin' Wonderland, which is about the future of organisations, and the co-author of the successful Dutch Basisboek Facility Management, of which the third edition was published in 2020. Bernard is now retired.

Acknowledgements

Photo's:
Hospitality Group, Amersfoort: p. 10, 113
Diane Levine: p. 12
Gom, Schiedam: p. 40
Lisa Hut, Amsterdam: p. 48
BUAS / Universiteit Breda, Breda: p. 52
Peter Ankerstjerne: p. 54
Kulturhus Sølvberget, Norway: p. 62
Shutterstock, New York: p. 64, 72, 73, 79, 82, 156, 224, 276, 291, 297, 300, 304, 316, 326, 331, 335, 341
TomTom: p. 66
Multiple generations in the workforce, kpcompanies.com: p. 72
European Co-working Hotspot Index, Cushman & Wakefield: p. 75
TIMAF, Utrecht: p. 77
The Future of Work in Figures, OECD: p. 81
Passive House Institute & Stichting Passief Bouwen, Eindhoven: p. 85
Joost Andela, Issy Les Moulineaux: p. 92
A.B. de Graaf / Deventer Ziekenhuis, Deventer: p. 96
Elizabeth Nelson: p. 98
Nicholas White: p. 98
Workplacefundi: p: 100
DSL Studio / Marco Cappelletti, Milaan: p. 109
Computer workstation ergonomics, http://www.safety.uwa.edu.au/topics/physical/ergonomics/workstation: p. 116
Hospice Het Vliethuys, Voorburg: p. 117
HappySmileys / Bee-Line B.V., Nieuwegein: p. 137 (adapted)
Employee Experience Journey Map, https://quod.lib.umich.edu/w/weave/images/12535642.0001.903-00000004.jpg: p.145
Loesje Ophuis: p. 150
Jens Schlüter, Berlin: p. 158
Saxion Hogeschool, Roadmap FSO: p. 171, 172, 173
Saxion Hogeschool, screendump Intranet: p.179
UNESCO: p. 120, 201, 204
FMN / NFC Index, Den Haag: p. 207
Davy van den Bosch, Strombeek-Bever (photo Matthijs Gallé): p. 218
Jacqueline Bakker: p. 226
Brigitte van Gompel: p. 270
Michiel Bakker, Sunnyvale (CA, USA): p. 278
Kadans, Haaren: p. 282
University Twente: p. 283
Industrial Park Kleefse Waard, Arnhem: p. 285
Merel de Ruijter, Amsterdam: p. 288
Nynke Visser, Utrecht: p. 294

Nick Steinbuch, Photogenick, Amsterdam (photo Thom Wernke): p. 301
Marcel van der Burg / Prima Beeld, Amsterdam: 306
Michel Tobé, Den Hoorn: p. 307
Charlotte de Block, Breda: p. 321
Susanne Balslev Nielsen: p. 328
Joop van Duren, Heerlen: p. 336
Anna Kimp, Amersfoort (photo Brenda Groen): p. 337
Frans Melissen, Breda: p. 339
Rogier Veldman, Hilversum (photo Mark Mobach): p. 340
Tender Services Group B.V., Den Haag (photo Thomas Klein Bramel): p. 342

Figures:
Figure 1.6: www.rocvantwente.nl/over-ons/bestuur-en-management.html
Figure 1.7: www.dz.nl/Organisatie
Figure 1.8: EN 15221 FM model
Figure 1.10: De Jong et al. (2013). *De facilitaire regieorganisatie in bedrijf*. Amsterdam: Vakmedianet (adapted)
Figure 1.11: Global FM Market Report (2018)
Figure 1.13: Jensen, P.A. (2009). The FM Value Map
Figure 3.2: Bolier, D.G.L. (1988). *Hotelmanagement, theorie en praktijk van frontoffice en housekeeping*, Zoetermeer: SVH Uitgeverij (adapted)
Figure 3.4: https://view.publitas.com/cushmanwakefeld/nederland-kantorenmarktmedio2018 (adapted)
Figure 3.7: Hospitality Group
Figure 3.17: Mood Maker, Hokkeling en De la Mar (2012)
Figure 3.18: Knol, E. (2009, 24-4). *Transparante facilitaire klantvriendelijkheid creëren. Een spiegel van het eigen gedrag* (adapted)
Figure 4.3: Landelijk opleidingsprofiel HBO-bachelor Facility Management, LOOFD, 2017 (adapted)
Figure 4.5: Facilitybeleving in Nederland 2018-2019, Integron (2019) (adapted)
Figure 4.8: Hillenius (2014). *De 13 gaps van facility management*. Deventer: SeLa Management (adapted)
Figure 4.10: Jaap Kloos
Figure 4.11: Van Berkel, Van Weele & Walhof (2017). *Inkoop in strategisch perspectief, shoppen voor professionals*. Amsterdam: Vakmedianet (adapted)
Figure 4.14: UNESCO
Figure 4.15: ISO 26000
Figure 4.16: Poiesz (1999)
Figure 4.17: ISO 31000
Figure 4.19: Hopkin (2010)
Figure 4.21: Ruigrok & Bosschers (2007)
Figure 4.22: FAMA Systems
Figure 5.4: www.klachtenmanagement.nl (adapted)
Figure 5.5: Focusvision
Figure 5.8: Mawson (2002). Cited in Jensen & Van der Voordt (2017)
Figure 5.9: Business Model Canvas, Osterwalder & Pigneur (2010)
Figure 5.11: EFQM (2019)
Figure 5.12 and 5.13: Dienst Facilitaire Bedrijven, University of Tilburg (2006) (adapted)
Figure 5.17: EN-15221-7

Technical drawings:
Integra, Pondicherry, India